CRASH COURSE

Anatomy

SECOND EDITION

Series editor
Daniel Horton-Szar
BSc (Hons), MBBS (Hons)
GP Registrar
Northgate Medical Practice
Canterbury
Kent

Faculty advisor
Ian Whitmore
MD MBBS LRCP MRCS
Professor of Anatomy
Stanford University School of Medicine
Stanford
California, USA

Anatomy

SECOND EDITION

Michael Dykes
BSc (Hons) Anantomy and Cell Biology
University of Sheffied Medical School
Royal Hallamshire Hospital
Sheffied

First Edition Author
Phillip Ameerally

Mosby

London • Edinburgh • New York • Philadelphia • St Louis • Sydney • Toronto 2002

MOSBY
An affiliate of Elsevier Science Limited

Commissioning Editor	**Alex Stibbe**
Project Manager	**Colin Arthur**
Project Development Manager	**Ruth Swan**
Designer	**Andy Chapman**
Illustration Management	**Mick Ruddy**

First edition 1998
Second edition 2002

ISBN 0723432473

British Library Cataloguing in Publication Data
A catalogue record for this book is available from the British Library

Library of Congress Cataloging in Publication Data
A catalog record for this book is available from the Library of Congress

Note
Medical knowledge is constantly changing. As new information becomes available, changes in treatment, procedures, equipment and the use of drugs become necessary. The author, editors and the publishers have taken care to ensure that the information given in this text is accurate and up to date. However, readers are strongly advised to confirm that the information, especially with regard to drug usage, complies with the latest legislation and standards of practice.

Typeset by Kolam, Pondicherry, India
Printed in Spain by Graphycems

The publisher's policy is to use **paper manufactured from sustainable forests**

Preface

Anatomy, medical students either love it or loathe it but we all need a good understanding of it (particularly at exam time!). It can often be a daunting task trying to commit this new vocabulary to memory, especially to those of us who have never studied it before. However, this book has been written to help remove this burden.

This Second Edition builds upon the First to provide you with an ever concise presentation of this subject but with better aids to help you learn. These include mnemonics, new and improved figures, and more clinical information to help illustrate the underlying anatomy.

The final section content has been increased to 100 multiple-choice questions with explanations, 20 short-answer questions with model answers and 20 essay questions for you to attempt. I hope you find this book useful throughout your study of this fascinating subject and good luck in your exams.

Michael Dykes

Anatomy in the Mosby Crash Course series is atypical: it is a discipline orientated book in a fundamentally system orientated series. However, the book will be useful to all students studying anatomy within medicine and allied subjects because the information in the book is so accessible. Students of anatomy now have information presented in so many different ways in their courses that a very concise book encompassing the whole subject will enable them to gain an overall knowledge, and yet find the detail they need for a particular PBL problem or case.

This Second Edition has been carefully revised, building on the sound base of the First. It is more consistent, and we have removed inaccuracies and clarified confused descriptions. The content is more even, having more detail than many full size textbooks, but still presenting the information in pictures, lists, tables, hints and summaries. There are more practice questions, and answers are expanded for multiple-choice questions and short-answer questions. I am proud to have been the faculty advisor for this book.

Ian Whitmore
Faculty Advisor

In the six years since the First Editions were published, there have been many changes in medicine, and in the way it is taught. These Second Editions have been largely rewritten to take these changes into account, and keep Crash Course up to date for the twenty-first century. New material has been added to include recent

research and all pharmacological and disease management information has been updated in line with current best practice. We have listened to feedback from hundreds of students who have been using Crash Course and have improved the structure and layout of the books accordingly: pathology material has been closely integrated with the relevant basic medical science; there are more multiple-choice questions and the clarity of text and figures is better than ever.

The principles on which we developed the series remain the same, however. Medicine is a huge subject, and the last thing a student needs when exams are looming is to waste time assembling information from different sources, and wading through pages of irrelevant detail. As before, Crash Course brings you all the information you need, in compact, manageable volumes that integrate basic medical science with clinical practice. We still tread the fine line between producing clear, concise text and providing enough detail for those aiming at distinction. The series is still written by medical students with recent exam experience, and checked for accuracy by senior faculty members from across the UK.

I wish you the best of luck in your future careers!

Dr Dan Horton-Szar
Series Editor (Basic Medical Sciences)

Acknowledgements

I would like to thank everyone who has helped in the production of this book, especially Dr Ian Whitmore and Dan Horton-Szar for reading my work and for their suggestions.

Dedication

To Mum, Dad, Philip, Gran, Marilyn and Katt for their encouragement and support throughout my studies and writing of this book, without whom I would not be who or where I am today

Contents

ANATOMY

1. Basic Concepts of Anatomy

Descriptive anatomical terms

The anatomical position

This is a standard position used in anatomy and clinical medicine to allow accurate and consistent description of one body part in relation to another (Fig. 1.1):

- The head is directed forwards with eyes looking into the distance.
- The body is upright, legs together, and directed forwards.
- The palms are turned forward, with the thumbs laterally.

Regions of the body

Note: the upper limb is composed of the scapular region, the arm, the forearm, and the hand; the lower limb is composed of the gluteal region, the thigh, the leg, and the foot.

Anatomical planes

These comprise the following (Fig. 1.2):

- The median sagittal plane is the vertical plane passing through the midline of the body from the front to the back. Any plane parallel to this is termed paramedian or sagittal.
- Coronal (or frontal) planes are vertical planes perpendicular to the sagittal planes.
- Horizontal or transverse planes lie at right angles to both the sagittal and coronal planes.

Terms of position

The terms of position commonly used in clinical practice and anatomy are illustrated in Fig. 1.3.

Terms of movement

Various terms are used to describe movements of the body (Fig. 1.4):

- Flexion—movement in a sagittal plane which reduces the angle at the joint (except at the ankle joint), e.g. bending the elbow.
- Extension—increases the angle at joints in the same plane.
- Abduction—movement away from the median plane.

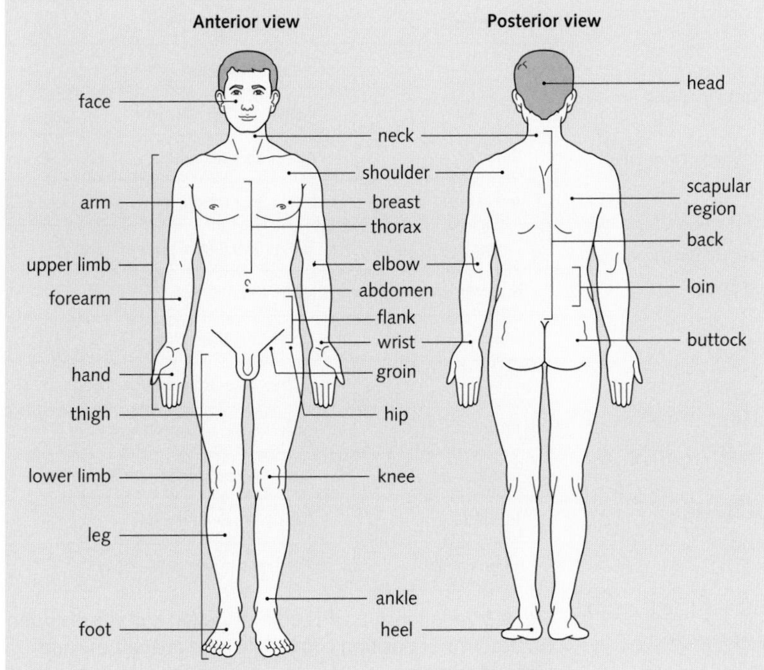

Fig. 1.1 Anatomical position and regions of the body.

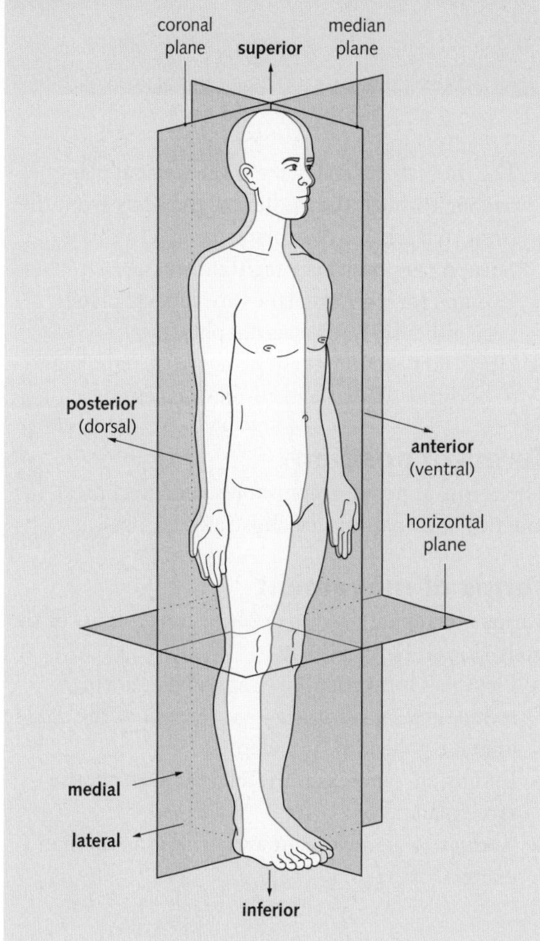

Fig. 1.2 Anatomical planes.

- Adduction—movement towards the median plane.
- Supination—movement, e.g. lateral rotation of the forearm, causing the palm to face anteriorly.
- Pronation—movement, e.g. medial rotation of the forearm, causing the palm to face posteriorly.
- Eversion—turning the sole of the foot outwards.
- Inversion—turning the sole of the foot inwards.
- Rotation—movement of part of the body around its long axis.
- Circumduction—a combination of flexion, extension, abduction, and adduction.

The terms used to describe movements of the thumb are perpendicular to the movements of the body (Fig. 1.5).

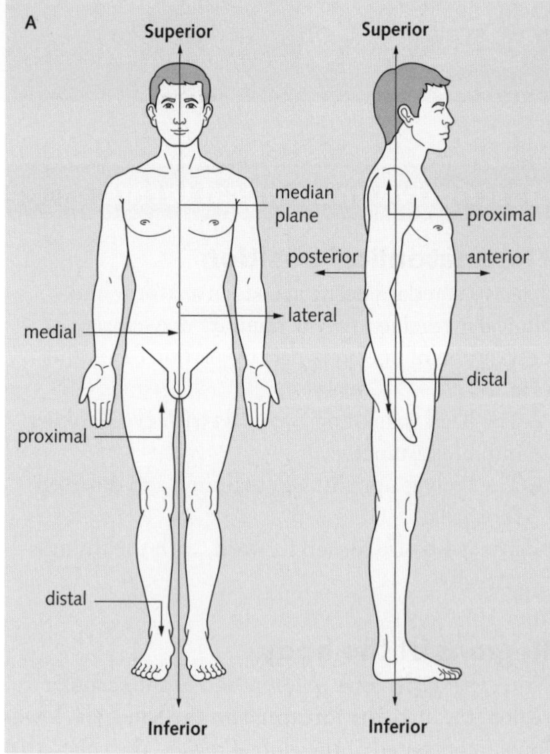

B	Classification of terms commonly used in anatomy and clinical practice
Position	**Description**
Anterior	In front of another structure
Posterior	Behind another structure
Superior	Above another structure
Inferior	Below another structure
Deep	Further away from body surface
Superficial	Closer to body surface
Medial	Closer to median plane
Lateral	Further away from median plane
Proximal	Closer to the trunk or origin
Distal	Further away from the trunk or origin
Ipsilateral	The same side of the body
Contralateral	The opposite side of the body

Fig. 1.3 Relationship and comparison (A) and classification (B) of terms of position commonly used in anatomy and clinical practice.

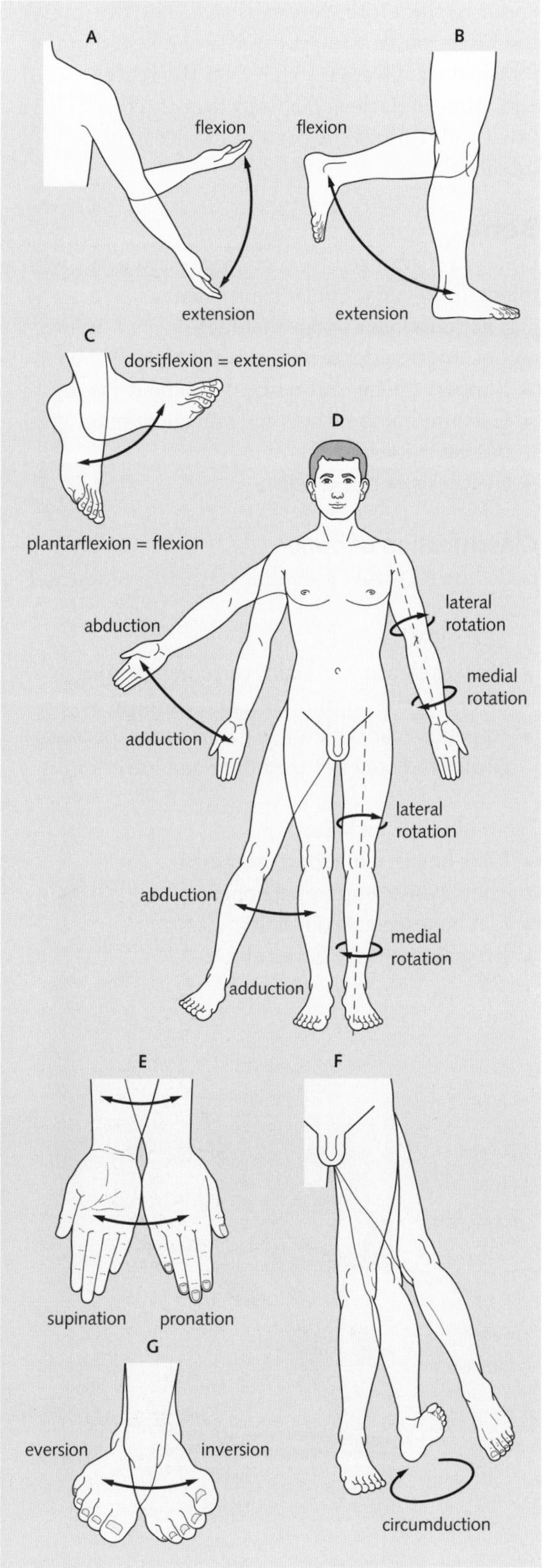

Fig. 1.4 Terms of movement.
(A) Flexion and extension of forearm at elbow joint.
(B) Flexion and extension of leg at knee joint.
(C) Dorsiflexion and plantarflexion of foot at ankle joint.
(D) Abduction and adduction of right limbs and rotation of left limbs at shoulder and hip joints, respectively.
(E) Pronation and supination of forearm at radioulnar joints.
(F) Circumduction (circular movement) of lower limb at hip joint.
(G) Inversion and eversion of foot at subtalar and transverse tarsal joints.

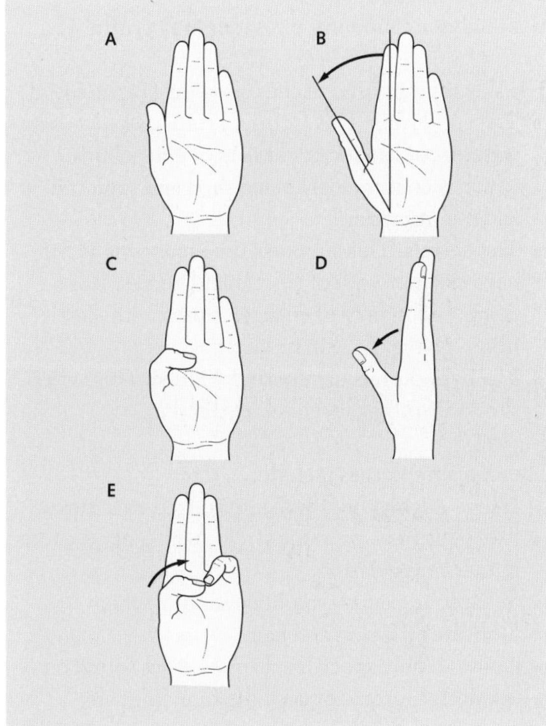

Fig. 1.5 (new) Terms of movement for the thumb. (Adapted from *Crash Course: Musculoskeletal System* by SV Biswas and R Iqbal. Mosby.)
(A) Neutral hand position.
(B) Extension.
(C) Flexion.
(D) Abduction.
(E) Opposition.

To differentiate supination from pronation remember that you hold a bowl of soup with a supinated forearm.

Basic structures of anatomy

Skin

The skin completely covers the body surface and is the largest organ of the body. The functions of the skin include:

- Protection from ultraviolet light and mechanical, chemical, and thermal insults.
- Sensations including pain, temperature, touch, and pressure.
- Thermoregulation.
- Metabolic functions, e.g. vitamin D synthesis.

The skin is composed of the following (Fig. 1.6):

- The epidermis. This forms a protective waterproof barrier. It consists of epithelium, which is continuously being shed and replaced, and it is avascular.
- The dermis. This supports the epidermis and it has a rich network of vessels and nerves. It is composed mainly of collagen fibres with elastic fibres, giving the skin its elasticity.
- The hypodermis or superficial fascia. This acts as a shock-absorbing layer below the skin.

The skin appendages include:

- Hairs—highly modified, keratinized structures.
- Sweat glands—produce sweat, which plays a role in thermoregulation.
- Sebaceous glands—produce sebum, which lubricates the skin and hair.
- Nails—highly specialized appendages found on the dorsal surface of each digit i.e. finger or toe.

Fascia

The fascia of the body may be divided into superficial and deep layers.

The superficial fascia (subcutaneous tissue) consists of loose areolar tissue that unites the dermis to the deep fascia. It contains cutaneous nerves, blood vessels, and lymphatics that travel to the dermis.

In some places sheets of muscle lie in the fascia, e.g. muscles of facial expression.

The deep fascia forms a layer of fibrous tissue around the limbs and body and the deep structures. Intermuscular septa extend from the deep fascia, attach to bone, and divide limb musculature into compartments. The fascia has a rich nerve supply and it is, therefore, very sensitive. The thickness of the fascia varies widely: e.g. it is thickened in the iliotibial tract but very thin over the rectus abdominis muscle and absent over the face. The fascia determines the pattern of spread of infection.

Bone

Bone is a specialized form of connective tissue with a mineralized extracellular component.

The functions of bone include:

- Locomotion (by serving as a rigid lever).
- Support (giving soft tissue permanent shape).
- Calcium homoeostasis and storage of other inorganic ions.
- Synthesis of blood cells.

Classification of bone

Bones are classified according to their position and shape.

The position can be:

- Axial skeleton, e.g. skull, vertebral column including the sacrum, ribs, and sternum.
- Appendicular skeleton, e.g. hip bones, pectoral girdle, and bones of the upper and lower limbs.

Types of shape include:

- Long bones, e.g. femur, humerus.
- Short bones, e.g. carpal bones.
- Flat bones, e.g. skull vault.
- Irregular bones, e.g. vertebrae.

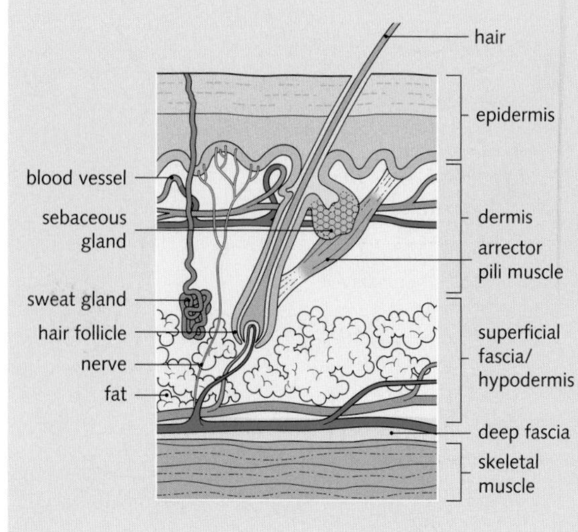

Fig. 1.6 Structure of skin and subcutaneous tissue.

General structure of bone

Bone is surrounded by a connective-tissue membrane called the periosteum (Fig. 1.7). This is continuous with muscle attachments, joint capsules, and the deep fascia. There is an outer fibrous layer and an inner cellular layer. The inner layer is vascular, and it provides the underlying bone with nutrition. It also contains progenitor cells, which may differentiate into osteoblasts when required.

After a fracture, the cells of the periosteum differentiate into osteoblasts and form a cuff of bone around the fracture site. This helps to stabilize the bone while it heals from the inside.

Bone includes the following components:

- The outer compact layer or cortical bone provides great strength and rigidity.
- The cancellous or spongy bone consists of a network of trabeculae arranged to resist external forces.
- The medullary cavity of long bones and the interstices of cancellous bone are filled with red (haematopoietic) or yellow (fatty) marrow. At birth virtually all the bone marrow is red, but this is replaced by yellow marrow—only the ribs, sternum, vertebrae, clavicle, pelvis, and skull bones contain red marrow in adult life.
- The endosteum is a single-cellular osteogenic layer lining the inner surface of bone.

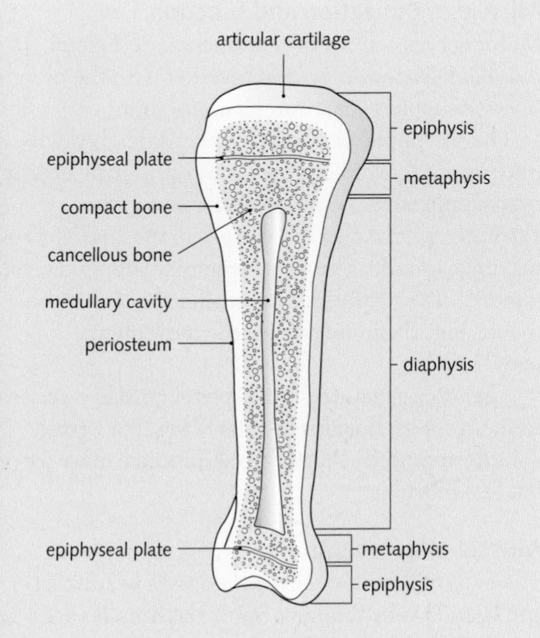

Fig. 1.7 Long bone and its components.

Blood supply of bones

There are two main sources of blood supply to bone:

- A major nutrient artery that supplies the marrow.
- Vessels from the periosteum.

The periosteal supply to bone assumes greater importance in the elderly. Extensive stripping of the periosteum, e.g. during surgery or following trauma, may result in bone death.

Joints

These are unions between bones (Fig. 1.8).

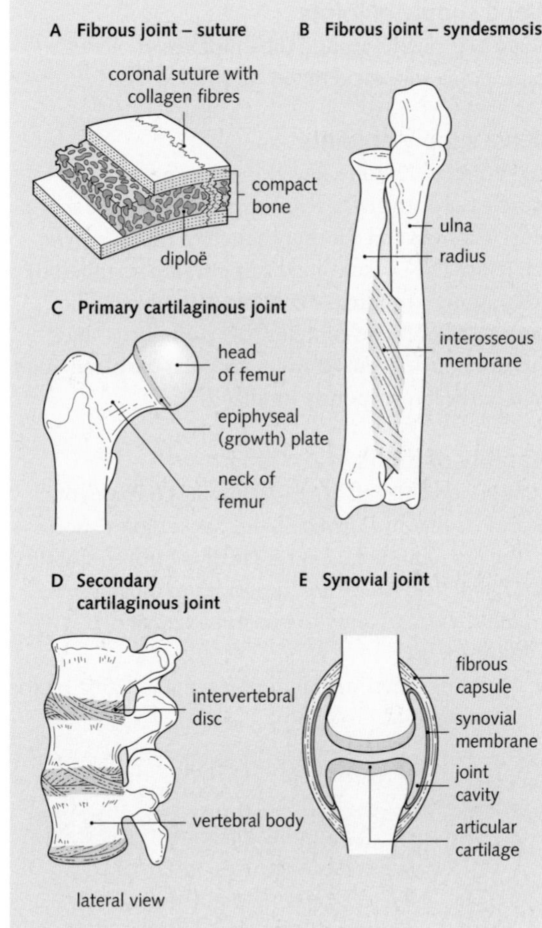

Fig. 1.8 Types of joints.
(A) Fibrous joint—sutural (bones are united by fibrous tissue, as in sutures of the skull).
(B) Fibrous joint—syndesmosis (bones are joined by a sheet of fibrous tissue).
(C) Primary cartilaginous joint (where bone and hyaline cartilage meet).
(D) Secondary cartilaginous joint (articular surfaces are covered by a thin lamina of hyaline cartilage; the hyaline laminae are united by fibrocartilage).
(E) Synovial joint.

Synovial joints

These have the following features:

- The bone ends are covered by hyaline articular cartilage.
- The joint is surrounded by a fibrous capsule.
- A synovial membrane lines the inner aspect of the joint, except where there is cartilage, and it contains synovial fluid, which lubricates the joint and transports nutrients, especially to the cartilage.
- Some synovial joints, e.g. the temporomandibular joints, are divided into two cavities by an articular disc.

Blood supply of joints

A vascular plexus around the epiphysis provides the joint with a very good blood supply.

Nerve supply of joints

According to Hilton's law, the motor nerve to a muscle tends to give a branch to the joint that the muscle moves and another branch to the skin over the joint. The capsule and ligaments are supplied by afferent nerve endings including pain fibres. The synovial membrane contains few pain fibres, and there are no afferent fibres in articular cartilage; joint pain is, therefore, poorly localized.

Stability of joints

Stability is achieved by the following components:

- Bony—e.g. in a firm ball-and-socket joint such as the hip joint, bony contours contribute to stability.
- Ligaments—these are important in most joints, and they act mainly to prevent excessive movement.
- Muscles—these are an important stabilizing factor in most joints.

In general, if a joint is very stable it has a reduced range of movement, e.g. the stable hip joint compared with the less stable shoulder joint; the latter has a greater range of movement.

Muscles and tendons

Skeletal muscles are aggregations of contractile fibres that move large structures such as the skeleton.

Muscles are usually joined to bone by tendons at their origin and insertion.

Muscle action

Muscles can be classified according to their action:

- Prime mover—the muscle is the major muscle responsible for a particular movement, e.g. brachialis is the prime mover in flexing the elbow.
- Antagonist—any muscle that opposes the action of the prime mover: it relaxes, but in a controlled manner, to assist the prime mover, e.g. triceps in flexion of the elbow.
- Fixator—prime mover and antagonist acting together to 'fix' a joint, e.g. muscles holding the scapula steady when deltoid moves the humerus.
- Synergist—prevents unwanted movement in an intermediate joint, e.g. extensors of the carpus contract to fix the wrist joint, allowing the long flexors of the fingers to function effectively.

Muscle design

Muscle fibres may be either parallel or oblique to the line of pull of the whole muscle.

Parallel fibres allow maximal range of mobility. These muscles may be quadrangular, fusiform, or strap. Examples include sartorius and sternocleidomastoid.

Oblique fibres increase the force generated at the expense of reduced mobility. These muscles may be unipennate, bipennate, multipennate, or triangular. Examples include flexor pollicis longus, dorsal interossei, deltoid, and temporalis.

Muscle organization and function

Motor nerves control the contraction of skeletal muscle. Each motor neuron together with the muscle fibres it supplies constitutes a motor unit.

The size of motor units varies considerably: where fine precise movements are required, a single neuron may supply only a few muscle fibres, e.g. the extrinsic eye muscles; conversely, in the large gluteus maximus muscle, a single neuron may supply several hundred muscle fibres. The smaller the size of the motor unit, the more precise the movements possible.

The force generated by a skeletal muscle is related to the cross-sectional area of its fibres. For a fixed volume of muscle, shorter fibres produce more force but less shortening.

Muscle attachments

The ends of muscles are attached to bone, cartilage, and ligaments by tendons. Some flat muscles are attached by a flattened tendon, an aponeurosis or fascia.

When symmetrical muscle fibres unite at an angle, e.g. in mylohyoid muscle, a raphe is formed.

When tendons cross joints, they are often enclosed in a synovial sheath, a layer of connective tissue lined by a synovial membrane and lubricated by synovial fluid.

Bursae are sacs of connective tissue filled with synovial fluid, which lie between tendons and bony areas, acting as cushioning devices.

Nerves

The nervous system is divided into the central nervous system and the peripheral nervous system: the central nervous system is composed of the brain and spinal cord; the peripheral nervous system consists of the cranial and spinal nerves, and their distribution. The nervous system may also be divided into the somatic and autonomic nervous systems.

The conducting cells of the nervous system are termed neurons. A typical motor neuron consists of a cell body, which contains the nucleus and gives off a single axon and numerous dendrites (Fig. 1.9). The cell bodies of most neurons are located within the central nervous system, where they aggregate to form nuclei. Cell bodies in the peripheral nervous system aggregate in ganglia.

Axons are the nerve fibres, and they conduct action potentials generated in the cell body, to influence other neurons or affect organs. They may be myelinated or non-myelinated.

Most nerves in the peripheral nervous system are bundles of motor, sensory, and autonomic axons. The region of the head is largely supplied by the 12 cranial nerves. The remainder of the trunk and the limbs are supplied by the segmental spinal nerves.

Autonomic nerves are either sympathetic or parasympathetic. Sympathetic preganglionic fibres arise from the thoracic and lumbar segments of the spinal cord. They synapse in a sympathetic chain ganglion, from which a postganglionic fibre can either enter a spinal nerve or innervate viscera. Some preganglionic fibres pass through the chain and synapse in prevertebral ganglia e.g. coeliac ganglion. Parasympathetic preganglionic fibres in cranial and sacral nerves synapse in ganglia associated with organs e.g. a pulmonary ganglion to form postganglionic fibres. These innervate the organ.

Spinal nerves

There are 31 pairs of spinal nerves: 8 cervical, 12 thoracic, 5 lumbar, 5 sacral, and the coccygeal nerve. The spinal cord ends at the lower border of the first lumbar vertebra in the adult. Below this, the nerve roots of the cord form a vertical bundle: the cauda equina.

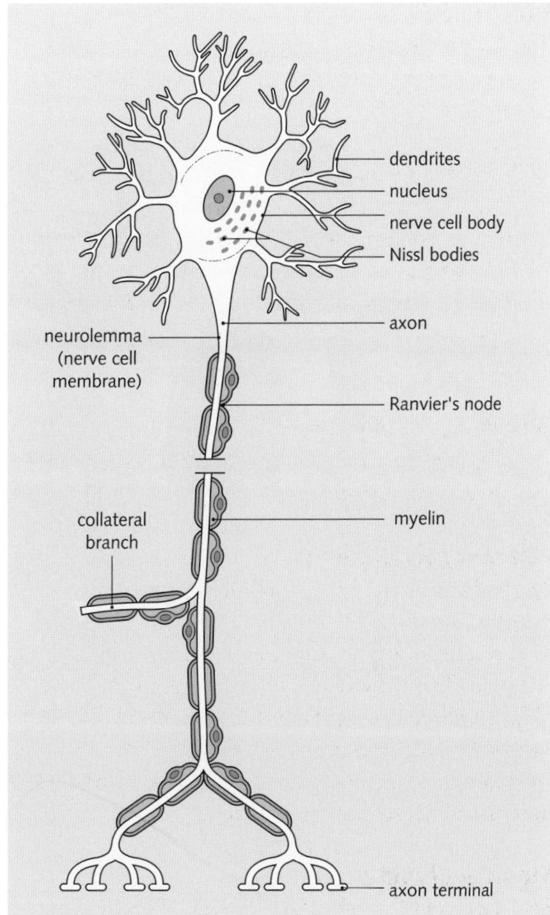

Fig. 1.9 Structure of a typical neuron.

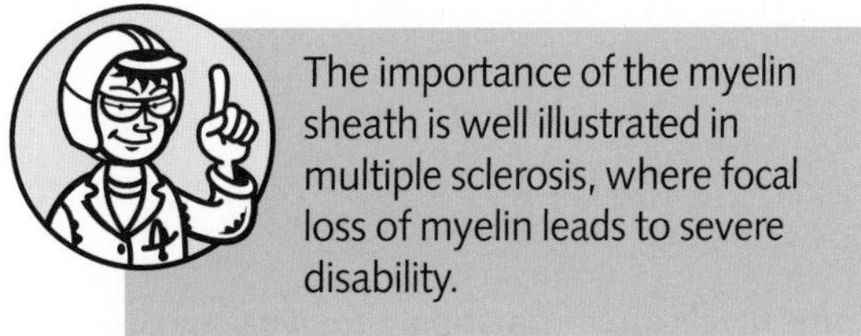

Each spinal nerve is formed by the union of the anterior and posterior roots (Fig. 1.10):

- The anterior root contains motor fibres for skeletal muscles. Those from T1 to L2 also contain sympathetic fibres; S2 to S4 also contain parasympathetic fibres.
- The posterior root contains sensory fibres whose cell bodies are in the posterior root ganglion.

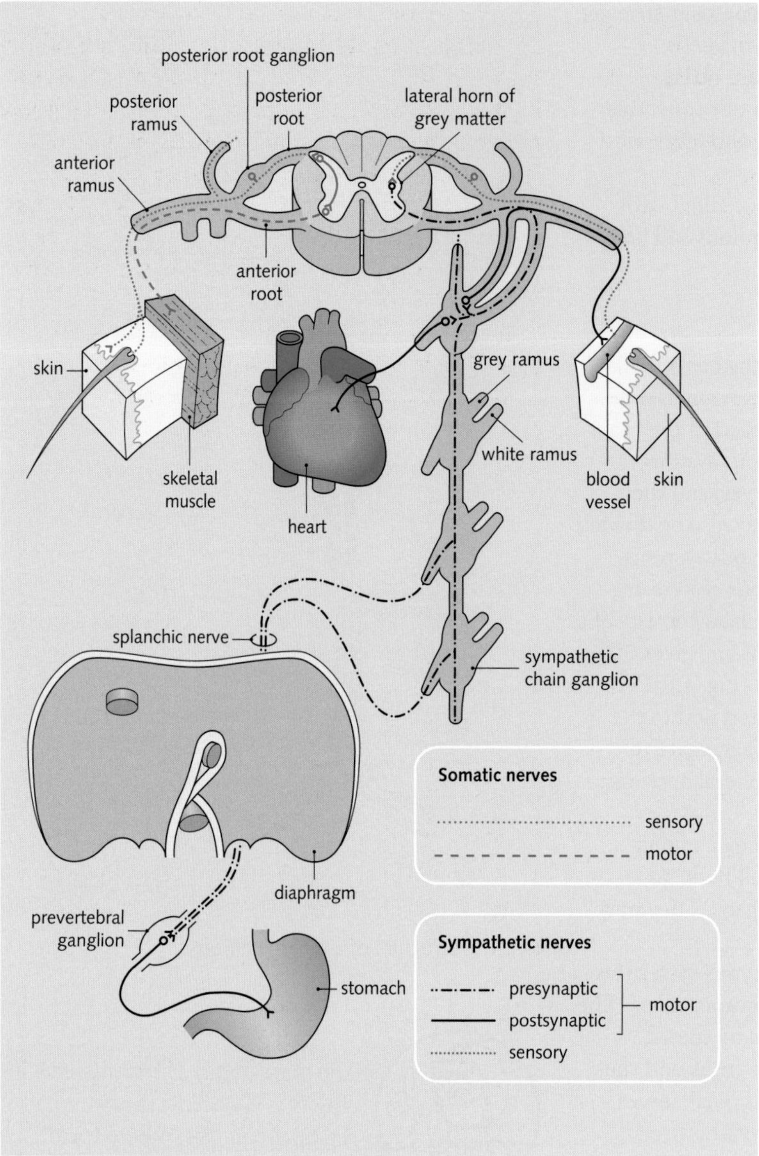

Fig. 1.10 Components of a typical spinal nerve.

Immediately after formation, the spinal nerve divides into anterior and posterior rami. The great nerve plexuses, e.g. the brachial, lumbar, and sacral, are formed by anterior rami.

Cardiovascular system

The cardiovascular system functions principally to transport oxygen and nutrients to the tissues and carbon dioxide and other metabolic waste products away from the tissues.

The right side of the heart pumps blood to the lungs via the pulmonary circulation. The left side of the heart pumps oxygenated blood through the aorta to the rest of the body via the systemic circulation.

Blood is distributed to the organs via the arteries and then arterioles, which branch to form capillaries where gaseous exchange occurs. Blood is eventually returned to the heart via the veins. Valves in the low-pressure venous system are required to prevent back-flow of blood. Loss of competence of the lower-limb valves results in varicose veins. However some veins have no true valves e.g. venae cavae, vertebral, pelvic, head, and neck veins.

Anastomosis

This is a communication between two vessels. Normally little flow of blood occurs through anastomoses; however, if an artery is occluded, the

anastomoses assume greater importance in helping to maintain the circulation to an organ. If an artery is slowly occluded, new vessels develop (collaterals), forming an alternative pathway.

When such communications are absent between arteries, the vessel is known as an end artery. Occlusion in these vessels leads to necrosis, e.g. in the central artery of the retina.

Lymphatics

Fig. 1.11 illustrates the lymphatic system in man.

Fluid moves into the tissues at the arterial end of the circulation and most is returned at the venous end. Excess fluid is drained into the lymphatic system as lymph. The lymphatics on the right side of the head, neck, upper limb and thorax drain into the right lymphatic duct. The rest of the body drains into the thoracic duct. Lymph is ultimately returned to the venous system and the right side of the heart.

Lymph also transports foreign materials to lymph nodes, initiating an immune response. Absorption of fats in the gastrointestinal tract is into the lymphatic system.

Lymphatics are found in all tissues except the central nervous system, eyeball, internal ear, cartilage, bone, and epidermis of the skin.

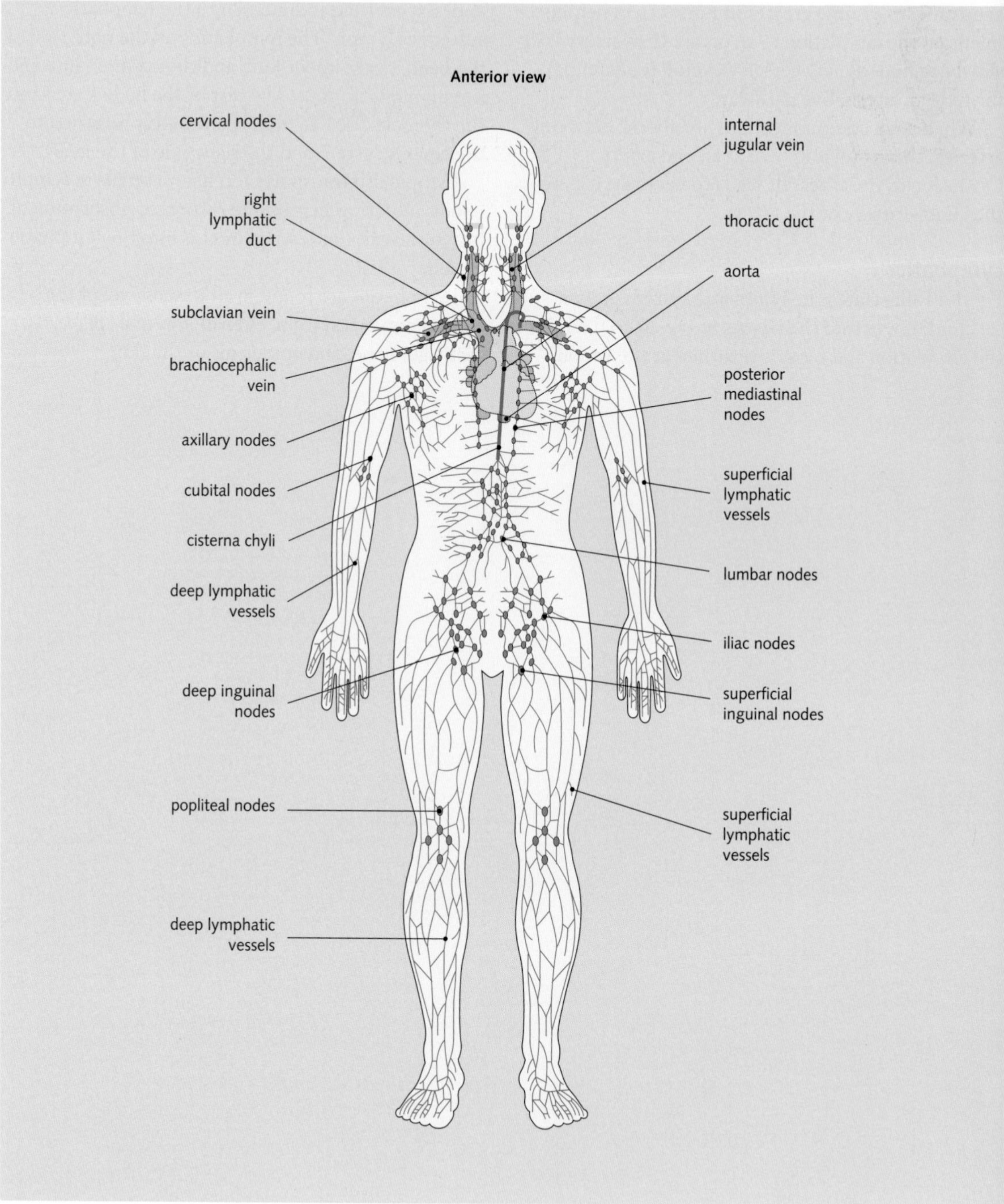

Anterior view

cervical nodes
right lymphatic duct
subclavian vein
brachiocephalic vein
axillary nodes
cubital nodes
cisterna chyli
deep lymphatic vessels
deep inguinal nodes
popliteal nodes
deep lymphatic vessels
internal jugular vein
thoracic duct
aorta
posterior mediastinal nodes
superficial lymphatic vessels
lumbar nodes
iliac nodes
superficial inguinal nodes
superficial lymphatic vessels

Fig. 1.11 The lymphatic system.

- Describe the anatomical position.
- What are the anatomical planes?
- Define the anatomical terms used in anatomy and clinical practice.
- Describe the terms of movement including the thumb.
- Discuss the structure of bone.
- What factors contribute to joint stability?
- What are the classifications of muscle according to their action?
- Describe the structure of muscle.
- Outline the components of a spinal nerve.
- Outline the layout of the cardiovascular and lymphatic systems.

2. The Upper Limb

Regions and components of the upper limb

The upper limb is joined to the trunk by the shoulder or pectoral girdle. The shoulder region is the area around the shoulder joint and girdle. The arm lies between the shoulder and the elbow. The forearm lies between the elbow and the wrist. The hand joins the forearm at the wrist.

The shoulder girdle is composed of the scapula and clavicle, which articulate at the acromioclavicular joint. The sternoclavicular joint is the only joint between the shoulder girdle and the axial skeleton, and connects them together. All the remaining attachments to the axial skeleton are muscular. The humerus lies in the arm, and it articulates with the scapula and with the ulna and radius. The radius articulates with the hand at the wrist joint.

The subclavian artery is the major arterial supply of the upper limb. It arises from the brachiocephalic trunk on the right side and directly from the aorta on the left side. It continues as the axillary artery and then as the brachial artery, which divides into the radial and ulnar arteries to supply the forearm and hand.

Blood is returned to the axillary vein, which becomes the subclavian vein. Superficial veins drain into the axillary vein.

The nerve supply to the upper limb is derived from the brachial plexus: the median, musculocutaneous, and ulnar nerves supply the anterior compartments; the posterior compartments are supplied by the radial nerve.

Surface anatomy and superficial structures

Surface anatomy

Surface anatomy of the shoulder region is shown in Fig. 2.1.

Scapula

The tip of the coracoid process can be felt on deep palpation in the lateral part of the deltopectoral triangle—a small depression situated below the outer third of the clavicle, bounded by the pectoralis major and deltoid muscles. The deltoid muscle forms the smooth round curve of the shoulder.

The acromion process is easily located in its subcutaneous position.

The crest of the scapula may be palpated and followed to its medial border. The inferior angle of the scapula can be palpated opposite the T7 vertebral spine.

Axilla and axillary folds

The anterior and posterior axillary folds may be palpated. The head of the humerus can be palpated through the floor of the axilla.

Elbow region

The medial and lateral epicondyles of the humerus and the olecranon process of the ulna can be palpated. The head of the radius can be palpated in a depression on the posterior aspect of the extended elbow, distal to the lateral epicondyle. The cubital fossa lies anterior to the elbow joint. The biceps brachii tendon is palpable as it enters the fossa.

The brachial artery may be palpated as it passes down the medial aspect of the arm.

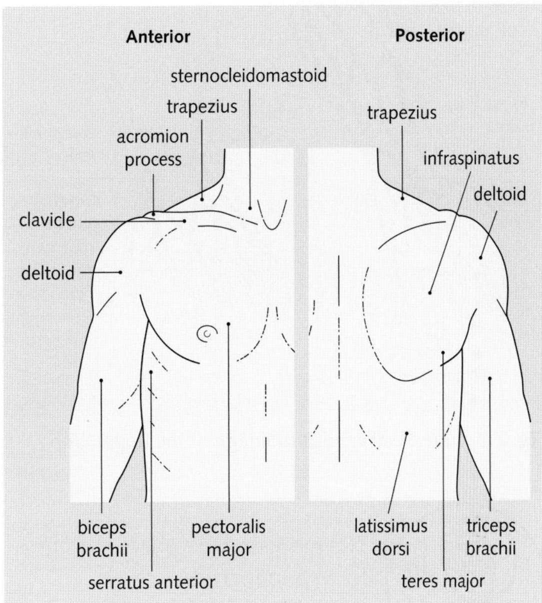

Fig. 2.1 Surface anatomy of the anterior and posterior views of the shoulder region.

The styloid processes of the radius and ulna may be palpated at the wrist.

Superficial venous drainage

The dorsal and the palmar veins drain into the dorsal venous network (Fig. 2.2). From this the medial basilic and lateral cephalic veins arise, and they ascend in the forearm to the arm. The basilic vein passes to halfway up the arm, pierces the deep fascia, and drains into the brachial vein. The cephalic vein passes onto the anterolateral aspect of the forearm, and it communicates with the basilic vein in the median cubital vein—this last vein is usually easy to identify and it is frequently used for venepuncture. The cephalic vein continues laterally up the arm to the deltopectoral groove and then to the infraclavicular fossa, where it drains into the axillary vein.

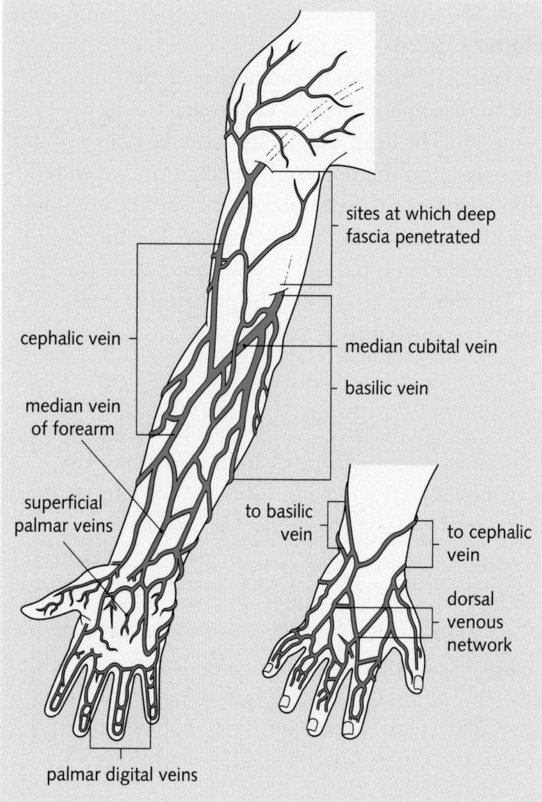

Fig. 2.2 Superficial venous drainage of the upper limb.

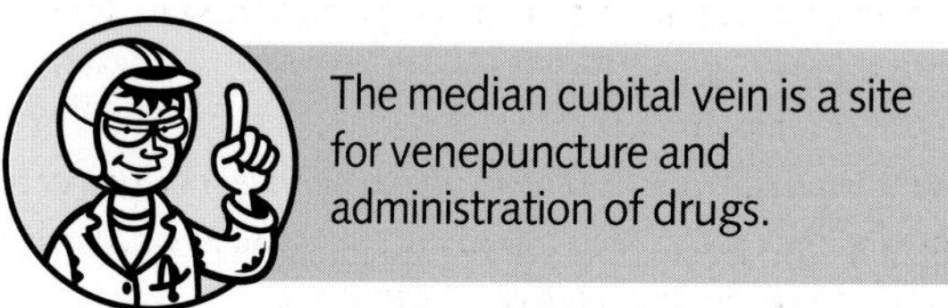

Lymphatic drainage

Lymphatics in the hand coalesce to form trunks that ascend the forearm and the arm with the cephalic and basilic veins and the deep veins. Vessels accompanying the cephalic vein drain into the infraclavicular nodes or the axillary nodes. Some vessels along the basilic vein are interrupted at the elbow by a supratrochlear node, but ultimately they all drain into the axillary nodes. Superficial vessels from the shoulder region drain into the axillary nodes.

Cutaneous innervation of the upper limb

Figs 2.3 and 2.4 illustrate the dermatomes and cutaneous innervation of the upper limb, respectively. This knowledge aids a differential diagnosis of determining the level of a spinal nerve injury or lesion from peripheral nerve damage and sensory loss.

The shoulder region and axilla

Pectoral girdle

The pectoral girdle (clavicle and scapula) suspends the upper limb whilst the clavicle holds the upper limb away from the trunk (Fig. 2.5). The girdle itself is suspended from the head and neck by the trapezius muscle. It articulates directly with the axial skeleton

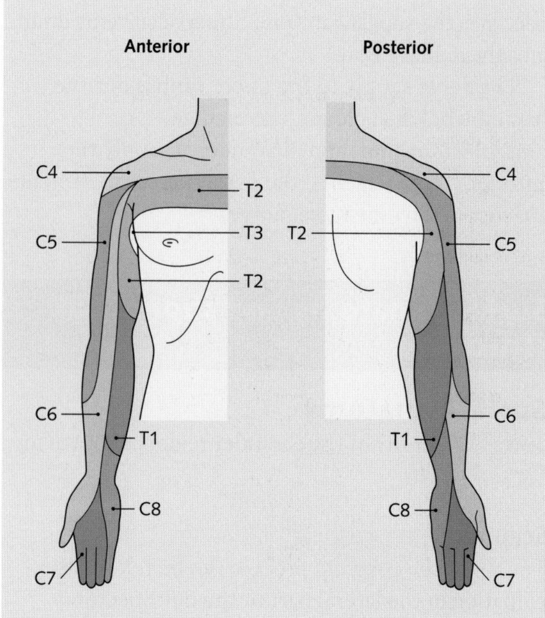

Fig. 2.3 Dermatomes of the upper limb.

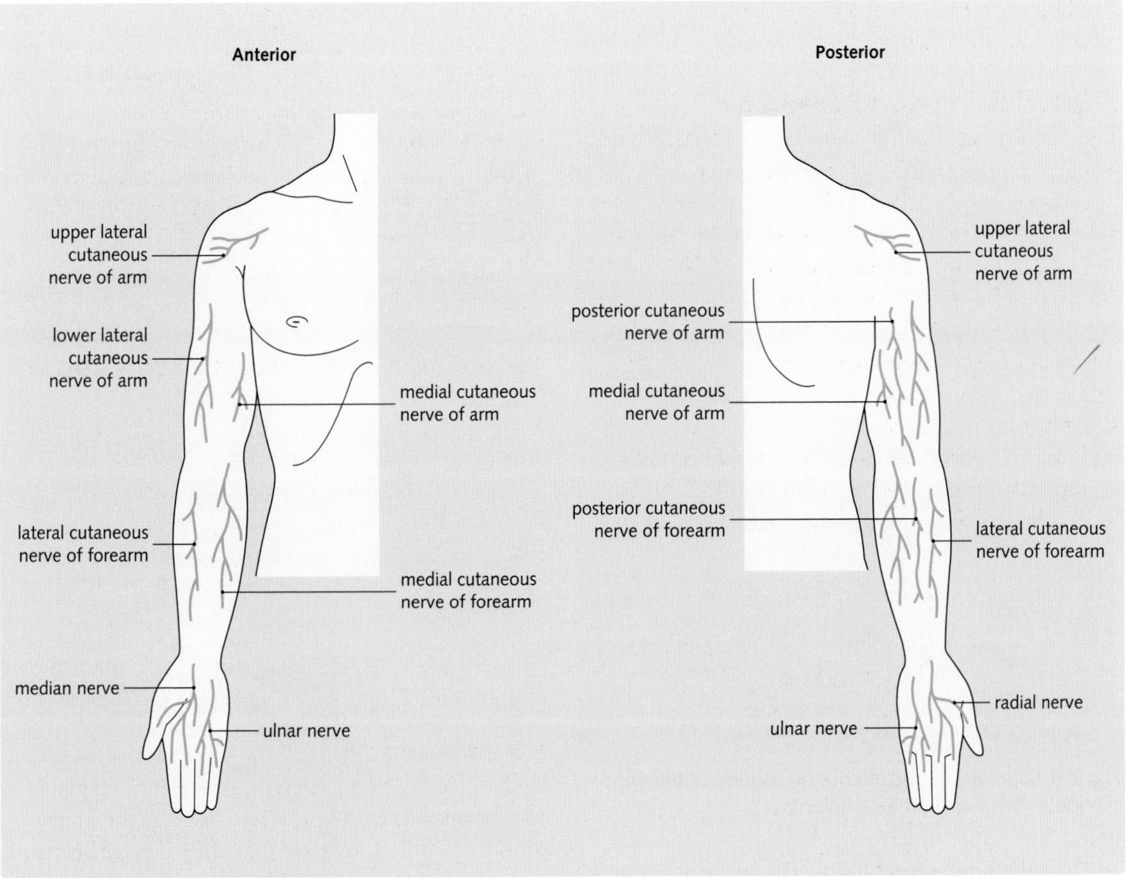

Fig. 2.4 Cutaneous innervation of the upper limb.

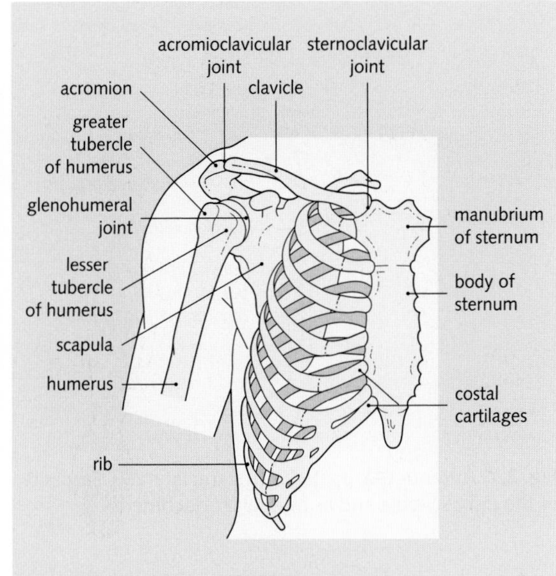

Fig. 2.5 Skeleton of the pectoral girdle.

only via the sternoclavicular joint; the remaining attachments are muscular. This partly accounts for the great mobility of the shoulder girdle.

Clavicle

The clavicle is subcutaneous, and it articulates with the sternum medially and with the acromion process of the scapula laterally (Fig. 2.6). Fracture of the clavicle is common, and it usually occurs at the junction of the outer and middle thirds.

Scapula

The scapula is a triangular flat bone lying on the posterior thoracic wall. It has superior, medial, and lateral borders, and superior and inferior angles (Fig. 2.7). The glenoid cavity articulates with the head of the humerus. The coracoid process projects upwards and forwards above the glenoid cavity, and it provides attachment for muscles and ligaments.

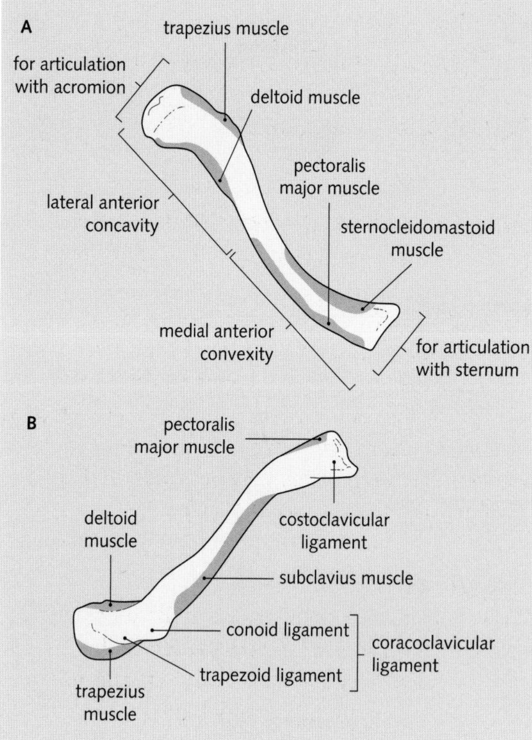

Fig. 2.6 Superior (A) and inferior (B) aspects of the right clavicle and its muscular attachments.

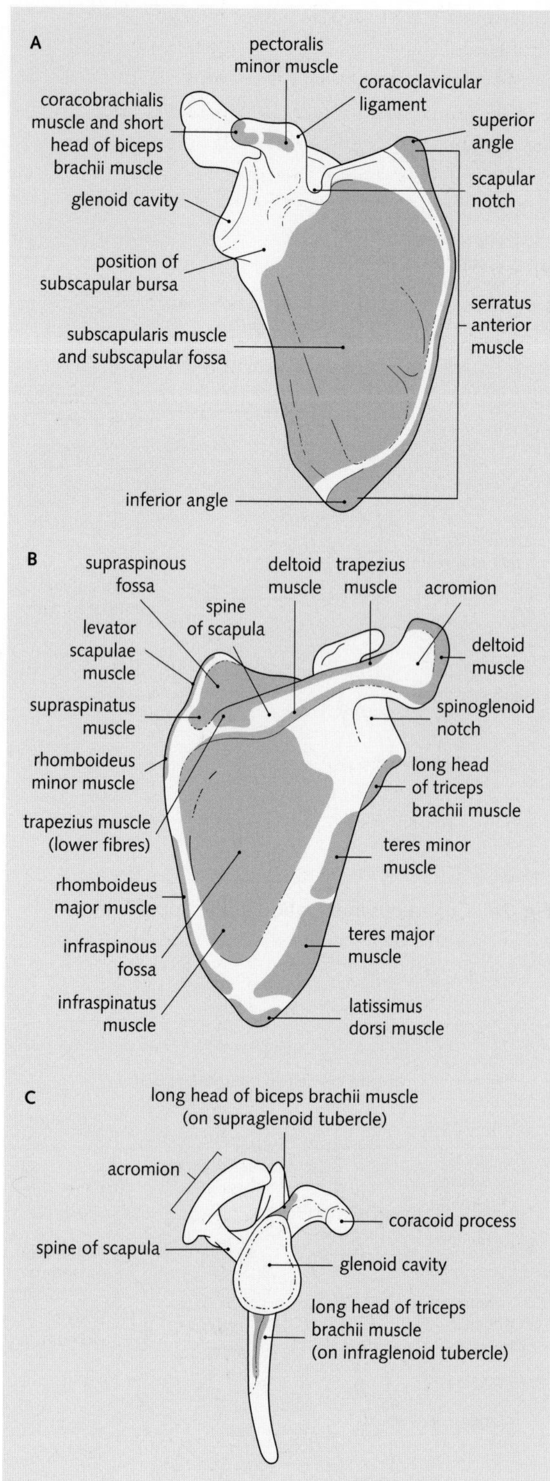

Fig. 2.7 Anterior (A), posterior (B), and lateral (C) aspects of the right scapula and its muscular attachments.

The subscapular fossa lies on the anterior surface; the supraspinous and infraspinous fossae are on the posterior surface, divided by the spine of the scapula, which expands laterally as the acromion.

Joints of the pectoral girdle

Sternoclavicular joint

This is an atypical synovial joint because the articular surfaces are covered by fibrocartilage, not hyaline cartilage. A capsule surrounds the joint and is reinforced by anterior and posterior sternoclavicular ligaments. The costoclavicular ligament also stabilizes the joint. An articular disc is attached to the capsule, dividing the joint into two cavities.

As the lateral end of the clavicle moves, its medial end moves in the opposite direction, moving around the costoclavicular ligament.

The joint is supplied by the medial supraclavicular nerve (C3–C4) from the cervical plexus.

Acromioclavicular joint

This is where the lateral end of the clavicle articulates with the medial border of the acromion. It is an atypical synovial joint, the articular surfaces being fibrocartilage.

A weak capsule surrounds the articular surfaces. It is reinforced by the acromioclavicular ligament

superiorly. The coracoclavicular ligament is very strong, and this is the major factor in joint stability.

Movements are passive as no muscle connects the bones to move the joint. Scapular movements involve movement at both ends of the clavicle.

The joint is supplied by the lateral supraclavicular nerve (C3–C4).

Humerus

The upper end of the humerus is shown in Fig. 2.8. The head articulates with the glenoid cavity of the scapula. The surgical neck is where fractures occur. The spiral groove of the humerus is related to the radial nerve.

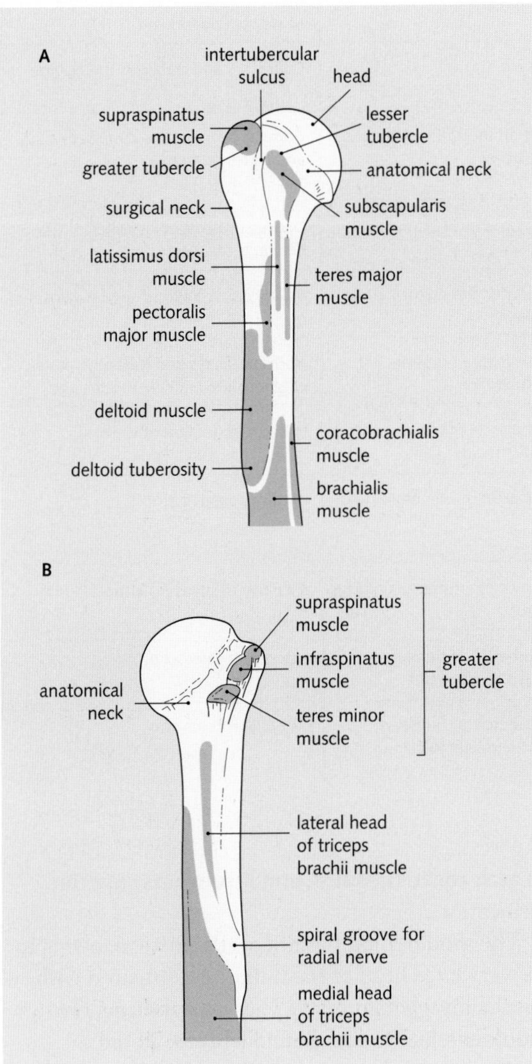

Fig. 2.8 (A) Anterior and (B) posterior views of the upper end of the humerus and its muscle attachments.

The surgical neck of the humerus is so called because it is a common site for fractures of the humerus.

Muscles of the upper limb

Fig. 2.9 outlines the major muscles of the upper limb.

Rotator cuff

The rotator cuff consists of the subscapularis, supraspinatus, infraspinatus, and teres minor. The tendons of these muscles surround the shoulder joint on all sides except inferiorly, and they blend with the capsule. They help to keep the large humeral head applied to the shallow glenoid cavity.

Clavipectoral fascia

This is a strong sheet of connective tissue that encloses the subclavius and is attached to the clavicle. Below, it splits to enclose pectoralis minor and continues as the suspensory ligament of the axilla.

The following structures pass through the clavipectoral fascia:

- Cephalic vein.
- Thoracoacromial artery.
- Lymphatic vessels from the infraclavicular nodes.
- Lateral pectoral nerve.

Quadrangular and triangular spaces

A number of spaces are formed by the muscles and bones in the axillary region (Fig. 2.10).

Shoulder joint

At the shoulder joint, there is articulation between the glenoid cavity of the scapula and the head of the humerus (Fig. 2.11). It is a multiaxial ball-and-socket synovial joint. A rim of fibrocartilage is attached to the margins of the glenoid cavity; it is called the glenoid labrum.

The capsule surrounds the joint, which is attached to the margins of the glenoid labrum and to the humerus around the anatomical neck. A gap in the anterior part of the capsule allows communication between the synovial cavity and subscapular bursa. The capsule is strong but loose, allowing great mobility. It is strengthened by the tendons of the rotator cuff. The long tendon of biceps brachii lies

Major muscles of the upper limb			
Name of muscle (nerve supply)	**Origin**	**Insertion**	**Action**
Latissimus dorsi (thoracodorsal nerve)	Iliac crest, lumbar fascia, spinal processes of lower six thoracic vertebrae, lower ribs, scapula	Floor of intertubercular sulcus of humerus	Extends, adducts, and medially rotates arm
Levator scapulae (C3 and C4 and dorsal scapular nerve)	Transverse processes of C1–C4	Medial border of scapula	Elevates scapula
Rhomboideus minor (dorsal scapular nerve)	Ligamentum nuchae, spines of C7 and T1	Medial border of scapula	Elevates and retracts medial border of scapula
Rhomboideus major (dorsal scapular nerve)	spines of T2–T5	Medial border of scapula	Elevates and retracts medial border of scapula
Trapezius (spinal part of XI nerve and C2 and C3)	Occipital bone, ligamentum nuchae, spinal processes of thoracic vertebrae	Lateral third of clavicle, acromion, spine of scapula	Elevates scapula, pulls scapula medially and pulls medial border of scapula downward
Subclavius (nerve to subclavius)	First costal cartilage	Clavicle	Depresses and stabilizes the clavicle
Pectoralis major (medial and lateral pectoral nerves)	Clavicle, sternum, upper six costal cartilages	Lateral lip of intertubercular sulcus of humerus	Adducts arm, rotates it medially, and flexes humerus
Pectoralis minor (medial pectoral nerve)	Third, fourth, and fifth ribs	Coracoid process of scapula	Depresses point of shoulder protracts shoulder
Serratus anterior (long thoracic nerve)	Upper eight ribs	Medial border and inferior angle of scapula	Pulls scapula forwards and rotates it
Deltoid (axillary nerve)	Clavicle, acromion, spine of scapula	Lateral surface of humerus (deltoid tubercle)	Abducts, flexes and medially rotates, extends, and laterally rotates arm
Supraspinatus (suprascapular nerve)	Supraspinous fossa of scapula	Greater tubercle of humerus, capsule of shoulder joint	Initiates abduction of the arm
Subscapularis (upper and lower subscapular nerves)	Subscapular fossa	Lesser tubercle of humerus	Medially rotates arm
Teres major (lower subscapular nerve)	Lateral border of scapula	Medial lip of intertubercular sulcus of humerus	Medially rotates and adducts arm
Teres minor (axillary nerve)	Lateral border of scapula	Greater tubercle of humerus, capsule of shoulder joint	Laterally rotates arm
Infraspinatus (suprascapular nerve)	Infraspinous fossa of scapula	Greater tubercle of humerus, capsule of shoulder joint	Laterally rotates arm

Fig. 2.9 Major muscles of the upper limb.

intracapsular and extrasynovial. The synovial membrane lines the capsule and part of the neck of the humerus. It communicates with the subscapular bursa and invests the long head of biceps brachii in a tubular sleeve.

The glenohumeral ligaments are three thickenings that slightly strengthen the capsule. The capsule is also reinforced by the strong coracohumeral ligament. The coracoacromial ligament forms an arch above the joint, and it prevents superior dislocation.

The shoulder joint is inherently unstable owing to the very large head of the humerus compared with the shallow glenoid cavity. Factors stabilizing the shoulder joint are the glenoid labrum, all the ligaments, and the muscles supporting the joint. The nerve supply to the joint is from the lateral pectoral nerve, the suprascapular nerve, and the axillary nerve.

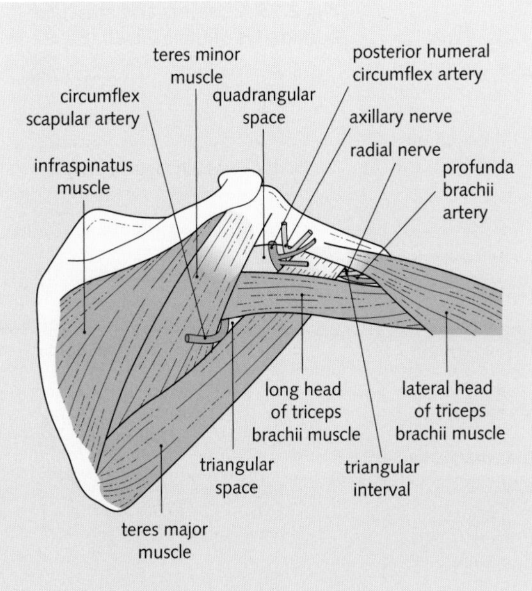

Fig. 2.10 Quadrangular and triangular spaces.

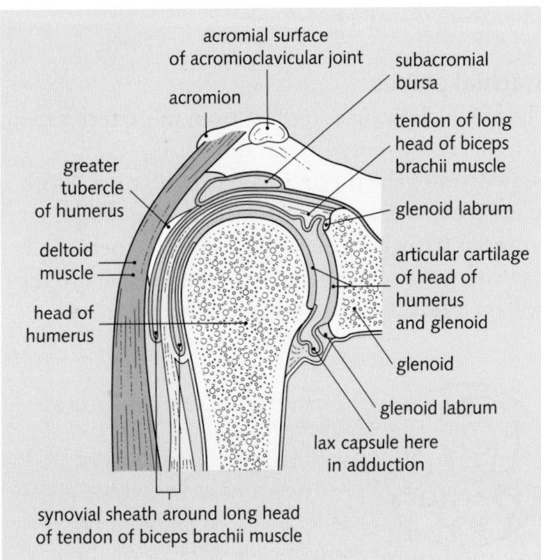

Fig. 2.11 Shoulder joint and related structures.

The movements at the shoulder joint, and the muscles performing them, are described in Fig. 2.12. The movement of abduction deserves special mention: a maximum of 120 degrees of abduction is possible at the glenohumeral joint. Further movement is obtained by rotating the inferior angle of the scapula laterally and anteriorly, turning the glenoid cavity upwards. This is achieved by serratus anterior and trapezius.

Movements of the shoulder joint and the muscles performing them

Movement	Muscles
Flexion	Pectoralis major, anterior fibres of deltoid
Extension	Posterior fibres of deltoid, latissimus dorsi, teres major
Abduction	Deltoid, supraspinatus
Adduction	Pectoralis major, latissimus dorsi, subscapularis, teres major, infraspinatus
Lateral rotation	Infraspinatus, teres minor, posterior fibres of deltoid
Medial rotation	Pectoralis major, anterior fibres of deltoid, latissimus dorsi, teres major, subscapularis
Circumduction	Varying combinations of flexion, extension, abduction, and adduction muscles

Fig. 2.12 Movements of the shoulder joint and the muscles performing them.

The capsule of the shoulder joint is weak inferiorly and the rotator cuff is also deficient inferiorly. This is where dislocation of the humeral head can occur.

Axilla

The major vessels of the upper limb leave the neck to enter the apex of the axilla. The boundaries of the axilla are:

- Anteriorly: clavipectoral fascia, pectoralis major, and minor muscles.
- Posteriorly: subscapularis, latissimus dorsi, and teres major muscles.
- Medially: the upper four ribs, intercostal, and serratus anterior muscles.
- Laterally: intertubercular sulcus, coracobrachialis, and biceps brachii muscles.
- The apex communicates with the root of the neck between the clavicle, first rib, and the superior border of the scapular.
- The base is composed of skin and axillary fascia.

The contents of the axilla are shown in Fig. 2.13 and include:

- Axillary artery.
- Axillary vein.

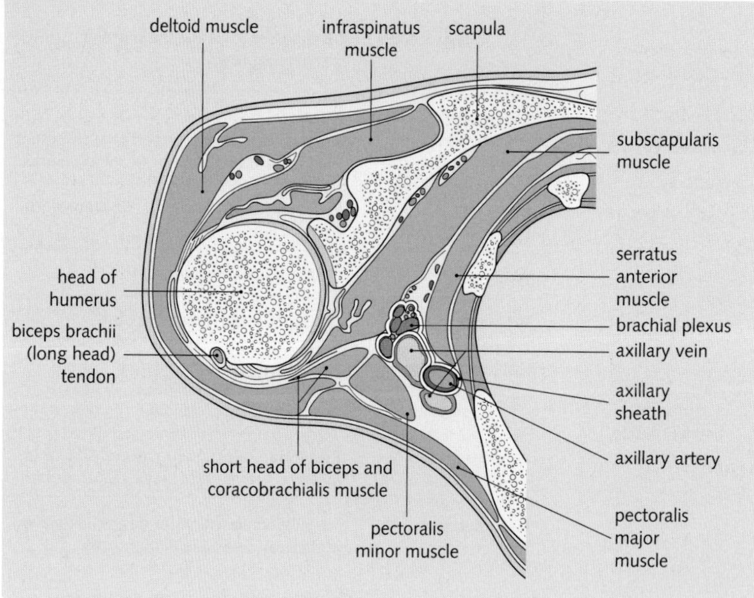

Fig. 2.13 Contents and muscular boundaries of the axilla.

- Brachial plexus.
- Axillary lymph nodes.

Axillary artery

The axillary artery is a continuation of the third part of the subclavian artery, and it commences at the outer border of the first rib. It is invested in fascia (axillary sheath), with the brachial plexus, derived from the prevertebral fascia. The axillary artery becomes the brachial artery at the lower border of teres major. It is divided into three parts by pectoralis minor (Fig. 2.14):

- The first part has one branch—the superior thoracic artery. This supplies both pectoral muscles and the thoracic wall.
- The second part has two branches—the thoracoacromial artery supplies the sternoclavicular joint, pectoral, and deltoid muscles. The lateral thoracic artery supplies serratus anterior, breast, and the axillary nodes.
- The third part has three branches—the subscapular artery supplies latissimus dorsi and forms part of a scapular anastomosis. The anterior and posterior circumflex humeral arteries supply the shoulder joint.

Axillary vein

The axillary vein is a continuation of the venae comitantes of the brachial artery, which are joined by the basilic vein. It commences at the lower border of teres major and ascends through the axilla medial to the axillary artery. At the outer border of the first rib, it becomes the subclavian vein.

Brachial plexus

The brachial plexus is formed from the anterior rami of spinal nerve roots C5–C8 and T1. Fig. 2.15 demonstrates the plexus is divided into roots (deep to the scalene muscles), trunks (found in the posterior triangle of the neck), divisions (behind the clavicle) and cords (named with their respect to the axillary artery. Fig. 2.16 tabulates the branches.

Notice, in Fig. 2.15, that the middle trunk of the brachial plexus as well as the lateral gives rise to an anterior division to form the lateral cord. All three trunks give a posterior division to form the posterior cord.

Axillary lymph nodes

These comprise (Fig. 2.17):

- Lateral group.
- Pectoral group.
- Subscapular group.
- Central group.
- Apical group.

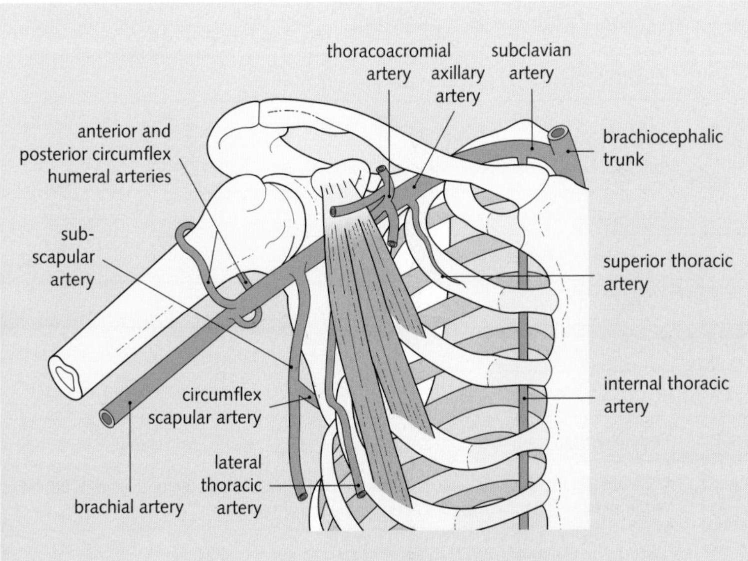

Fig. 2.14 Axillary artery and its branches.

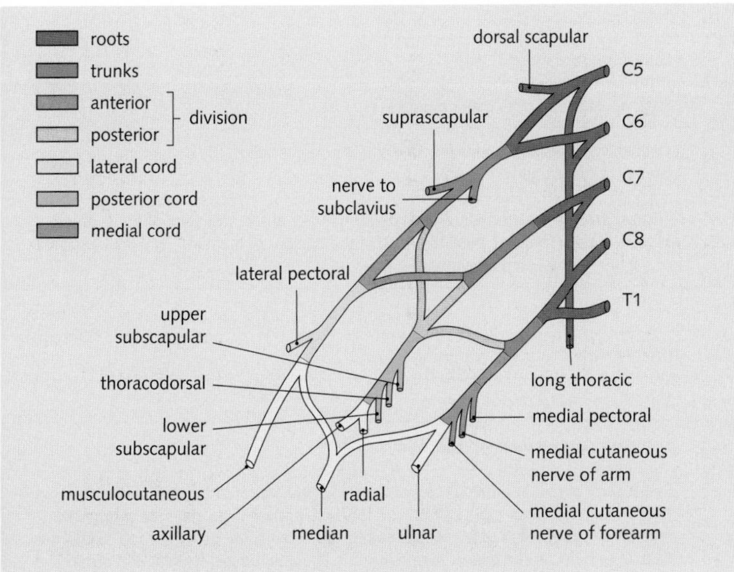

Fig. 2.15 Brachial plexus showing the trunks, divisions, cords, and branches.

The arm

The arm lies between the shoulder and elbow joint. It has anterior and posterior compartments separated by the medial and lateral intermuscular septa. These septa arise from the deep fascia surrounding the arm. The lateral intermuscular septum extends from the lateral lip of the intertubercular sulcus to the lateral epicondyle of the humerus. The medial intermuscular septum extends from the medial lip of the intertubercular sulcus to the medial epicondyle of the humerus.

Flexor compartment of the arm

The bony skeleton of the arm and the muscle attachments of the anterior compartment are shown in Fig. 2.18.

Muscles of the arm

The muscles of the arm are shown in Fig. 2.19.

Vessels of the arm

Brachial artery

The brachial artery is a continuation of the axillary artery, commencing at the lower border of teres

Branches of the brachial plexus and their distribution	
Branches	**Distribution**
Roots	
Dorsal scapular nerve (C5)	Rhomboid major, rhomboid minor, and levator scapulae muscles
Long thoracic nerve (C5–C7)	Serratus anterior muscle
Upper trunk	
Suprascapular nerve (C5, C6)	Supraspinatus and infraspinatus muscles
Nerve to subclavius (C5, C6)	Subclavius muscle
Lateral cord	
Lateral pectoral nerve (C5–C7)	Pectoralis major muscle
Musculocutaneous nerve (C5–C7)	Coracobrachialis, biceps brachii, brachialis muscles, and the skin along the lateral border of the forearm (lateral cutaneous nerve of the forearm)
Lateral root of median nerve (C5–C7)	Joins the medial root (C8, T1) to form the median nerve (see below)
Posterior cord	
Upper subscapular nerve (C5–C6)	Subscapularis muscle
Thoracodorsal nerve (C6–C8)	Latissimus dorsi muscle
Lower subscapular nerve (C5–C6)	Subscapularis and teres major muscles
Axillary nerve (C5–C6)	Deltoid and teres minor muscles. Skin over the lower half of the deltoid muscle (upper lateral cutaneous nerve of arm)
Radial nerve (C5–C8, T1)	Triceps, brachialis, anconeus, and posterior muscles of forearm. Skin of the posterior aspects of arm, forearm, the lateral half of the dorsum of the hand, and dorsal surface of the lateral three and a half digits.
Medial cord	
Medial pectoral nerve (C8, T1)	Pectoralis major and minor muscles
Medial cutaneous nerve of the arm (C8, T1)	Skin of the medial side of the arm
Medial cutaneous nerve of the forearm (C8, T1)	Skin of the medial side of the forearm
Ulnar nerve (C8, T1)	Flexor carpi ulnaris and medial half of flexor digitorum profundus (in forearm). Hypothenar, adductor pollicis, third and fourth lumbrical, interossei, palmaris brevis muscles (in hand). Skin of the medial half of the dorsum and palm of hand, skin of the palmar, and dorsal surfaces of the medial one and a half digits (palmar and dorsal digital cutaneous branches).
Median nerve (C5–C8, T1)	(In forearm) pronator teres, flexor carpi radialis, flexor digitorum superficialis (median nerve). Flexor pollicis longus, lateral half of flexor digitorum profundus and pronator quadratus (anterior interosseous branch). (In hand) thenar muscles, first two lumbricals (median nerve). Skin of lateral half of palm and palmar surface of lateral three and a half digits (palmar and digital cutaneous branches).

Fig. 2.16 Branches of the brachial plexus and their distribution.

major (Fig. 2.20). It terminates at the neck of the radius by dividing into the radial and ulnar arteries. This artery and its profunda brachii branch supply the anterior and posterior compartments of the arm respectively. Collateral branches anastomose with recurrent radial and ulnar branches to form a collateral circulation around the elbow. The artery is very superficial throughout its course, being covered by skin and fascia only. It lies just behind the medial border of biceps brachii.

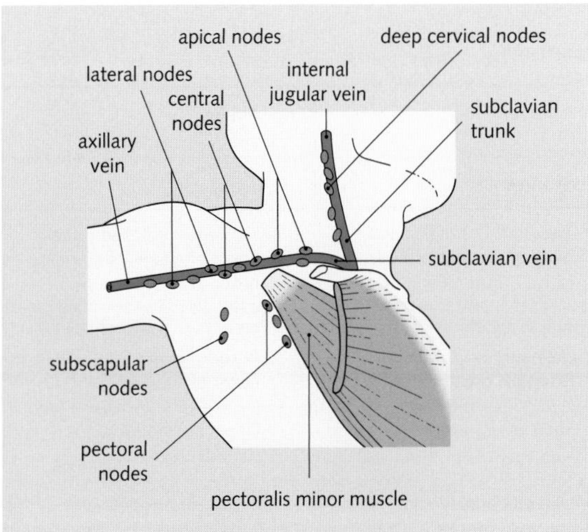

Fig. 2.17 Arrangement of axillary lymph nodes.

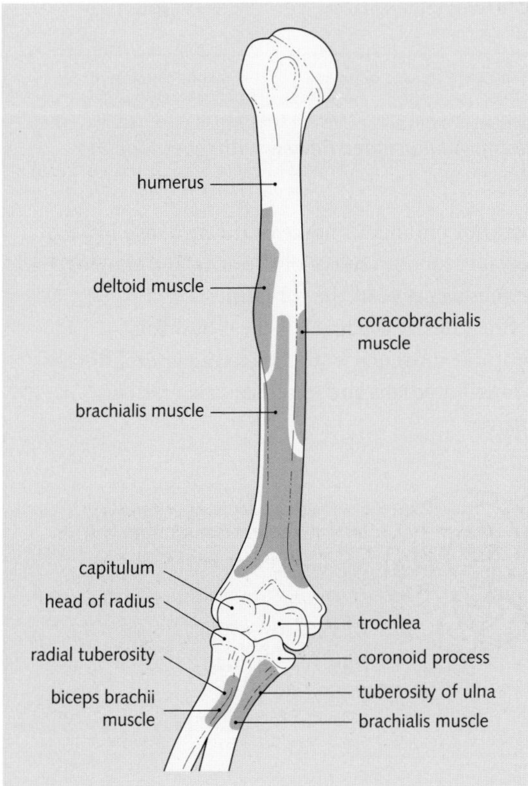

Fig. 2.18 Skeleton of the arm, showing sites of muscle attachment.

Brachial veins

Usually a pair of venae comitantes accompany the brachial artery. They are joined by tributaries that correspond to branches of the brachial arteries. The veins receive the basilic vein before becoming the axillary vein.

To remember the contents of the anterior arm compartment use the mnemonic B, B, C (biceps brachii, brachialis and coracobrachialis).

The brachial artery is easily palpated on the medial aspect of the arm under biceps brachii.

Nerves of the arm

Median nerve

The median nerve enters the arm on the lateral side of the brachial artery and crosses in front of the artery to be on its medial side. It continues to the elbow in this relationship, and it is crossed by the bicipital aponeurosis. The median nerve has no branches in the arm.

Ulnar nerve

The ulnar nerve passes down the arm medial to the brachial artery, and it pierces the medial intermuscular septum halfway down, accompanied by the superior ulnar collateral artery, to enter the posterior compartment. It continues between the medial intermuscular septum and the medial head of triceps, then on the posterior aspect of the medial epicondyle of the humerus; it enters the forearm between the heads of flexor carpi ulnaris muscle. Pressure on the nerve as it crosses the medial epicondyle results in a tingling sensation—this area is referred to as the 'funny bone'. The ulnar nerve has no branches in the arm.

Radial nerve

The radial nerve enters the posterior compartment of the arm by passing over the lower border of teres major through the triangular interval (see Axilla, p. 21). It runs, with the profunda brachii artery, in the spiral groove of the humerus. The nerve enters the anterior compartment of the arm and then passes into the forearm, deep to the brachioradialis muscle.

Branches in the axilla and arm comprise muscular branches to the triceps and anconeus, and the

Name of muscle (nerve supply)	Origin	Insertion	Action
Muscles of the upper arm			
Anterior fascial compartment			
Biceps brachii—long head (musculocutaneous nerve)	Supraglenoid tubercle of scapula	Tuberosity of radius and bicipital aponeurosis into deep fascia of forearm	Supinator of flexed forearm, flexor of elbow joint, weak flexor of shoulder joint
Biceps brachii—short head (musculocutaneous nerve)	Coracoid process of scapula	Tuberosity of radius and bicipital aponeurosis into deep fascia of forearm	Supinator of flexed forearm, flexor of elbow joint, weak flexor of shoulder joint
Coracobrachialis (musculocutaneous nerve)	Coracoid process of scapula	Shaft of humerus	Flexes and adducts shoulder joint
Brachialis (musculocutaneous nerve and radial nerve)	Front of humerus	Ulnar tuberosity and coronoid process	Flexes elbow joint
Posterior fascial compartment			
Triceps—long head (radial nerve)	Infraglenoid tubercle of scapula	Olecranon process of ulna	Extends elbow joint
Triceps—lateral head (radial nerve)	Posterior surface of humerus (upper part)	Olecranon process of ulna	Extends elbow joint
Triceps—medial head (radial nerve)	Posterior surface of humerus (lower part)	Olecranon process of ulna	Extends elbow joint

Fig. 2.19 Muscles of the upper arm. (Adapted from *Clinical Anatomy, An Illustrated Review with Questions and Explanations*, 2nd edn, by R S Snell. Little Brown & Co)

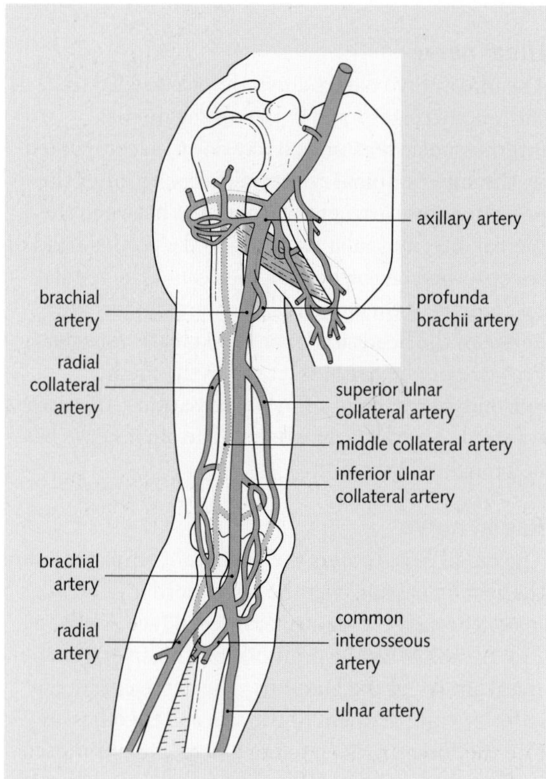

Fig. 2.20 Brachial artery and its branches.

posterior cutaneous nerve of the arm, the lower lateral cutaneous nerve of the arm, and the posterior cutaneous nerve of the forearm.

Branches at the level of the elbow joint comprise muscular branches to the lateral fibres of brachioradialis and extensor carpi radialis longus.

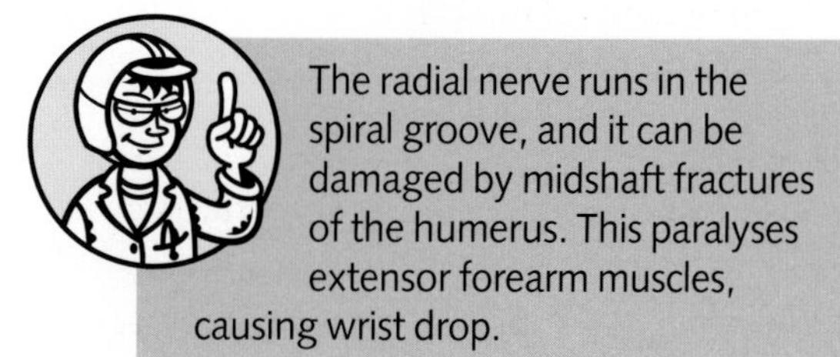

The radial nerve runs in the spiral groove, and it can be damaged by midshaft fractures of the humerus. This paralyses extensor forearm muscles, causing wrist drop.

Musculocutaneous nerve

After its formation, the musculocutaneous nerve passes through coracobrachialis and runs down between biceps and brachialis to reach the lateral aspect of brachialis. It supplies the muscles of the anterior compartment, and it terminates by piercing the deep fascia to become the lateral cutaneous nerve of the forearm.

The cubital fossa and elbow joint

Cubital fossa

The cubital fossa is a triangular region lying anterior to the elbow joint. Its boundaries are:

- Lateral—brachioradialis muscle.
- Medial—pronator teres.
- Base—an imaginary line drawn between the two epicondyles of the humerus.
- Floor—the supinator and brachialis muscles.
- Roof—skin, fascia, and the bicipital aponeurosis.

Fig. 2.21 shows the contents of the cubital fossa. Note:

- The lateral cutaneous nerve of the forearm emerges between biceps and brachialis.
- The radial nerve enters the anterior compartment deep to brachioradialis.
- The brachial artery crosses the tendon of biceps, and it lies deep to the bicipital aponeurosis.
- The median nerve lies medial to the brachial artery.
- The medial cutaneous nerve of the forearm lies superficial to pronator teres.

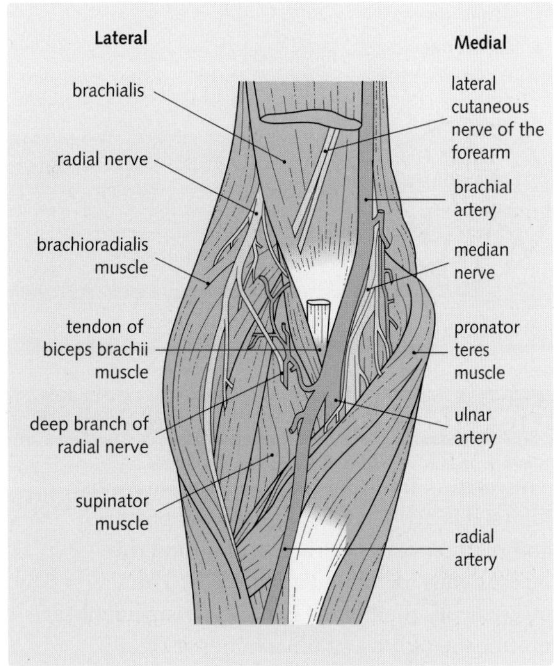

Fig. 2.21 Contents of the cubital fossa.

Elbow joint

The elbow joint is a synovial hinge joint between the lower end of the humerus and the upper end of the radius and the ulna (Fig. 2.22).

Articular surfaces comprise:

- Superior surface of the radius.
- Capitulum of the humerus laterally.
- Trochlear notch of the ulna.
- Trochlea of the humerus medially.

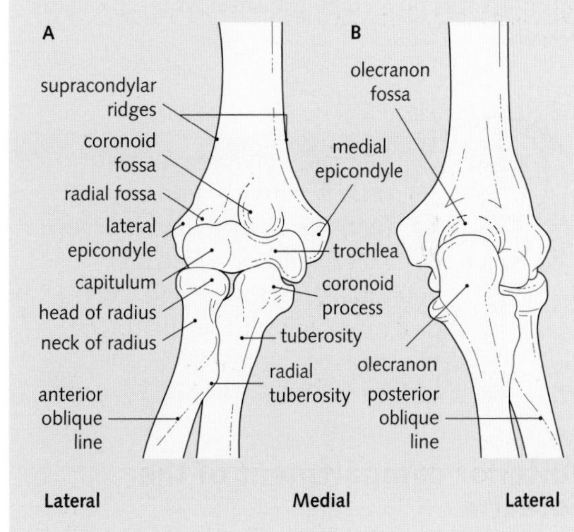

Fig. 2.22 Anterior (A) and posterior (B) aspects of the humerus and the upper end of the ulna and radius.

Movements of the elbow joint are limited to flexion and extension. Independent rotation of the radius is possible at the proximal radioulnar joint in the movements of pronation and supination of the forearm.

The capsule is lax anteroposteriorly, but it is strengthened medially and laterally by collateral ligaments.

Ligaments comprise:

- The ulnar collateral ligament—this triangular ligament consists of three bands, and it runs between the ulnar and humeral bones.
- The radial collateral ligament—this is a band joining the lateral epicondyle of the humerus to the anular ligament.
- The anular ligament—this is attached to the margins of the radial notch of the ulna. It clasps the head and neck of the radius in the proximal radioulnar joint.

The nerve supply of the elbow joint consists of:

- Musculocutaneous nerve.
- Median nerve.

- Ulnar nerve.
- Radial nerve.

Remember that flexion and extension occur at the elbow joint; rotation occurs at the proximal radioulnar joint.

The forearm

The forearm lies between the elbow and wrist joints. It is divided into anterior and posterior compartments by the radius, ulna, and the interosseous membrane (Fig. 2.23). The interosseous membrane is a thin strong membrane uniting the radius and ulnar bones at their interosseous borders. It provides attachments for muscles, and superiorly it is incomplete, allowing the posterior interosseus vessels to pass, whilst inferiorly it is pierced by the anterior interosseous vessels.

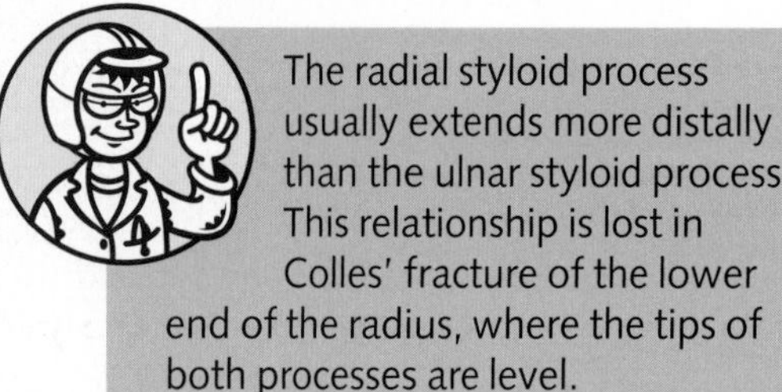

The radial styloid process usually extends more distally than the ulnar styloid process. This relationship is lost in Colles' fracture of the lower end of the radius, where the tips of both processes are level.

Anterior compartment of the forearm

Muscles of the anterior compartment

The muscles of the anterior compartment are divided into superficial and deep groups (Fig. 2.24). The superficial muscles arise from the medial supracondylar ridge and the epicondyle of the humerus (the common flexor origin).

Vessels of the anterior compartment

The brachial artery enters the forearm through the cubital fossa, and it divides into the radial and ulnar arteries (Fig. 2.25).

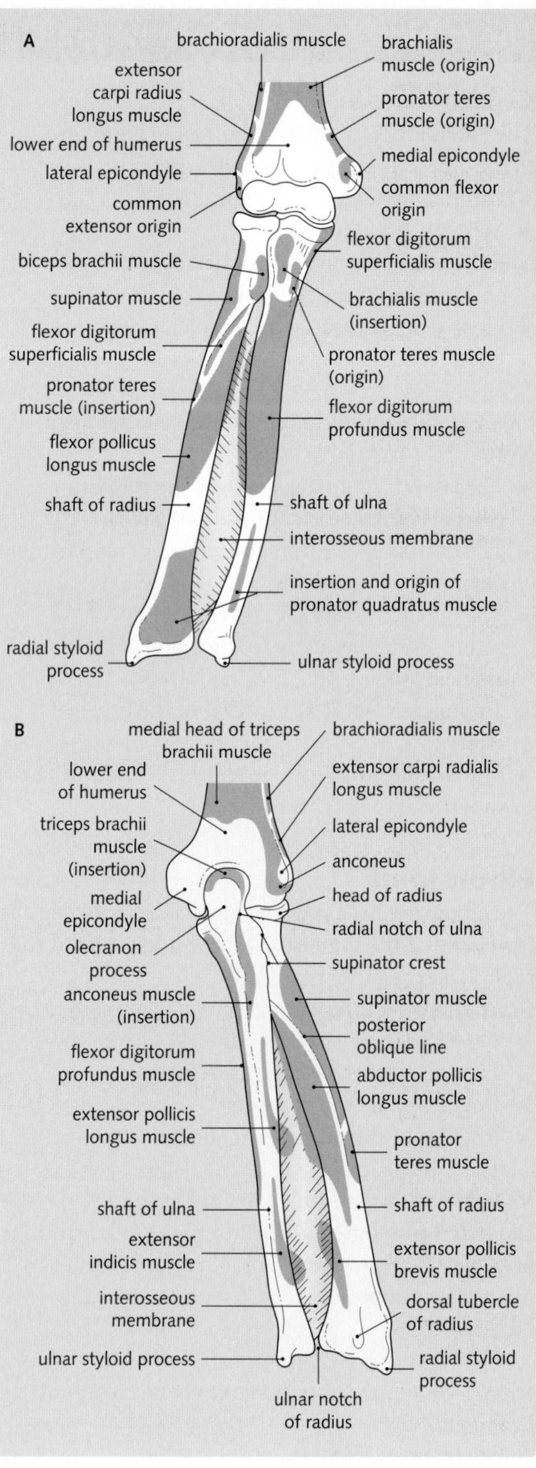

Fig. 2.23 Anterior (A) and posterior (B) aspects of the right radius and ulna, showing sites of muscular attachments.

Nerves of the anterior compartment

The median and ulnar nerves pass through the anterior compartment, and they supply all of the

Muscles of the anterior compartment of the forearm			
Name of muscle (nerve supply)	**Origin**	**Insertion**	**Action**
Superficial			
Flexor carpi radialis (median nerve)	Common flexor origin	Second and third metacarpal bones	Flexion and adduction of wrist joint
Flexor carpi ulnaris—humeral head (ulnar nerve)	Common flexor origin	Pisiform and through pisometacarpal ligament to fifth metacarpal bone	Flexion and adduction of wrist joint
Flexor carpi ulnaris—ulnar head (ulnar nerve)	Olecranon process and posterior border of ulna	Pisiform and through pisometacarpal ligament to fifth metacarpal bone	Flexion and adduction of wrist joint
Flexor digitorum superficialis—humeroulnar head (median nerve)	Common flexor origin and coronoid process of ulna	Middle phalanges of medial four digits	Flexion of PIP and MCP joints of the medial four digits and wrist joint
Flexor digitorum superficialis—radial head (median nerve)	Anterior oblique line of radius	Middle phalanges of medial four digits	Flexion of PIP and MCP joints of the medial four digits and wrist joint
Pronators teres—humeral head (median nerve)	Common flexor origin and medial supracondylar ridge	Lateral aspect of shaft of radius	Pronation of forearm and flexion of elbow
Pronator teres—ulnar head (median nerve)	Coronoid process of ulna	Lateral aspect of shaft of radius	Pronation of forearm
Palmaris longus (median nerve)	Common flexor origin	Palmar aponeurosis	Flexion of wrist joint
Deep			
Pronator quadratus (anterior interosseous nerve)	Anterior surface of ulna	Anterior surface of radius	Pronation of forearm
Flexor pollicis longus (anterior interosseous nerve)	Anterior surface of radius and interosseous membrane	Distal phalanx of thumb	Flexion of interphalangeal and MCP joints
Flexor digitorum profundus (medial half by ulnar nerve and lateral half by anterior interosseous nerve)	Anterior surface of ulna and interosseous membrane	Distal phalanges of medial four digits	Flexion of DIP, PIP, MCP, and wrist joint

Fig. 2.24 Muscles of the anterior compartment of the forearm. (DIP, distal interphalangeal; PIP, proximal interphalangeal; MCP, metacarpophalangeal)

muscles in this compartment. The superficial branch of the radial nerve runs part of its course in this compartment.

Median nerve

This enters the forearm between the heads of pronator teres. It crosses the ulnar artery and then runs deep to flexor digitorum superficialis until just above the wrist, where it appears between the tendon of this muscle and the tendon of flexor carpi radialis.

Branches of the median nerve in the forearm comprise:

- Anterior interosseous nerve. This leaves the median nerve as it passes through pronator teres. It joins the anterior interosseous artery and passes down the forearm on the anterior surface of the interosseous membrane between flexor pollicis longus and flexor digitorum profundus. It supplies flexor pollicis longus, the lateral part of flexor digitorum profundus, and pronator quadratus, and it has articular branches to the distal radioulnar, wrist, and carpal joints.
- Muscular branches to the superficial muscles except flexor carpi ulnaris.
- Palmar cutaneous nerve. This is given off just above the wrist joint. It supplies the skin over the thenar eminence and the central part of the palm of the hand.

- Articular branches to the elbow joint and proximal radioulnar joint.

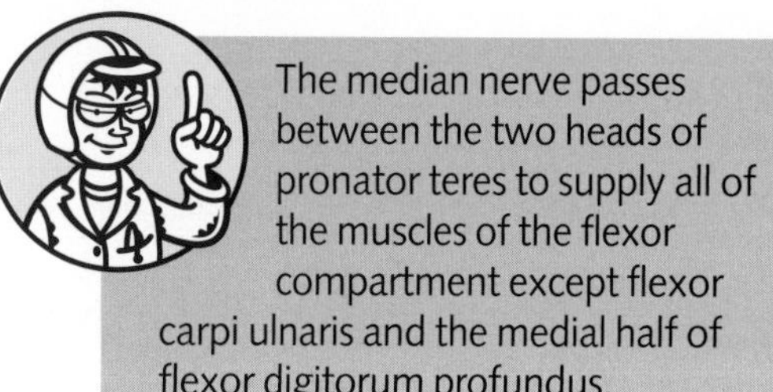

The median nerve passes between the two heads of pronator teres to supply all of the muscles of the flexor compartment except flexor carpi ulnaris and the medial half of flexor digitorum profundus.

Ulnar nerve

This enters the anterior compartment by passing between the heads of flexor carpi ulnaris. It runs with the ulnar artery between flexor carpi ulnaris and flexor digitorum profundus. In the lower half of the forearm both artery and nerve become superficial on the lateral side of flexor carpi ulnaris.

Branches of the ulnar nerve comprise:

- Muscular branches to flexor carpi ulnaris and the medial half of flexor digitorum profundus.
- A palmar cutaneous branch, which supplies the skin over the medial part of the palm.
- A dorsal branch, which passes deep to flexor carpi ulnaris to reach the dorsal aspect of the hand.

Radial nerve

This enters the forearm deep to the brachioradialis muscle. It immediately divides into a superficial and a deep branch. The deep branch passes laterally around the radius between the layers of supinator to enter the extensor compartment as the posterior interosseous nerve. The superficial branch continues down the forearm deep to brachioradialis, and it is joined by the radial artery. Both pass onto the dorsum of the hand.

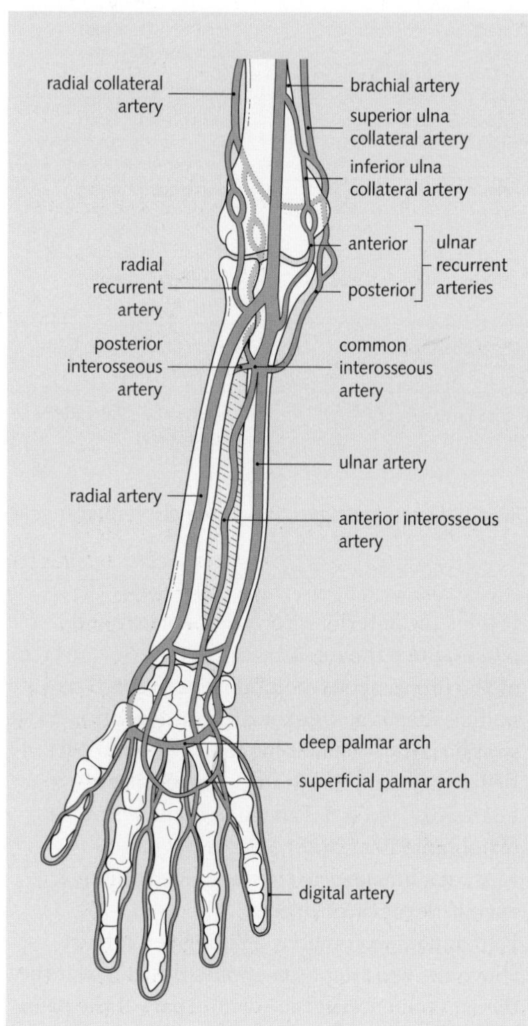

Fig. 2.25 Arteries of the forearm.

Remember, the radial nerve supplies the posterior compartment of the arm then pierces the supinator muscle to supply the extensor forearm muscles.

Posterior compartment of the forearm

Muscles of the posterior compartment

The superficial group of muscles arise from the lateral epicondyle (the common extensor origin) and the supracondylar ridge of the humerus (Figs 2.23 and 2.26).

Vessels of the posterior compartment

The ulnar artery gives off the common interosseous artery near its origin. The latter divides into the anterior and posterior interosseous arteries, which both contribute to the supply for the extensor compartment.

The anterior interosseous artery passes down the anterior surface of the interosseous membrane. Branches pierce the membrane to supply the underlying muscles. At the superior border of pronator quadratus the artery pierces the membrane to anastomose with the posterior interosseous artery, and then it continues to the wrist to join the dorsal carpal arch.

Muscles of the posterior compartment of the forearm			
Name of muscle (nerve supply)	**Origin**	**Insertion**	**Action**
Brachioradialis (radial nerve)	Lateral supracondylar ridge of humerus	Styloid process of radius	Flexes elbow and rotates forearm
Extensor carpi radialis longus (radial nerve)		Base of second metacarpal bone	Extends and abducts hand at wrist joint
Extensor carpi radialis brevis (posterior interosseous nerve)	Common extensor origin	Base of third metacarpal bone	Extends and abducts hand at wrist joint
Extensor digitorum (posterior interosseous nerve)		Extensor expansion of middle and distal phalanges of the medial four digits	Extends the medial four fingers and hand at wrist joint
Extensor digiti minimi (posterior interosseous nerve)		Extensor expansion of little finger	Extends little finger
Extensor carpi ulnaris (posterior interosseous nerve)		Base of fifth metacarpal bone	Extends and adducts hand at the wrist
Anconeus (radial nerve)		Olecranon process and shaft of ulna	Extends and stabilizes the elbow joint
Supinator (posterior interosseous nerve)	Common extensor origin and supinator crest of ulna	Neck and shaft of radius	Supination of forearm
Abductor pollicis longus (posterior interosseous nerve)	Shafts of radius and ulna and interosseous membrane	Base of first metacarpal bone	Abducts thumb
Extensor pollicis brevis (posterior interosseous nerve)	Shaft of radius and interosseous membrane	Base of proximal phalanx of thumb	Extends metacarpophalangeal joint of thumb
Extensor pollicis longus (posterior interosseous nerve)	Shaft of ulna and interosseous membrane	Base of distal phalanx of thumb	Extends thumb
Extensor indicis (posterior interosseous nerve)		Extensor expansion of index finger	Extends index finger

Fig. 2.26 Muscles of the posterior compartment of the forearm.

The posterior interosseous artery passes posteriorly above the upper border of the interosseous membrane, and it accompanies the posterior interosseous nerve to supply the deep muscles of the extensor compartment.

Nerves of the posterior compartment

Brachioradialis and extensor carpi radialis longus are supplied directly by the radial nerve. All the other extensor muscles are supplied by the posterior interosseous nerve.

This nerve emerges from supinator and runs on the interosseous membrane as far as the wrist joint, which it supplies.

Anatomical snuffbox

This is a depression proximal to the base of the first metacarpal and overlying the scaphoid and trapezoid bones when the thumb is actively extended. The anterior margin is formed by the tendons of abductor pollicis longus and extensor pollicis brevis.

The posterior margin is formed by the tendon of extensor pollicis longus. The radial artery runs through the anatomical snuffbox on its course to the dorsum of the hand. It lies on the scaphoid bone here.

Acute tenderness in this region following a fall indicates a probable fracture of the scaphoid bone.

Radioulnar joints

The movements of pronation and supination occur at the proximal and distal radioulnar joints. In the proximal radioulnar joint, the head of radius rotates in an osseofibrous ring formed by the radial notch of the ulna and the anular ligament. The radius also rotates around the ulna in the distal radioulnar joint.

Biceps brachii and supinator cause supination, whereas pronator teres and quadratus are responsible for pronation.

The wrist and hand

Fig. 2.27 shows the skeleton of the wrist and hand.

Radiocarpal (wrist) joint

This is a synovial joint where the distal end of the radius articulates with the scaphoid, lunate, and triquetral bones. An articular disc separates the joint cavity from the distal radioulnar joint (Fig. 2.28). The joint is strengthened by radiocarpal, ulnocarpal, and collateral ligaments. The nerve supply to the joint is from anterior and posterior interosseous nerves.

Movements of the wrist joint are inseparable, functionally, from those at the midcarpal joint (synovial joints between the proximal and distal rows of carpal bones):

- The midcarpal joint participates mainly in flexion and abduction.
- The radiocarpal joint contributes mainly to extension and adduction.

All the joints in the wrist and hand are synovial joints.

Dorsum of the hand

The skin on the dorsum of the hand is thin and loose and the dorsal venous network of veins is usually visible. The veins drain into the cephalic and basilic veins.

The long extensor tendons of the forearm lie beneath the superficial veins. As the tendons cross the wrist joint, they are surrounded by synovial sheaths and bound down by the extensor retinaculum. The extensor retinaculum attachments run from the radius to the pisiform and triquetral bones. The dorsal interossei are the only intrinsic muscles of the dorsum of the hand. The extensor tendons divide into three slips over the proximal phalanx: a central part that inserts into the middle phalanx, and two collateral bands that insert into the distal phalanx. These collateral bands receive a strong attachment from the interossei and lumbrical tendons. It is this attachment, which forms the extensor expansion (Fig. 2.29).

Nerve supply to the dorsum of the hand

Fig. 2.30 shows the cutaneous innervation of the hand. Note the fingertips are supplied by palmar digital branches of the median and ulnar nerves.

Vessels of the dorsum of the hand

Fig. 2.31 shows the blood supply to the dorsum of the hand.

Palm of the hand

Skin

The skin is thick and hairless. Flexure creases and papillary ridges occupy the entire flexor surface, improving the gripping ability of the hand. Fibrous bands bind the skin down to the palmar aponeurosis and divide the subcutaneous fat into loculi, forming a cushion capable of withstanding pressure. A small palmaris brevis muscle attaches the dermis of the skin to the palmar aponeurosis and flexor retinaculum. Its action of wrinkling the skin over the hypothenar eminence improves the ability to grip objects.

The palmaris brevis muscle wrinkles the skin medially to aid the gripping of objects.

Palmar aponeurosis

The palmar aponeurosis is a tough sheet lying between the thenar and hypothenar eminences, where it is continuous with the deep fascia. Distally it separates into four slips, which are joined by the superficial transverse metacarpal ligaments. At the base of each finger the four slips divide and fuse with the fibrous flexor sheaths, the capsule of the metacarpophalangeal joint, and the proximal phalanx.

Contraction of the palmar aponeurosis leads to fixed flexion of the digits (Dupuytren's contracture), which affects most severely the ring and little fingers.

Flexor retinaculum

The flexor retinaculum is a strong band that runs from the scaphoid and trapezium to the pisiform and

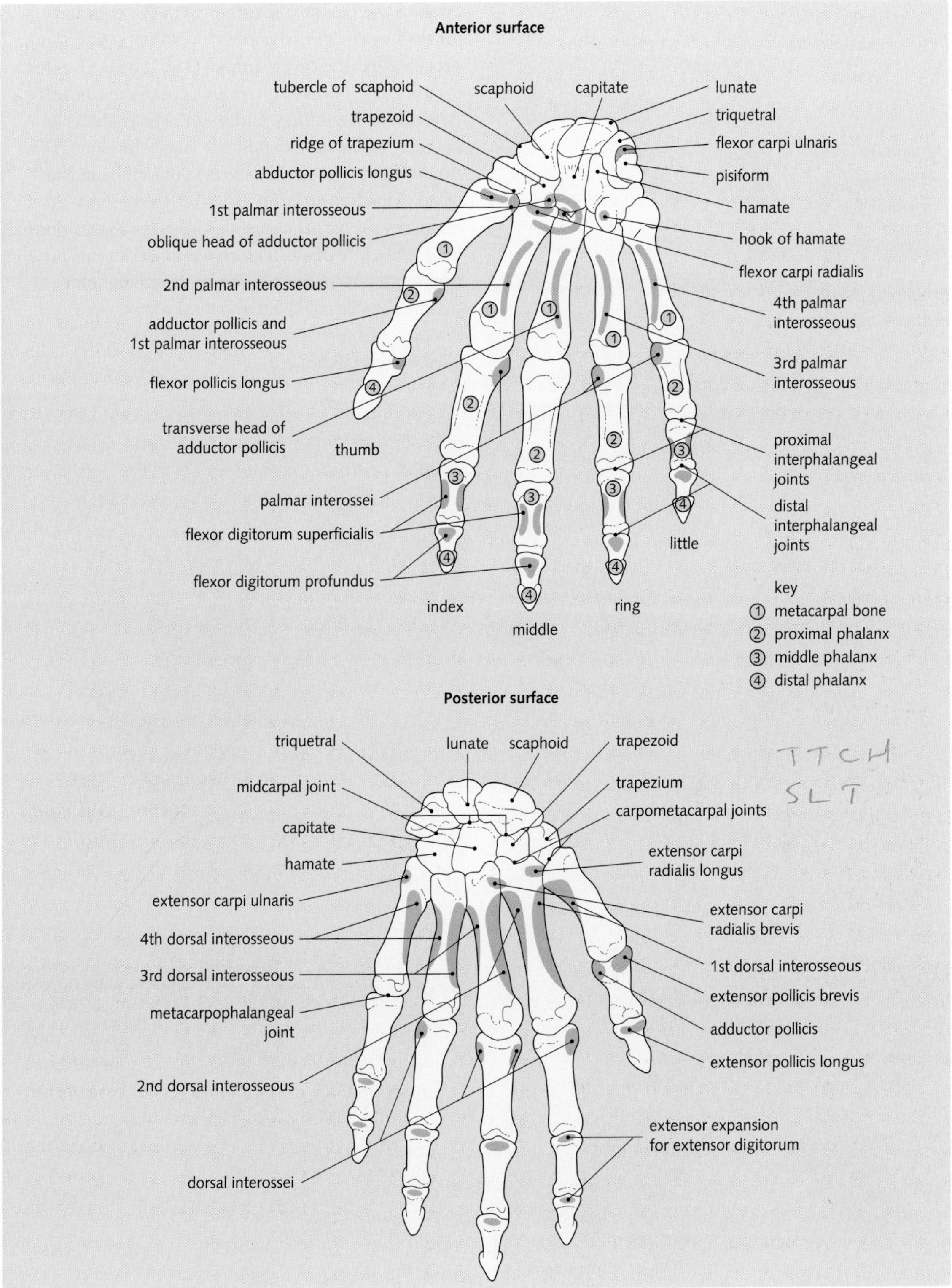

Anterior surface

tubercle of scaphoid
trapezoid
ridge of trapezium
abductor pollicis longus
1st palmar interosseous
oblique head of adductor pollicis
2nd palmar interosseous
adductor pollicis and 1st palmar interosseous
flexor pollicis longus
transverse head of adductor pollicis
thumb
palmar interossei
flexor digitorum superficialis
flexor digitorum profundus
index
middle
ring
little
scaphoid
capitate
lunate
triquetral
flexor carpi ulnaris
pisiform
hamate
hook of hamate
flexor carpi radialis
4th palmar interosseous
3rd palmar interosseous
proximal interphalangeal joints
distal interphalangeal joints

key
① metacarpal bone
② proximal phalanx
③ middle phalanx
④ distal phalanx

Posterior surface

triquetral
midcarpal joint
capitate
hamate
extensor carpi ulnaris
4th dorsal interosseous
3rd dorsal interosseous
metacarpophalangeal joint
2nd dorsal interosseous
dorsal interossei
lunate
scaphoid
trapezoid
trapezium
carpometacarpal joints
extensor carpi radialis longus
extensor carpi radialis brevis
1st dorsal interosseous
extensor pollicis brevis
adductor pollicis
extensor pollicis longus
extensor expansion for extensor digitorum

Fig. 2.27 Bones and joints of the wrist and hand.

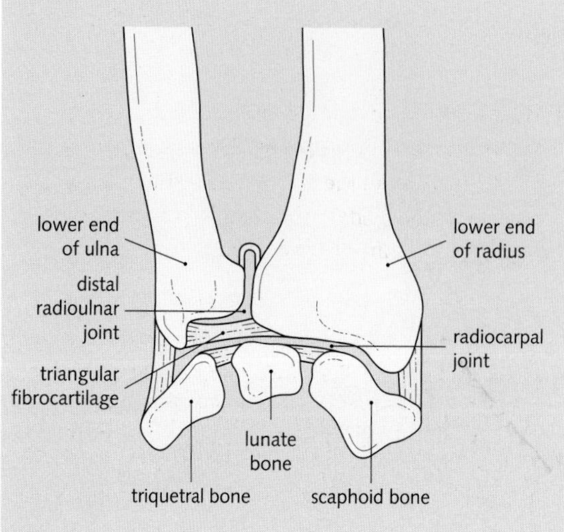

Fig. 2.28 Relationship of the distal radioulnar joint to the radiocarpal joint.

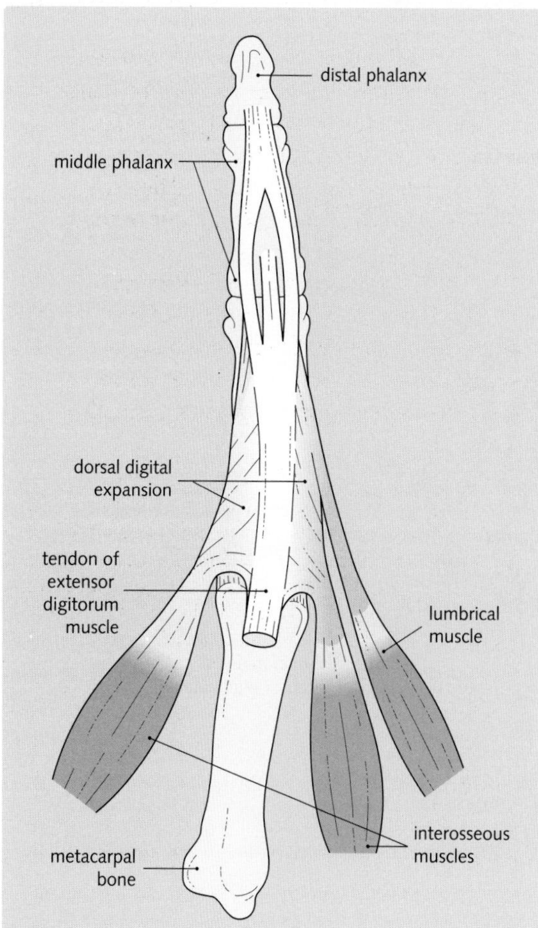

Fig. 2.29 Dorsal digital expansion and extensor tendon.

hook of the hamate. Together with the concavity created by the carpal bones it forms an osseofibrous canal, called the carpal tunnel (Fig. 2.32). The flexor digitorum superficialis tendons enter the tunnel in two rows. The middle and ring finger tendons lie anterior to the index and little finger tendons. The flexor digitorum profundus tendons all lie in the same plane beneath the superficialis tendons. At the distal row of carpal bones the superficialis tendons all lie in the same plane. The remaining contents are the flexor pollicis longus tendon and median nerve.

The muscles of the thenar and hypothenar eminences arise from the flexor retinaculum and adjacent carpal bones.

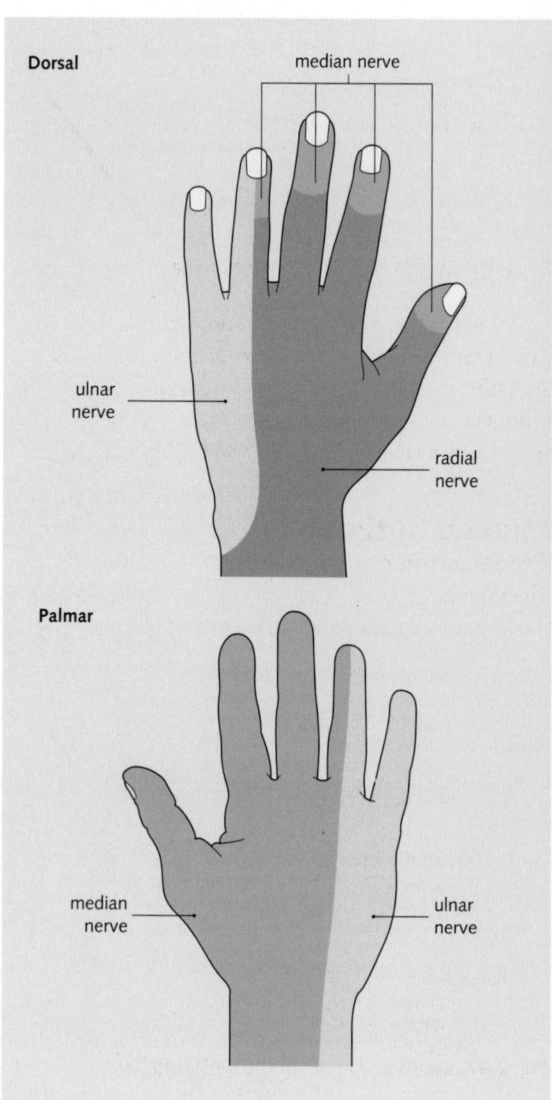

Fig. 2.30 Cutaneous innervation of the dorsal and palmar surfaces of the hand.

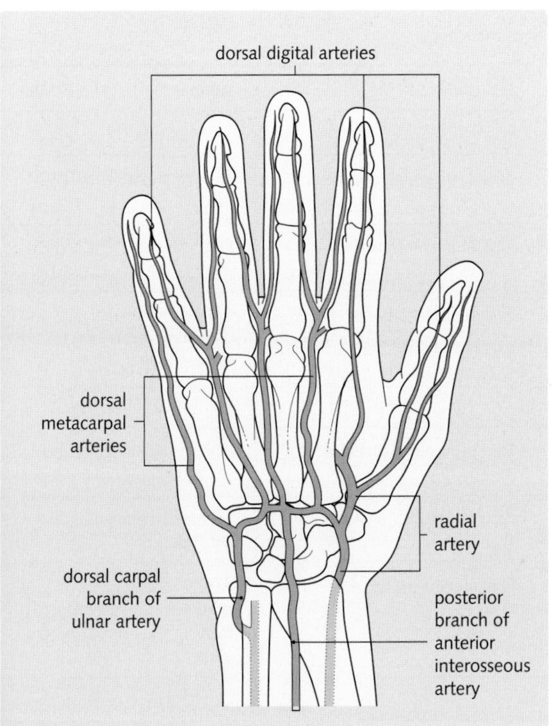

Fig. 2.31 Vessels of the dorsum of the hand.

Compression of the median nerve in the carpal tunnel results in wasting of the thenar muscles and anaesthesia of the lateral three-and-a-half digits. This is known as carpal tunnel syndrome. Division of the retinaculum relieves the pressure and symptoms.

Muscles of the hand

Thenar eminence

The thenar eminence is the prominent region between the base of the thumb and the wrist. It is composed of three muscles: abductor pollicis brevis, flexor pollicis brevis, and opponens pollicis (Fig. 2.33).

Hypothenar eminence

The hypothenar eminence lies between the base of the small finger and the wrist. It consists of the abductor digiti minimi, flexor digiti minimi, and opponens digiti minimi (Fig. 2.33).

The median nerve supplies the thenar eminence; the ulnar nerve supplies the hypothenar eminence.

Other intrinsic hand muscles

This includes adductor pollicis, lumbrical, palmaris brevis, dorsal and palmar interosseous muscles (Fig. 2.33).

When dorsal and palmar interossei contract together their individual actions, abduction and adduction respectively, cancel each other out. Instead they, along with the lumbrical muscles, extend the interphalangeal joints and flex the metacarpophalangeal joints. This is due to the interossei insertion to the proximal phalanx causing flexion at the metacarpophalangeal joint. The interossei and lumbricals pull on their insertion into the extensor expansion causing extension at the interphalangeal joints. This is a simple explanation to a rather complex action.

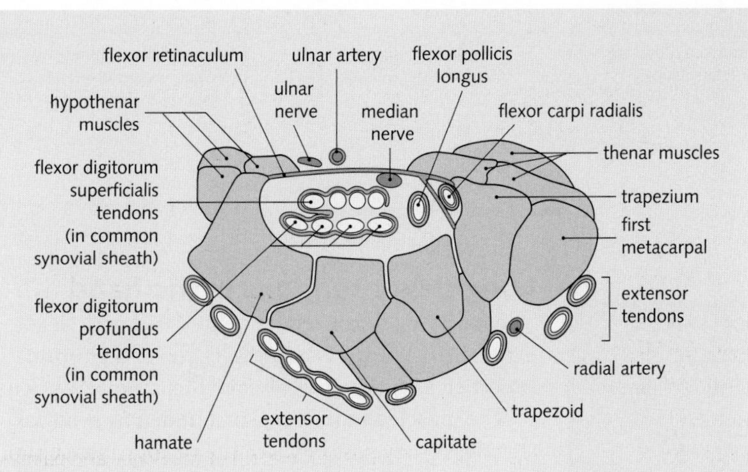

Fig. 2.32 Cross-section of the right carpal tunnel at the distal row of carpal bones showing its contents. (Adapted from *Gray's Anatomy*, 38th edn, edited by L H Bannister et al. Harcourt Brace and Co.)

The intrinsic muscles of the hand			
Name of muscle (nerve supply)	**Origin**	**Insertion**	**Action**
Thenar eminence muscles			
Abductor pollicis brevis (recurrent branch of median nerve)	Scaphoid, trapezium and flexor retinaculum	Base of proximal phalanx	Abducts thumb at the MCP joint
Flexor pollicis brevis (recurrent branch of median nerve)	Trapezium and flexor retinaculum	Base of proximal phalanx	Flexes thumb at the MCP joint
Opponens pollicis (recurrent branch of median nerve)	Trapezium and flexor retinaculum	First metacarpal bone	Rotates metacarpal at carpometacarpal joint to oppose thumb
Hypothenar eminence muscles			
Abductor digiti minimi (deep branch of ulnar nerve)	Pisiform and flexor retinaculum	Base of proximal phalanx	Abducts little finger at the MCP joint
Flexor digiti minimi (deep branch of ulnar nerve)	Hook of hamate and flexor retinaculum	Base of proximal phalanx	Abducts little finger at the MCP joint
Opponens digiti minimi (deep branch of ulnar nerve)	Hook of hamate and flexor retinaculum	Fifth metacarpal bone	Assists in flexing the carpometacarpal joint, cupping the palm to assist gripping
Other intrinsic hand muscles			
Lumbricals (first and second: median nerve; third and fourth: deep branch of ulnar nerve)	Tendons of flexor digitorum profundus	Extensor expansion of the medial four digits	Extends the DIP and PIP joints of medial four digits. Flexes the MCP joint of the medial four digits
Palmar interossei (deep branch of ulnar nerve)	First, second, fourth and fifth metacarpal bones	Base of the proximal phalanx and extensor expansion	Adduct the digits away from the middle finger. Flexes digit at MCP and extends interphalangeal joints
Dorsal interossei (deep branch of ulnar nerve)	Adjacent sides of the five metacarpal bones	Base of proximal phalanx and extensor expansion	Abduct the digits towards the middle finger. Flexes digit at MCP and extends interphalangeal joints
Adductor pollicis (deep branch of ulnar nerve)	Oblique head: capitate, trapezoid and second and third metacarpals. Transverse head: distal part of third metacarpal	Base of proximal phalanx	Adducts thumb
Palmaris brevis (superficial branch of the ulnar nerve)	Palmar aponeurosis and flexor retinaculum	Dermis of the skin on medial border of hand	Wrinkles the skin over the hypothenar eminence and improve the grip of the hand

Fig. 2.33 The intrinsic muscles of the hand.

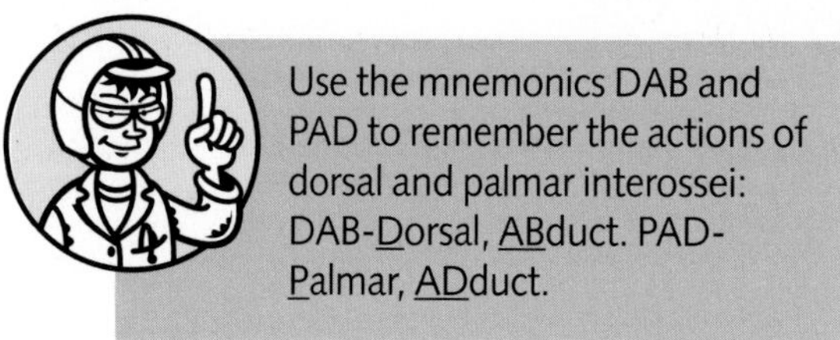

Long flexor tendons in the hand

The following flexor tendons enter the hand: flexor carpi ulnaris, flexor carpi radialis, flexor digitorum superficialis and profundus, and flexor pollicis longus.

The muscle tendons are surrounded by synovial sheaths as they pass beneath the flexor retinaculum, and as they enter the digits.

Synovial flexor tendon sheaths

There is a common synovial sheath for flexor digitorum superficialis and profundus, which is incomplete on its radial side. This sheath extends from above the flexor retinaculum to the palm around all digitorum tendons except for digitus minimus (the little finger). Here the sheath continues to the distal phalanx. For the second, third, and fourth digit tendons, there is a bare area before the synovial sheath encloses them again before entering the digits. From this area the lumbrical muscles gain origin.

The sheath around the flexor pollicis longus tendon extends from just proximal to the flexor retinaculum to the distal phalanx of the pollex (the thumb). See Fig. 2.34.

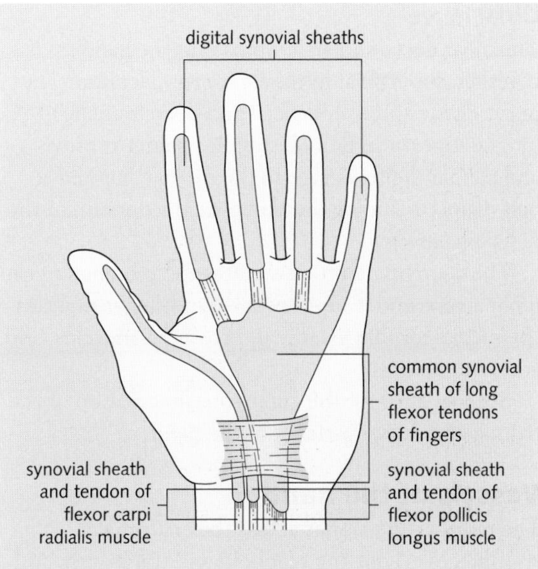

Fig. 2.34 Flexor synovial sheaths in the hand.

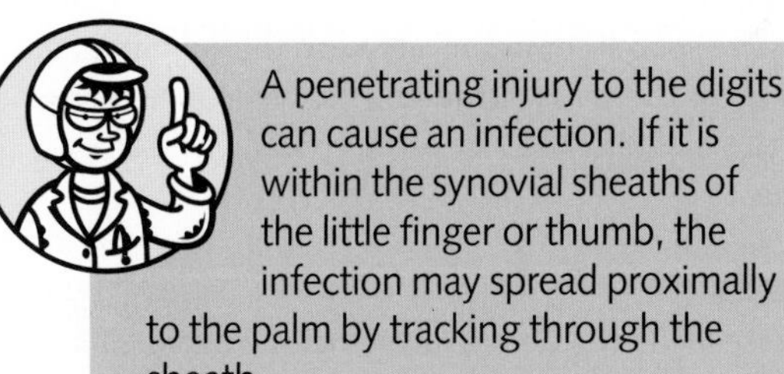

A penetrating injury to the digits can cause an infection. If it is within the synovial sheaths of the little finger or thumb, the infection may spread proximally to the palm by tracking through the sheath.

Long flexor tendons in the digits

The tendon of flexor digitorum superficialis bifurcates into two slips before inserting into the middle phalanx, whilst that of flexor digitorum profundus runs between these slips to insert into the terminal phalanx. The tendons are attached to their sheaths via bands called vincula. Fibrous flexor sheaths bind the tendons down to the finger, which are strong and thick over the phalanges but weak and loose over digit joints to allow movement (Fig. 2.35).

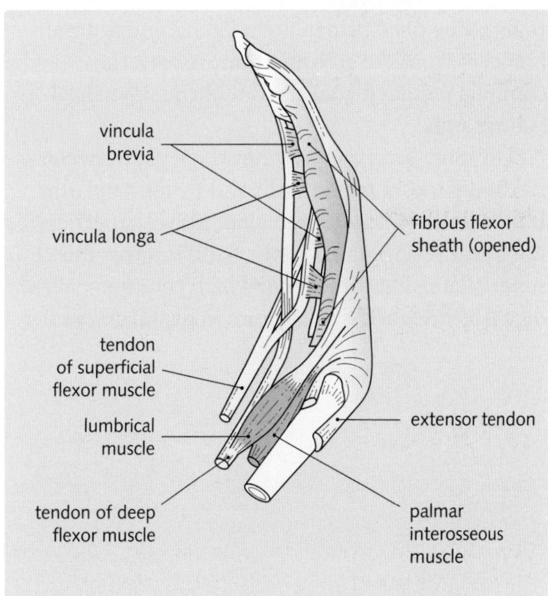

Fig. 2.35 Long flexor tendons of a finger and their vincula.

Nerves in the hand

Median nerve

The median nerve emerges from the carpal tunnel to enter the palm. Branches of the nerve in this region include:

- Muscular branches to the muscles of the thenar eminence.
- Palmar digital nerves providing sensory innervation to the lateral three-and-a-half digits (including the nail bed and skin on the dorsum of the digit over the terminal phalanx) and motor supply to the first and second lumbricals.

Damage to the median nerve, e.g. trauma to the wrist, results in loss of sensation to the lateral three-and-a-half digits and loss of function in the thumb. This is a very incapacitating injury that renders the entire upper limb almost functionless.

Ulnar nerve

The ulnar nerve and artery pass into the hand together, superficial to the flexor retinaculum. The nerve divides into superficial and deep branches.

The superficial branch supplies palmaris brevis and palmar digital nerves to the medial one-and-a-half digits (including the skin over the dorsum of the distal phalanx).

The deep branch runs with the deep branch of the ulnar artery, and it supplies the hypothenar muscles, the medial lumbricals, the interosseous muscles, and adductor pollicis.

See Fig. 2.30 for the cutaneous innervation of the palmar and dorsal surfaces of the hand.

Vessels of the hand

The radial artery slopes across the anatomical snuffbox overlying the scaphoid and trapezium, and it passes between the two heads of the first dorsal interosseous. It appears deep in the palm of the hand, emerging between the two heads of adductor pollicis to form the deep palmar arch. Palmar metacarpal branches from the arch anastomose with the common palmar digital arteries of the superficial palmar arch.

The ulnar artery approaches the wrist between flexor digitorum superficialis and flexor carpi ulnaris. It enters the wrist with the ulnar nerve, superficial to the flexor retinaculum. In the hand it forms the superficial palmar arch from which common palmar digital arteries arise. The common digital arteries divide to form (proper) palmar digital arteries that enter the digits to supply the joints and phalanges.

The radial pulse can be taken at the wrist, laterally, or in the anatomical snuffbox.

Palmar spaces

The intermediate palmar septum from the palmar aponeurosis to the third metacarpal divides the central part of the palm into two fascial spaces: the thenar space lies laterally and the midpalmar space lies medially.

The spaces communicate with the subcutaneous tissue of the webs of the fingers. Deep infections of the midpalmar space often spread to these sites.

Nails

These lie on the dorsal surface of the distal phalanges, and they are formed from modified skin tissue.

- Describe the superficial veins of the upper limb.
- Describe the lymphatic drainage of the upper limb.
- Describe the bones and joints of the pectoral girdle.
- Discuss the anatomy of the shoulder joint.
- What are the components and actions of the rotator cuff muscles?
- List the movements of the shoulder joint and the muscles causing them.
- List the boundaries and contents of the axilla.
- Describe the formation of the brachial plexus between the neck and axilla.
- Name the muscles and their nerve supply that are involved in elbow flexion and extension.
- Describe the vessels of the arm and their branches.
- List the boundaries and contents of the cubital fossa.
- Describe the elbow joint.
- Discuss the muscles involved in wrist and digit movement.
- Describe the branches of the median nerve in the forearm.
- Outline the boundaries and contents of the anatomical snuffbox.
- Describe a dorsal digital expansion.
- Outline the cutaneous nerve supply of the hand.
- Discuss the synovial sheaths of the flexor tendons.
- Outline the carpal tunnel and the structures passing through it.
- Discuss the muscles that move the thumb.

3. The Thorax

Regions and components of the thorax

The thorax lies between the neck and the abdomen.

The thoracic cavity contains the heart, lungs, great vessels, trachea, and oesophagus. The lungs lie laterally, and the other structures lie in the mediastinum.

The thoracic cage is formed by the thoracic vertebrae, the ribs, the costal cartilages, and the sternum. It protects the contents of the thoracic cavity and some abdominal contents, e.g. the liver and spleen.

Superiorly the thorax communicates with the neck through the thoracic inlet. Inferiorly the diaphragm separates the thorax from the abdominal cavity.

Surface anatomy and superficial structures

Bony landmarks

The bony landmarks of the thorax are illustrated in Fig. 3.1.

The bony sternum can be palpated in the midline of the thorax anteriorly. The manubrium sterni joins the body of the sternum at the manubriosternal joint (the angle of Louis). The manubriosternal angle is important clinically as it is level with the second costal cartilage. Inferiorly, the body of the sternum articulates with the xiphoid process.

Ribs 1–7 articulate with the manubrium and sternum laterally via the costal cartilages. The costal margin runs from the xiphisternum and comprises the costal cartilages of ribs 7–10.

Posteriorly the spinous processes of the thoracic vertebrae are palpable inferior to the vertebra prominens—the spinous process of C7 vertebra.

The manubriosternal joint is where the second costal cartilages join it. This can be used as a reference point for counting the ribs.

Trachea, lungs, and pleurae

The trachea is palpable in the midline, above the suprasternal notch. It bifurcates behind the manubriosternal angle (at the level of T4) into the right and left main bronchi.

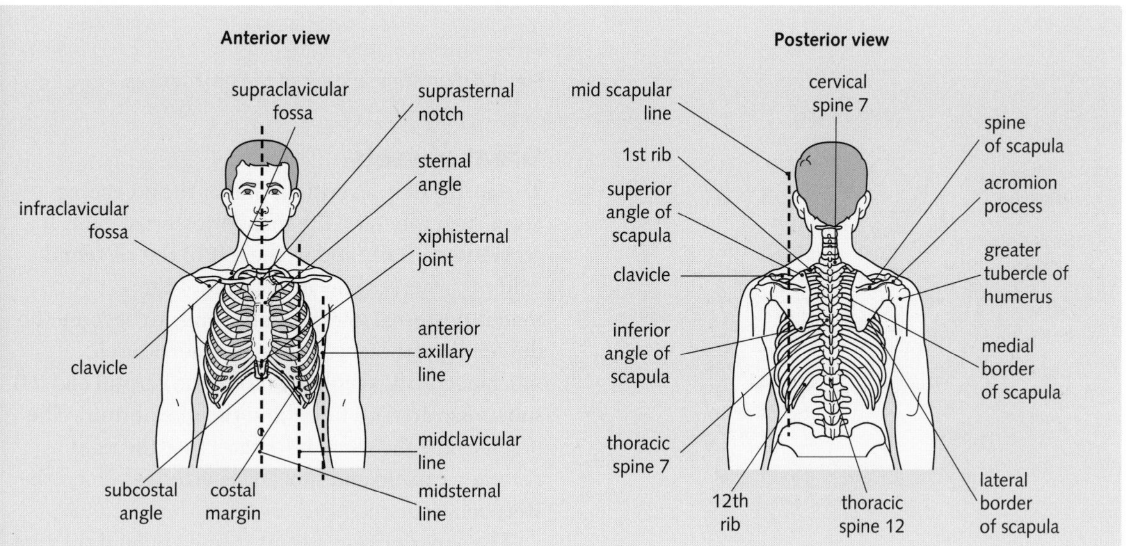

Fig. 3.1 Surface markings of the anterior and posterior thoracic walls.

The dome of the pleura extends about 2.5 cm above the medial end of the clavicle (Fig. 3.2). The anterior border of the right pleura passes down behind the sternal angle to the xiphisternal joint. The anterior border of the left pleura follows a similar course, but at the level of the 4th costal cartilage it deviates laterally to form the cardiac notch. The lower border follows a curved line, being at the level of the 8th rib in the midclavicular line, the 10th rib in the midaxillary line, and the 12th rib adjacent to the vertebral column.

Note, trauma to the root of the neck or at the 11th or 12th ribs can easily penetrate the pleura, causing a pneumothorax.

The surface markings of the lungs in mid-inspiration are similar to those of the pleurae except inferiorly, where they lie at the level of the 6th, 8th, and 10th ribs in the midclavicular line, midaxillary line, and adjacent to the vertebral column, respectively.

The space between the lower border of the pleura and lung is the costodiaphragmatic recess. In life, it is filled by the lungs in full inspiration.

Heart

The surface markings of the heart are outlined in Fig. 3.3. The apex of the heart is taken as lying approximately in the midclavicular line of the fifth intercostal space. This point takes into account differences in stature and children. The surface markings of the heart valves are illustrated in Fig. 3.4.

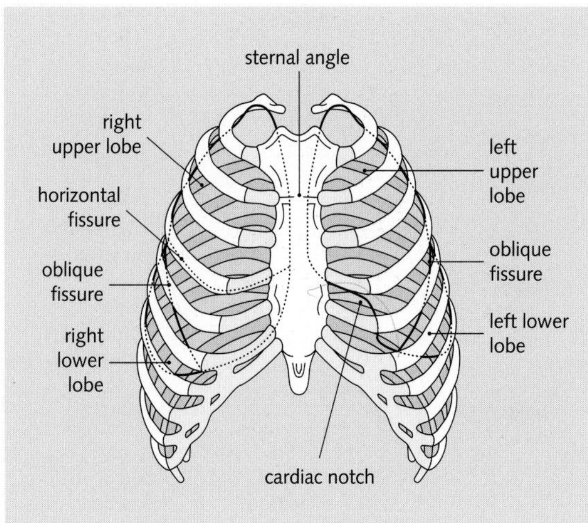

Fig. 3.2 Surface markings of the lungs.

Outline of the surface landmarks of the heart

Border	Area covered
Superior border	From the second left costal cartilage to the third right costal cartilage
Right border	From the third right costal cartilage to the sixth right costal cartilage
Left border	From the second left costal cartilage to the apex of the heart
Inferior border	From the sixth right costal cartilage to the apex
Apex	Lies in the fifth intercostal space, in the midclavicular line

Fig. 3.3 Outline of the surface landmarks of the heart.

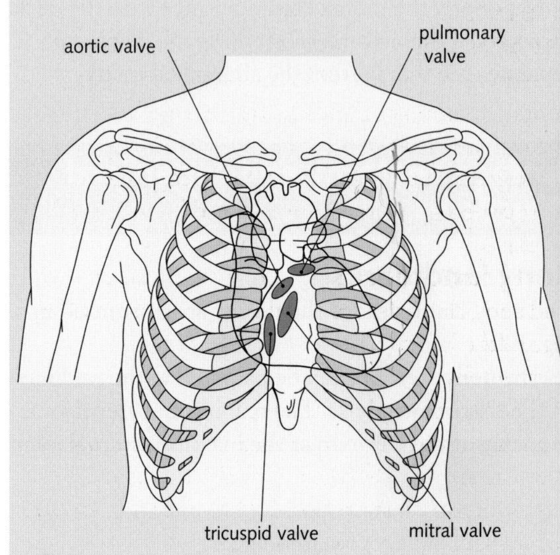

Fig. 3.4 Surface markings of the heart valves.

Great vessels

The aortic arch, a continuation of the ascending aorta, begins behind the manubriosternal joint. It arches posteriorly and to the left of the vertebral column, where, again at the level of the manubriosternal joint, the arch ends to become the descending (thoracic) aorta. The aortic arch, brachiocephalic trunk, left common carotid and left subclavian arteries lie behind the manubrium. The brachiocephalic trunk bifurcates into the right common carotid and subclavian arteries behind the sternoclavicular joint.

The superior vena cava is formed behind the first right costal cartilage by the union of the

brachiocephalic veins. The superior vena cava runs inferiorly along the right sternal border to enter the right atrium behind the third costal cartilage. The azygos vein joins the superior vena cava at the level of the second right costal cartilage.

The internal thoracic vessels run vertically downwards, posterior to the costal cartilages and 1 cm lateral to the sternal edge, as far as the 6th intercostal space. The arteries now divide into musculophrenic and superior epigastric arteries.

Diaphragm

The central tendon of the diaphragm lies behind the xiphisternal joint. In mid-respiration the right dome arches upwards to the upper border of the 5th rib in the midclavicular line; the left dome reaches only the lower border of the 5th rib.

Breasts

The breasts lie in the superficial fascia, mainly superficial to the pectoralis major muscle. They contain mammary glandular tissue, which drains through lactiferous ducts into the nipple. The nipple is the greatest prominence of the breast, and it is surrounded by a circular pigmented area called the areola.

The base of the breast usually lies between the 2nd and 6th rib vertically, and from the midaxillary line to the lateral border of the sternum horizontally.

The blood supply to the breast is derived from branches of the internal thoracic artery, the lateral thoracic and the thoracoacromial arteries, and the posterior intercostal arteries. Venous drainage is into the axillary and internal thoracic veins.

Lymph drains into the subareolar plexus and then into either axillary nodes, internal thoracic nodes, or the other breast.

Axillary nodes receive most of the lymphatic drainage of the superior and lateral parts of the breast. Lymphatics from the inferior and medial part of the breast drain into lymph nodes along the internal thoracic vessels and then via the bronchomediastinal lymph trunk into the lymphatics at the root of the neck.

Lymphatics may communicate with vessels from the opposite breast.

Carcinoma of the breast—the most common cancer affecting women—spreads via the lymphatics in many cases. A detailed knowledge of this drainage is required for appropriate treatment.

In breast carcinoma, skin dimpling occurs due to the tumour pulling and thus shortening the suspensory ligaments.

The thoracic wall

The thoracic skeleton is formed by the sternum, the ribs and costal cartilages, and the thoracic vertebrae (Fig. 3.5).

Sternum

The sternum (breast bone) has three components:

- The manubrium is the upper part of the sternum. It articulates with the clavicles and with the 1st and upper part of the 2nd costal cartilages.
- The body of the sternum articulates with the manubrium at the manubriosternal joint superiorly, and with the xiphisternum inferiorly. These are fibrocartilaginous joints. Laterally the sternal body articulates with the 2nd to 7th costal cartilages via synovial joints.
- The xiphoid process is the lowest part of the sternum.

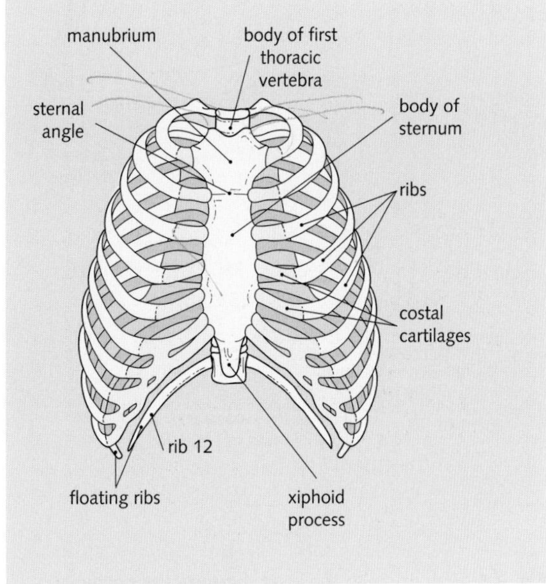

Fig. 3.5 Thoracic skeleton.

Ribs and costal cartilages

There are 12 pairs of ribs. They all articulate with a costal cartilage anteriorly. The upper seven ribs articulate directly with the sternum via their own costal cartilage. The 8th to 10th ribs have costal cartilages that are attached to each other and to the 7th anteriorly, and thence to the sternum.

The floating 11th and 12th ribs have no anterior attachment for their costal cartilages.

Typical ribs

A typical rib has the following features (Fig. 3.6):

- It is a long curved flattened bone with a rounded superior border and a sharp thin inferior border forming the costal groove.
- It has a head with two demifacets for articulation with the numerically similar vertebral body and that of the vertebra immediately above.
- The neck separates the head and the tubercle.
- The tubercle has a facet for articulation with the transverse process of the corresponding vertebra.
- The shaft is thin, flat, and curved, with an angle at its point of greatest change in curvature.

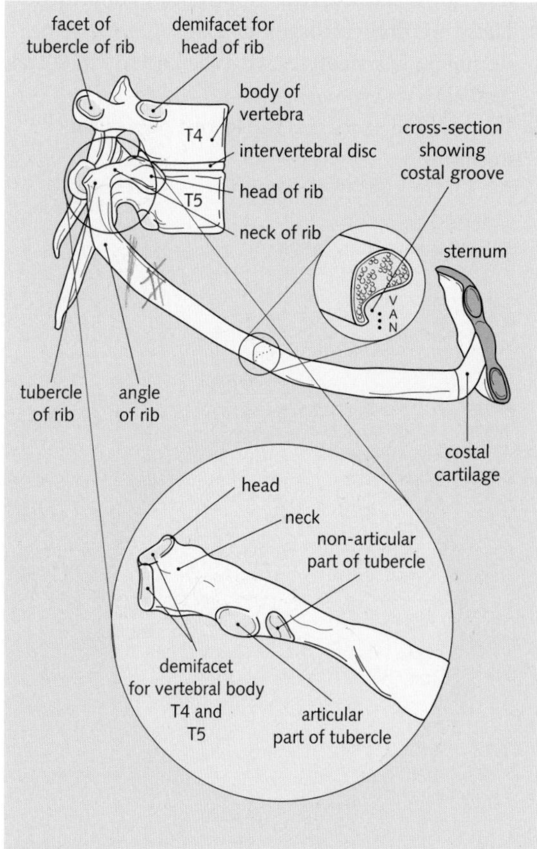

Fig. 3.6 Fifth rib, with the inset showing the posterior surface of the rib.

The weakest part of the rib is anterior to the angle. Fractures may injure thoracic structures e.g. the pleura, causing a pneumothorax.

Atypical ribs

Fig. 3.7 shows the 1st rib and its relations in the thoracic inlet. This is the broadest, shortest, and most sharply curved rib. It has a tubercle on the upper surface for attachment of the scalenus anterior. The subclavian artery and vein pass posterior and anterior to the tubercle, respectively. The artery forms the subclavian groove on the rib.

The 10th, 11th, and 12th ribs have one facet on the head for articulation with their own vertebra.

A cervical rib may compress the subclavian artery or the inferior trunk of the brachial plexus, causing ischaemic pain and numbness along the medial border of the forearm.

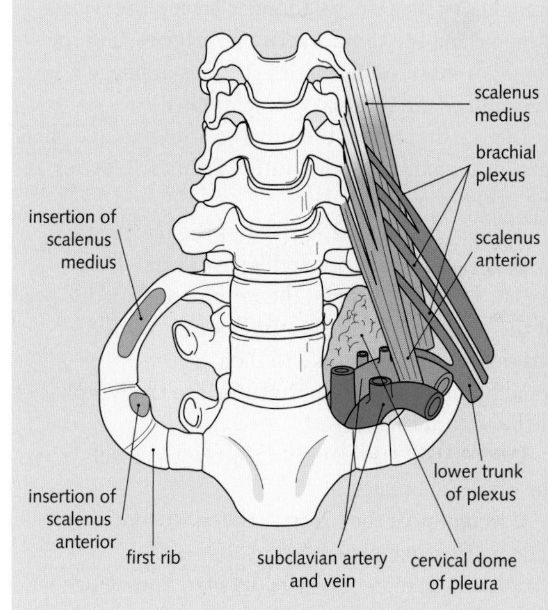

Fig. 3.7 First rib and its relations to the thoracic inlet.

Thoracic vertebrae

There are 12 thoracic vertebrae with their intervening intervertebral discs (Fig. 3.8).

Openings into the thorax

Thoracic inlet (superior thoracic aperture)

The thoracic cavity communicates with the root of the neck via the thoracic inlet (see Fig. 3.7). The margins of the inlet include:

- T1 vertebra.
- Medial border of the 1st rib and its costal cartilage.
- Superior border of the manubrium.

The oesophagus, trachea, and the apices of the lungs, together with various vessels and nerves, pass through the inlet.

Thoracic outlet (inferior thoracic aperture)

The thoracic outlet lies between the thorax and the abdomen. It is bounded posteriorly by the 12th thoracic vertebra and the 12th pair of ribs, anteriorly by the costal margin and xiphoid process. It is covered mainly by the diaphragm. Numerous structures pass through this opening (see Fig. 3.12).

Intercostal spaces

The intercostal spaces lie between the ribs.

Below the skin and superficial fascia lie the three intercostal muscles (Fig. 3.9). The innermost muscle layer is lined by the endothoracic fascia and the parietal pleura.

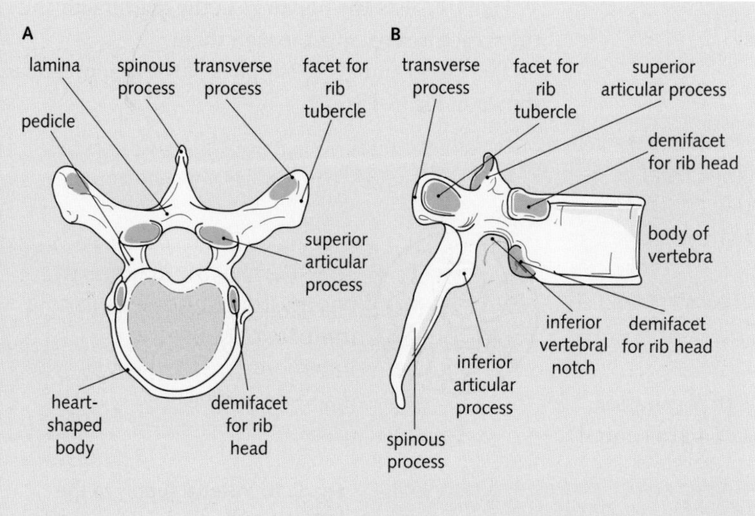

Fig. 3.8 Superior (A) and lateral (B) surfaces of a thoracic vertebra.

Muscles of the thorax			
Name of muscle (nerve supply)	**Origin**	**Insertion**	**Action**
External intercostal (intercostal)	Inferior border of rib above	Superior border of rib above	All the intercostal muscles assist in both inspiration and expiration during ventilation
Internal intercostal (intercostal)	Inferior border of rib above	Superior border of rib above	
Innermost intercostal (intercostal)	Adjacent ribs	Adjacent ribs	
Diaphragm (phrenic)	Xiphoid process, lower sixth costal cartilages, L1–L3 vertebrae by crura, and medial and lateral arcuate ligaments	Central tendon	Important muscle of inspiration; increases vertical diameter of thorax by pulling down central tendon

Fig. 3.9 Muscles of the thorax.

Intercostal nerves and vessels

The intercostal nerves are the anterior rami of the upper eleven thoracic spinal nerves. The 12th nerve is the subcostal nerve.

Each intercostal nerve runs in a plane between the parietal pleura and the posterior intercostal membrane, and it passes forward in the costal groove of the corresponding rib, between the two deeper intercostal muscles. The nerve has a collateral branch that runs along the upper border of the lower rib.

All the intercostal nerves supply the skin and parietal pleura and the intercostal muscles of their respective spaces. Most of the first intercostal nerve passes to the brachial plexus. The 7th to 11th nerves also pass to the anterior abdominal wall to supply the skin, peritoneum, and anterior abdominal wall muscles. The subcostal nerve lies in the abdominal wall.

Fig. 3.10 shows the arterial supply to the thoracic wall.

When inserting an intercostal needle, introduce it into the middle of the space, avoiding the collateral branches of the neurovascular bundle and the structures in the costal groove.

The intercostal neurovascular bundle running in the costal groove is remembered as VAN (vein, artery, nerve) from superior to inferior.

Diaphragm

The diaphragm is the primary muscle of respiration. It consists of a peripheral muscular part and a central tendon, and it separates the thoracic and abdominal cavities. As viewed from the front, the diaphragm curves up into two domes, the right dome being higher than the left (Fig. 3.11). When viewed from the side, it assumes an inverted-J shape.

Fig. 3.12 lists the openings in the diaphragm and the structures passing through them.

The blood and nerve supplies of the diaphragm are shown in Fig. 3.13.

To remember the diaphragm has the phrenic nerve as its sole motor supply, use the mnemonic: C3, 4, 5 keeps the diaphragm alive.

Arterial supply to the thoracic wall

Artery	Origin	Distribution
Anterior intercostal (spaces 1–6)	Internal thoracic artery	Intercostal spaces and parietal pleura
Anterior intercostal (spaces 7–9)	Musculophrenic artery	
Posterior intercostal (spaces 1–2)	Superior intercostal artery (from costocervical trunk of subclavian artery)	
Posterior intercostal (all other spaces)	Thoracic aorta	
Internal thoracic	Subclavian artery	Runs down lateral to the sternum and terminates by dividing into the superior epigastric and musculophrenic arteries
Subcostal	Thoracic aorta	Abdominal wall

Fig. 3.10 Arterial supply to the thoracic wall.

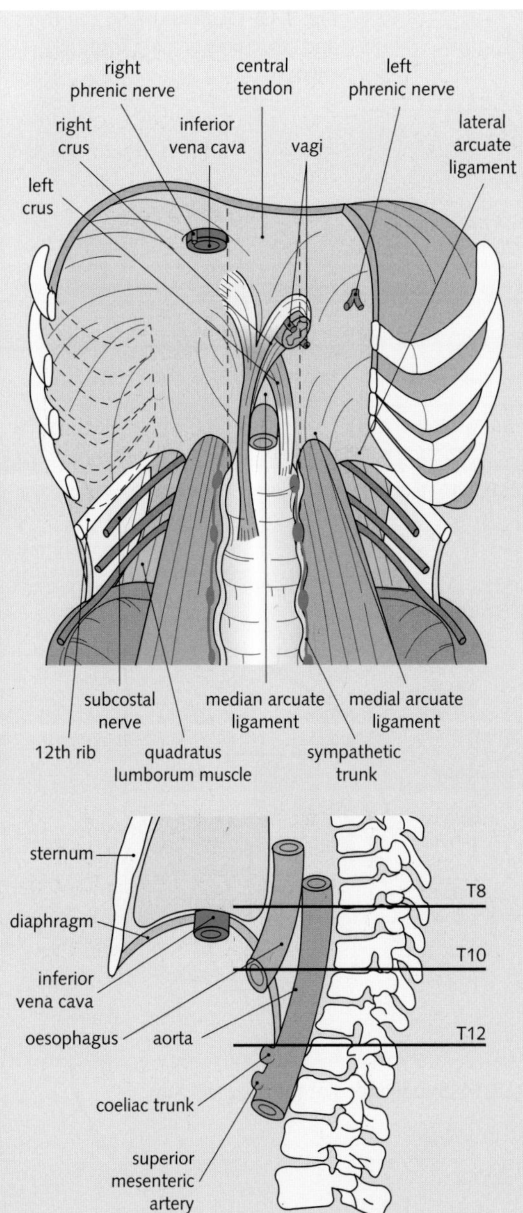

Fig. 3.11 Diaphragm as seen from below (the anterior portion of the right side has been removed) and as a sagittal section.

The thoracic cavity

The thoracic cavity is filled laterally by the lungs and the pleural cavities. The median partition separating the lungs and pleurae is the mediastinum. This may be divided by a plane passing through the sternal angle (of Louis) and T4 vertebra (Fig. 3.14). The superior mediastinum lies above this plane and contains the thymus, great vessels, trachea, oesophagus, thoracic duct, and sympathetic trunk. The inferior mediastinum lies below the plane, and it is further subdivided into:

- The anterior mediastinum, lying between the pericardium and the sternum, which contains lymph nodes.
- The middle mediastinum, which contains the pericardium and heart.
- The posterior mediastinum, lying between the pericardium and the vertebral column, which contains the oesophagus, thoracic duct, sympathetic trunk, and descending aorta.

Fig. 3.15 shows the left and right sides of the mediastinum.

The mediastinum

Pericardium

The pericardium is a double-walled fibroserous sac that encloses the heart and the roots of the great vessels. It is divided into the fibrous pericardium and the two layers of the serous pericardium: parietal and visceral. The fibrous pericardium is a strong layer that limits the movement of the heart. It is attached to the central tendon of the diaphragm, the sternum, and the tunica adventitia of the great vessels.

The nerve supply to the fibrous and parietal layer of the pericardium is by the phrenic nerve. However, the visceral layer has no somatic innervation, and so it is insensitive to pain. Therefore, pain from inflammation of the sac (pericarditis) is transmitted through the phrenic nerve and referred to the shoulder through C3 and C4 spinal nerves.

The pericardium may be affected in many diseases, including tuberculosis and other infections. Also, trauma to the heart may result in bleeding into the pericardium, which may compress the heart (cardiac tamponade); this is an emergency—immediate drainage is needed if the patient is to survive.

Pericardial sinuses

The reflection of the serous pericardium around the large veins forms a recess called the oblique sinus. The reflection around the aorta and the pulmonary trunk, together with the reflection around the great veins, form the transverse sinus. Both sinuses lie on the posterior surface of the heart (see Fig. 3.16).

Diaphragmatic apertures and structures passing through them	
Opening	**Structures**
Aortic opening (behind the diaphragm at the level of T12)	Aorta, thoracic duct, azygos and hemiazygos veins
Oesophageal opening (in the muscle of the diaphragm at the level of T10)	Oesophagus, vagus nerves, oesophageal branches of the left gastric vessels and lymphatics
Vena caval opening (in the central tendon at the level of T8)	Inferior vena cava, right phrenic nerve
Structures passing behind the diaphragm	Splanchnic nerves and sympathetic trunk
Structures passing anteriorly to the diaphragm	Superior epigastric vessels and lymphatics
Structures passing laterally to the diaphragm	Lower six intercostal vessels and nerves.

Fig. 3.12 Diaphragmatic apertures and structures passing through them.

Nerves and vessels of the diaphragm	
Innervation	Motor supply: phrenic nerves (C3–C5); Sensory supply: centrally by phrenic nerves (C3–C5), peripherally by intercostal nerves (T5–T11) and subcostal nerve (T12)
Arterial supply	Superior phrenic arteries; musculophrenic arteries; inferior phrenic arteries; pericardiacophrenic arteries
Venous drainage	Musculophrenic and pericardiacophrenic veins drain into internal thoracic vein; inferior phrenic veins
Lymphatic drainage	Diaphragmatic lymph nodes drain to posterior mediastinal nodes eventually; superior lumbar lymph nodes; lymphatic plexuses on superior and inferior surfaces communicate freely

Fig. 3.13 Nerves and vessels of the diaphragm.

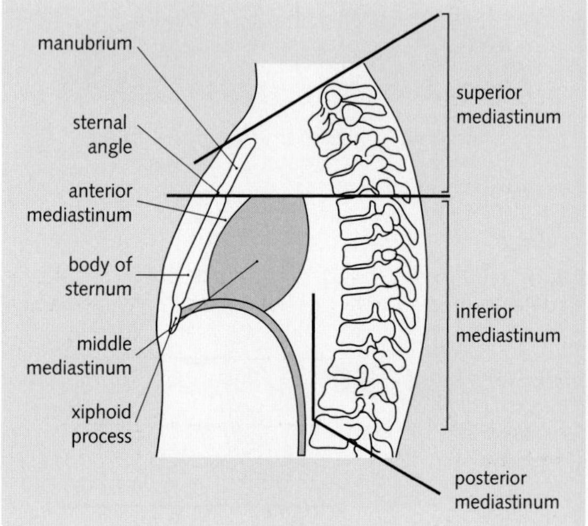

Fig. 3.14 Subdivisions of the mediastinum.

Heart

This is the muscular organ responsible for pumping blood throughout the body (Fig. 3.17). It lies free in the pericardium, connected only at its base to the great vessels.

The walls of the heart consist mainly of heart muscle (myocardium), lined internally by the endocardium and externally by the epicardium (visceral serous pericardium).

The heart has four chambers: two atria and two ventricles. The right side of the heart pumps blood to the lungs while the left side propels blood throughout the remainder of the body.

Chambers of the heart

Right atrium

This chamber consists of the right atrium proper and an atrial appendage, the right auricle (Fig. 3.18). The two parts are separated externally by a groove—the sulcus terminalis—and internally by a ridge—the crista terminalis. The area posterior to the crista terminalis is smooth walled, while the region anterior

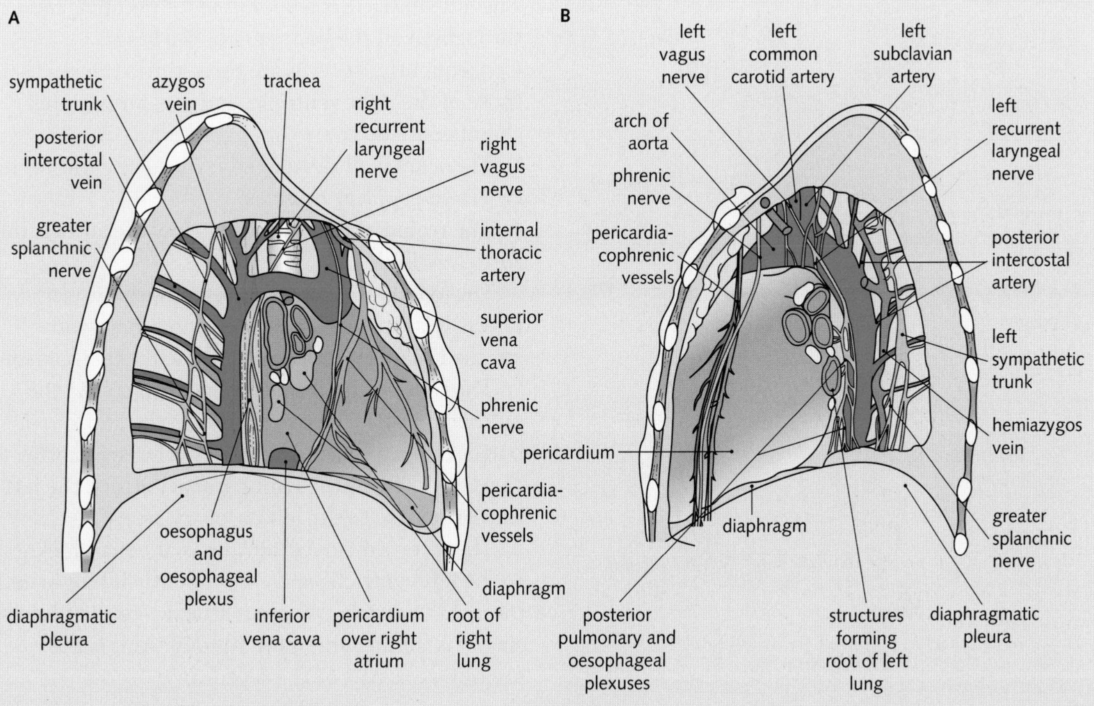

Fig. 3.15 (A) Right surface of the mediastinum and the right posterior thoracic wall. (B) Left surface of the mediastinum and the left posterior thoracic wall. Mediastinal pleura has been removed.

to it is ridged by muscle fibres – the musculi pectinati.

Interatrial septum

This forms the posterior wall of the right atrium. In the lower part of the septum is a depression, called the fossa ovalis. This represents the foramen ovale in the fetal heart. The anulus ovalis forms the crescentic upper margin of the fossa. It is the remnant of the lower edge of the septum secundum of the fetal heart.

Right ventricle

This chamber communicates with the right atrium via the tricuspid valve (see Fig. 3.18), and with the pulmonary artery through the pulmonary valve. The ventricle becomes funnel shaped as it approaches the pulmonary orifice—this region is known as the infundibulum.

The tricuspid valve has three cusps (anterior, posterior, and septal), the bases of which are attached to the fibrous ring of the skeleton of the heart.

The three cusps of the pulmonary valve are attached to the arterial wall at their lower margins. At the root of the pulmonary trunk are three dilations called the sinuses, one external to each cusp.

The ventricular wall has many muscular elevations, called trabeculae carneae, of which there are three types:

- Papillary muscles—attached to the ventricular wall and to the cusps of the tricuspid valve via fibrous cords, the chordae tendineae.
- Moderator band—forms part of the conducting system of the heart and runs from the septal wall to the anterior wall.
- Prominent muscular ridges.

Left atrium

This consists of a main cavity and an atrial appendage, the left auricle. The interior is smooth, but the auricle is ridged. The left atrium forms most of the base of the heart. The four pulmonary veins open into the posterior wall. The left atrioventricular orifice is protected by the bicuspid mitral valve.

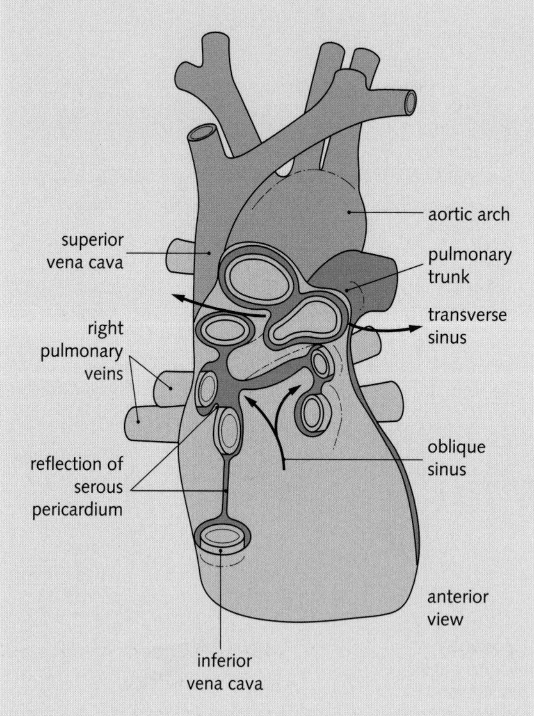

Fig. 3.16 Pericardial sinuses and reflections (the heart has been removed). Anterior view. (Adapted from *Gray's Anatomy*, 38th edn, edited by L H Bannister et al. Harcourt Brace and Co.)

Left ventricle

The left ventricle is responsible for pumping blood throughout all the body except the lungs. Consequently, its walls are three times thicker than those of the right ventricle, and pressures in this chamber are up to six times greater.

There are well-developed trabeculae carneae and two large papillary muscles.

The region of the ventricle below the aorta is the aortic vestibule.

The mitral valve guards the left atrioventricular orifice. It is bicuspid (anterior and posterior), with attached chordae tendineae similar to the tricuspid valve.

The three-cusped aortic valve is similar to the pulmonary valve. Above each cusp the aortic wall bulges to form the aortic sinuses. The right aortic sinus gives rise to the right coronary artery; the left sinus gives rise to the left coronary artery.

The interventricular septum is of equal thickness to the left ventricle, and consequently it bulges into the right ventricle. Where it attaches to the fibrous ring, it is thinner and more fibrous—this is the membranous part of the septum.

Skeleton of the heart

The two atria and two ventricles are attached to a pair of fibrous rings around the atrioventricular orifice. The fibrous rings separate the muscle

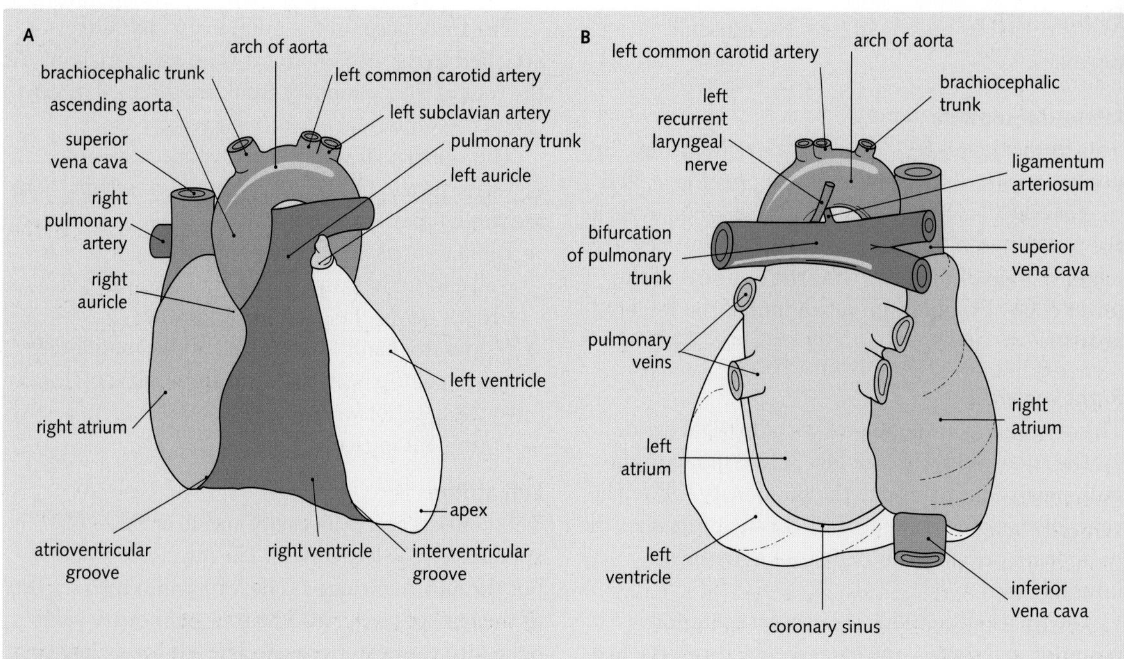

Fig. 3.17 Anterior (A) and posterior (B) surfaces of the heart.

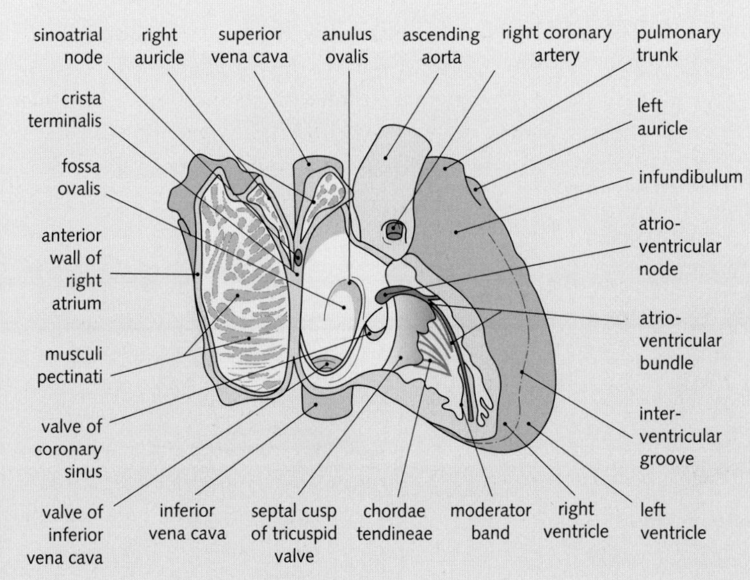

Fig. 3.18 Interior of right atrium and right ventricle.

fibres of the atria and ventricles, with the atrioventricular conducting system forming the only physiological connection between the atria and ventricles.

The bases of the cusps of the atrioventricular valves and the membranous part of the atrioventricular septum are also attached to the fibrous skeleton.

Blood supply of the heart

Fig. 3.19 illustrates the blood supply of the heart. The ascending aorta has two branches, which arise from aortic sinuses. These branches are the right and left coronary arteries. Both run in the coronary groove.

The right coronary artery in 60% of individuals supplies the sinoatrial node and in 90% of individuals supplies the atrioventricular node. The atrioventricular bundle and its right terminal branch are supplied by the right coronary artery. The left terminal branch is supplied by both right and left coronary arteries.

Left and right dominance of the heart is a reference to the coronary artery that the posterior interventricular branch arises from. Right dominance is more common: the posterior interventricular branch arises from the right coronary artery.

Narrowing of the coronary arteries (ischaemic heart disease), usually due to atheromatous deposition in their walls, gives rise to conditions such as angina pectoris and myocardial infarction.

Ischaemic heart disease is the most common cause of death in the Western world. The anterior interventricular artery is most commonly affected.

In a heart coronary bypass operation the internal thoracic artery or the saphenous vein can be used as a graft.

Venous drainage of the heart

Most of the venous blood in the heart drains into the coronary sinus, which lies in the posterior atrioventricular groove. It is a continuation of the great cardiac vein, and it opens into the right atrium.

The small and middle cardiac veins empty into the coronary sinus.

The remaining blood is returned to the heart via the anterior cardiac veins to the right atrium and other small veins (venae cordis minimae) that open directly into the heart chambers.

Conducting system of the heart

The heart contracts rhythmically at about 70 beats per minute.

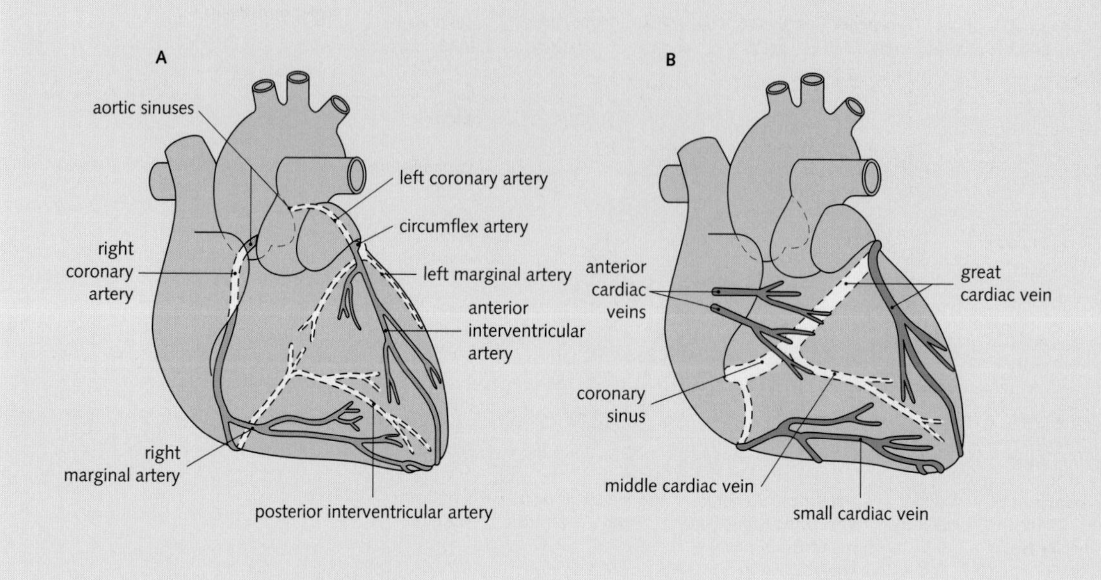

Fig. 3.19 Blood supply of the heart: (A) coronary arteries; (B) cardiac veins.

The sinoatrial node (pacemaker) lies in the upper part of the sulcus terminalis. Impulses generated here are transmitted throughout the atria, causing them to contract, and to the atrioventricular node, which lies in the atrial septum just above the attachment of the septal cusp of the tricuspid valve. The atrioventricular node conducts impulses to the atrioventricular bundle.

The atrioventricular bundle conducts the impulses to the inferior part of the interventricular septum. At this point the bundle divides into two branches, one for each ventricle:

- The right branch conducts the impulse to the apex of the right ventricle and, via the moderator band, to the anterior wall. The impulses are then distributed throughout the muscle.
- The left branch pierces the septum and passes down on the left side beneath the endocardium. It divides into two branches.

Referred pain from a myocardial infarction is transmitted through the cardiac branches of the cervical sympathetic trunk. Afferent fibres enter the spinal cord of the upper five thoracic nerves, and pain is referred to dermatomes these spinal nerves supply.

Fig. 3.20 details the nerve supply of the heart.

Great vessels of the thorax

Fig. 3.21 outlines the thoracic aorta and its branches.

Fig. 3.22 shows the aortic arch and pulmonary trunk.

Pulmonary trunk

This vessel transports deoxygenated blood from the right ventricle to the lungs.

Note the phrenic nerve passes anterior to the lung root. The vagus nerve passes posterior to the lung root.

The ligamentum arteriosum (see Fig. 3.22) is a fibrous band that connects the bifurcation of the pulmonary trunk to the aortic arch. It is the remnant of the ductus arteriosus, which in the fetus conducts blood from the pulmonary trunk to the aorta, bypassing the lungs.

Outline of the nerve supply of the heart		
Nerve type	**Origin**	**Action**
Sympathetic nerves	Cervical and upper thoracic part of the sympathetic trunk via the cardiac plexuses	Increase the rate and force of contraction
Parasympathetic nerves	Vagus nerves via the cardiac plexuses	Reduce the rate and force of contraction

Fig. 3.20 Outline of the nerve supply of the heart.

Aorta and its branches in the thorax		
Artery	**Course and origin**	**Branches**
Ascending aorta	Originates from the left ventricle, ascends and becomes the aortic arch at the level of the sternal angle	Right and left coronary arteries
Aortic arch	Arches posteriorly to the left of the trachea and the oesophagus, and continues as the descending aorta	Brachiocephalic trunk, left common carotid artery, left subclavian artery
Descending aorta	Descends to the left of the vertebral column, moves anterior to reach the midline, and leaves the thorax by passing behind the diaphragm at T12	Posterior intercostal arteries, subcostal arteries and visceral branches
Bronchial	Arises from the descending aorta	To bronchi and visceral pleura
Oesophageal	Arises from the descending aorta	To oesophagus
Superior phrenic	Arises from the descending aorta	To diaphragm
Posterior intercostal	Arises from the descending aorta	To intercostal muscles and thoracic wall

Fig. 3.21 Aorta and its branches in the thorax.

Superior vena cava

The superior vena cava drains the upper limbs and the head and neck directly. It receives the azygos vein (before entering the pericardium), which receives intercostal, pericardial, bronchial, mediastinal, oesophageal, and vertebral venous plexus tributaries. Thus, the superior vena cava drains thoracic structures indirectly. The superior vena cava drains into the right atrium.

Inferior vena cava

The vein pierces the central tendon of the diaphragm opposite T8 vertebra and almost immediately it enters the right atrium.

The azygos, accessory hemiazygos, and hemiazygos veins offer an alternative venous blood flow route from thoracic, abdominal, and back regions.

Fig. 3.23 shows the superior and inferior venae cavae and their main tributaries.

Pulmonary veins

Two pulmonary veins leave each lung, carrying oxygenated blood to the left atrium of the heart.

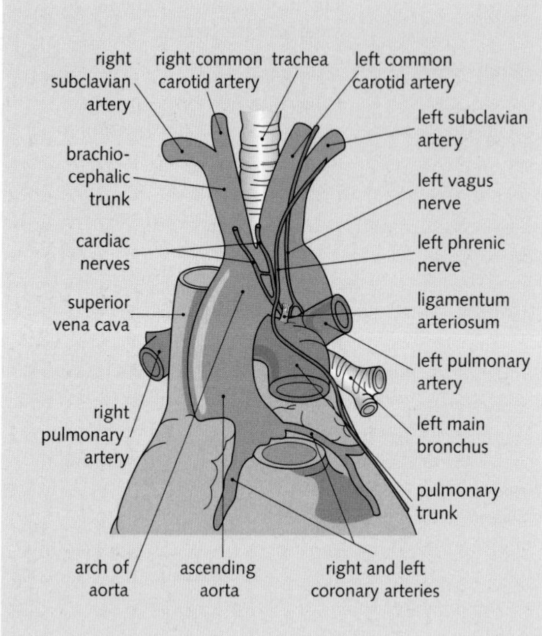

Fig. 3.22 Aorta and pulmonary trunk.

Azygos system of veins

Fig. 3.24 shows the azygos system of veins and the thoracic duct.

Nerves of the thorax

Fig. 3.25 lists the main nerves of the thorax (see also Fig. 3.15).

Thoracic sympathetic trunk

The sympathetic trunks follow a paravertebral course. Superiorly the trunks lie over the heads of the ribs, but inferiorly they lie over the body of the vertebrae. They leave the thorax by passing behind the medial arcuate ligaments.

There is usually a ganglion for each thoracic spinal nerve, but the first ganglion usually merges with the inferior cervical ganglion to form the stellate ganglion.

White and grey rami communicantes communicate with the thoracic spinal nerves.

The sympathetic trunk gives rise to the greater, lesser, and least splanchnic nerves. The greater nerve is formed by branches from the 5th to 9th ganglia. The lesser nerve is formed by branches from the 10th to 11th ganglia, and the least nerve is a branch of the 12th ganglion. These pass into the abdomen behind the diaphragm to relay in the coeliac ganglion and supply abdominal structures.

Lymphatic drainage of the thorax

Fig. 3.26 shows the lymphatic drainage of the thoracic cavity.

The thoracic duct (see Fig. 3.24) receives lymph from:

- The lower half of the body.
- The left posterior intercostal nodes.
- The left side of the head and neck, and the left upper limb, via the left jugular and subclavian lymph trunks, respectively.

The right lymph duct receives lymph from the posterior right thoracic wall. The right side of the head and neck, and the right upper limb drain into the right jugular and subclavian trunks, respectively. These vessels open into the great veins either independently or as a single trunk, the right lymph trunk.

Thymus

This is the major organ responsible for the maturation of T lymphocytes. It is a large bilobed organ at birth, lying in the superior and anterior divisions of the mediastinum, anterior to the great vessels. The gland involutes after puberty.

Trachea

The trachea (Fig. 3.27) commences in the neck, at the lower border of the cricoid cartilage. It passes into the superior mediastinum, anterior to the oesophagus and close to the midline. At the level of T4 vertebra, the trachea bifurcates into the right and left main bronchi. The right main bronchus is shorter, wider, and more vertical than the left—foreign bodies are therefore more likely to lodge in this bronchus or one of its branches.

Oesophagus

This organ commences as a continuation of the laryngopharynx at the level of C6 vertebra. In the thorax the oesophagus lies between the trachea and the vertebral column. It then passes through the diaphragm at the level of T10 vertebra and after 1–2 cm enters the stomach. Fibres from the right crus of the diaphragm form a sling around the oesophagus.

The upper oesophageal sphincter lies at the level of cricopharyngeus. This is the narrowest part of the oesophagus. Other constrictions in the oesophagus

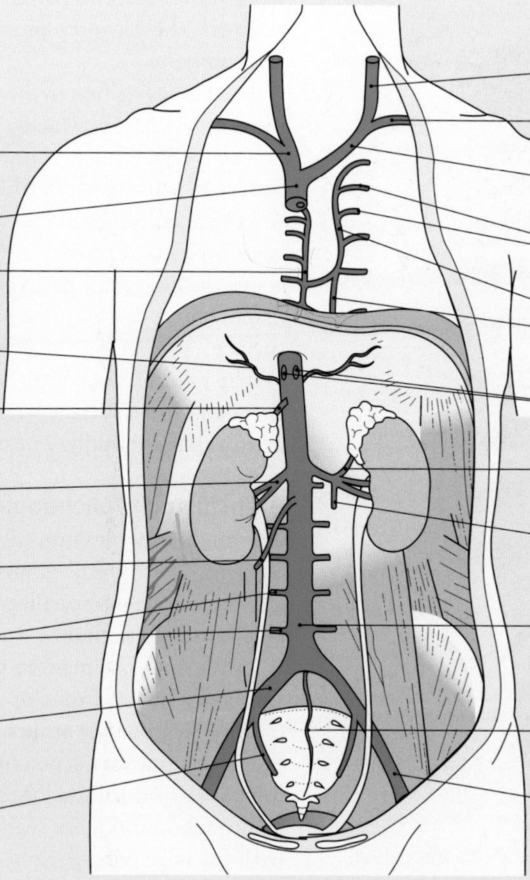

right brachiocephalic vein
superior vena cava
azygos vein
inferior phrenic vein
right suprarenal vein
right renal vein
right gonadal vein
right lumbar veins
right common iliac vein
right internal iliac vein
left internal jugular vein
left subclavian vein
left brachiocephalic vein
posterior intercostal veins
hemiazygos veins
hepatic veins
left suprarenal vein
left gonadal vein
inferior vena cava
median sacral vein
right external iliac vein

Fig. 3.23 Superior and inferior venae cavae and their main tributaries.

are found where it is crossed by the aortic arch and left main bronchus, and where it pierces the diaphragm.

Fig. 3.28 outlines the blood supply, lymphatic drainage, and nerve supply of the oesophagus.

Pleurae and lungs

Pleurae

The pleurae surround the lungs. The pleura is divided into two layers: the parietal and visceral pleurae (Fig. 3.29).

The parietal pleura lines the thoracic wall (costal pleura), the thoracic surface of the diaphragm (diaphragmatic pleura), and the lateral aspect of the mediastinum (mediastinal pleura). At the thoracic inlet, it arches over the lungs as the cervical pleura. It lies above the clavicle at this point.

The visceral pleura completely invests the outer surface of the lung and invaginates into the fissures of the lungs. Consequently, it is firmly adherent to the lungs.

The two layers become continuous with each other at the root of the lung. Here, there is a double layer of pleura that hangs down as the pulmonary ligament.

The pleural cavity lies between the two pleural layers. It is a potential space and, in health, contains a small quantity of clear pleural fluid.

Where the parietal pleura is reflected off the diaphragm onto the thoracic wall, a recess is formed that is not filled with lung except in deep inspiration. This is the costodiaphragmatic recess. A similar recess is formed between the thoracic wall and mediastinum (costomediastinal recess).

The nerve supply of the pleura is described in Fig. 3.30. The visceral pleura has no somatic

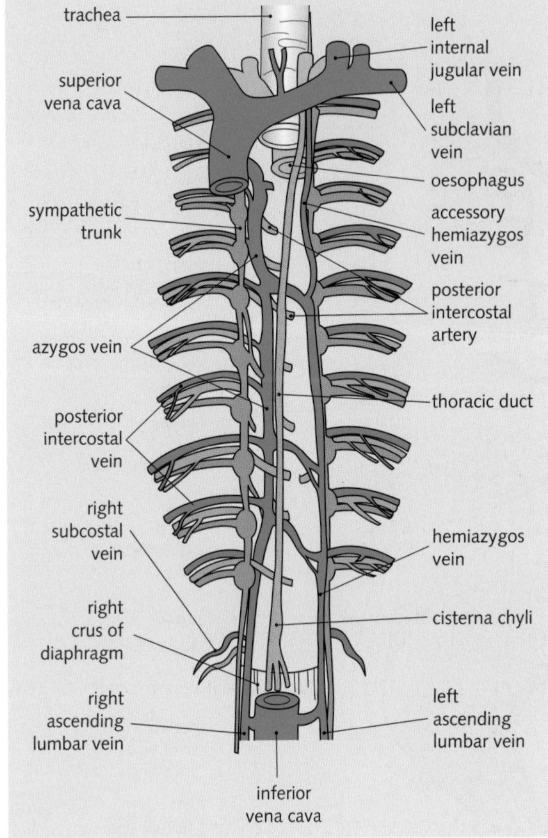

Fig. 3.24 Azygos system of veins and the thoracic duct.

innervation, and it is, therefore, insensitive to pain.

Blood supply of the pleurae

The parietal pleura is supplied by the intercostal arteries and branches of the internal thoracic artery. Venous drainage and lymphatic drainage are similarly shared.

The visceral pleura is supplied by the bronchial arteries. It shares venous and lymphatic drainage with the lung.

Endothoracic fascia

This layer of loose connective tissue separates the parietal pleura from the thoracic wall. It includes the suprapleural membrane, which covers the dome of the parietal pleura where it projects into the root of the neck.

Lungs

The lungs in life are light, spongy, and elastic. The surface changes from a pink colour at birth to a mottled darker colour in later life because of deposition of carbon particles from atmospheric pollution. This is more pronounced in city-dwellers and smokers.

Each lung lies free in the pleural cavity except at its root, where it is attached to the mediastinum.

Surfaces and borders of the lungs

These are indicated in Figs 3.31 and 3.32. Each lung has an apex that projects into the neck about 1 cm above the clavicle, a base that lies against the diaphragm, a costal surface, and a mediastinal surface.

The hilum (root) of the lung (see Fig. 3.32) lies on the mediastinal surface where the bronchi and neurovascular bundles enter the lungs.

Bronchi and bronchopulmonary segments

The trachea divides into the right and left main bronchi. These are fibromuscular tubes reinforced by incomplete rings of cartilage. They are lined by respiratory epithelium.

In the lung, the main bronchus divides into secondary (lobar) bronchi, which in turn divide into tertiary (segmental) bronchi. The latter supply the bronchopulmonary segments, which are the functional units of the lungs (see Fig. 3.33).

Bronchopulmonary segments are wedge shaped, with the base lying peripherally and the apex lying towards the root of the lungs. There are 10 segments for each lung. Each segment has its own segmental bronchus, segmental artery, lymphatic vessels, and autonomic nerves. The veins lie both in and between segments. Within each segment the segmental bronchus repeatedly subdivides and reduces in size until the cartilage in the walls disappears. The airway is now called a bronchiole. These subdivide and they eventually give rise to the alveoli, which are thin-walled sacs with no muscle in the wall. These subdivisions greatly increase the surface area of the lungs, allowing efficient exchange of gases.

The basal superior bronchopulmonary segment is at 90 degrees to the bronchial tree. In supine patients, bronchial secretions drain into this segment and cause pneumonia.

Nerves of the thorax	
Nerve (origin)	**Course and distribution**
Vagus, X (medulla oblongata)	Enters superior mediastinum posterior to sternoclavicular joint and brachiocephalic vein to supply pulmonary plexus, oesophageal plexus, and cardiac plexus
Phrenic (anterior rami of C3–C5)	Enters thorax and runs between mediastinal pleura and pericardium to supply motor and sensory innervation to diaphragm and sensory innervation to mediastinal pleura, pericardium, diaphragmatic pleura, and peritoneum
Intercostal nerves (anterior rami of T1–T11)	Run between internal and innermost layers of intercostal muscles and supply skin, muscles, and parietal pleura of intercostal spaces. Lower intercostal nerves also supply skin, muscles, and peritoneum of anterior abdominal wall
Subcostal (anterior ramus of T12)	Follows inferior border of 12th rib and passes into abdominal wall
Recurrent laryngeal (X)	Loops around subclavian artery on right and arch of aorta on left, and ascends in tracheoesophageal groove; supplies intrinsic muscles of larynx (except cricothyroid) and sensation inferior to level of vocal folds
Cardiac plexus (X and sympathetic trunks)	Fibres pass along coronary arteries to sinoatrial node; parasympathetic fibres reduce heart rate and force of contraction, sympathetic fibres increase rate and contraction force
Pulmonary plexus (X and sympathetic trunks)	Plexus forms on root of lung and extends along branches of bronchi; parasympathetic fibres constrict bronchioles, sympathetic fibres dilate them
Oesophageal (X and sympathetic trunks)	Vagus and sympathetic nerves form a plexus around oesophagus to supply muscle and glands of oesophagus

ig. 3.25 Nerves of the thorax.

Jerve supply of the lungs

ympathetic and parasympathetic nerves from the ulmonary plexuses, which lie anterior and posterior to he lung roots, supply the smooth muscle of the ronchial tree, the vessels, and the mucous nembrane.

Blood supply of the lungs

'he lungs have a dual blood supply:

- Bronchial arteries from the thoracic aorta supply the bronchi, the connective tissue of the lung, and the visceral pleura. Bronchial veins drain into the azygos system of veins.
- Pulmonary arteries transport deoxygenated blood to the alveolar capillaries. The pulmonary veins return oxygenated blood to the lungs.

'here are anastomoses between the pulmonary and ronchial circulations.

Pulmonary thromboembolism is a common ause of morbidity and mortality in the UK. pproximately 10% of hospital patients develop a ulmonary embolus (PE); although most cases are lent, in severe cases this can lead to pulmonary nfarction or sudden death. The source of the mbolus is often from the deep veins of the legs.

Lymphatic drainage of the lungs

Fig. 3.34 illustrates the lymphatic drainage of the lungs.

Bronchial carcinoma is the most common lethal cancer in the UK and USA. It usually spreads via the lymphatics, and it has a poor prognosis. An understanding of the anatomy and lymphatic drainage is essential for planning treatment and assessing the prognosis of these tumours.

In bronchocarcinoma, enlarging of the tracheobronchial nodes can widen the tracheal bifurcation, and also it may compress the left recurrent laryngeal nerve, paralysing the left laryngeal muscles and causing voice hoarseness and weakness.

Mechanics of respiration

Respiration consists of an inspiratory and an expiratory phase (Fig. 3.35).

Patients with severe obstruction of the airways, e.g. severe asthma or chronic obstructive airways

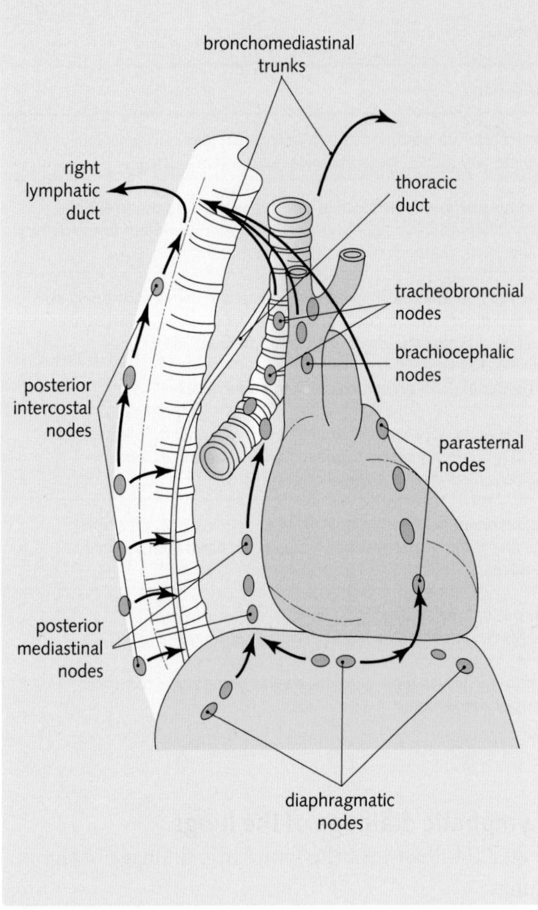

Fig. 3.26 Lymphatic drainage of the thoracic cavity.

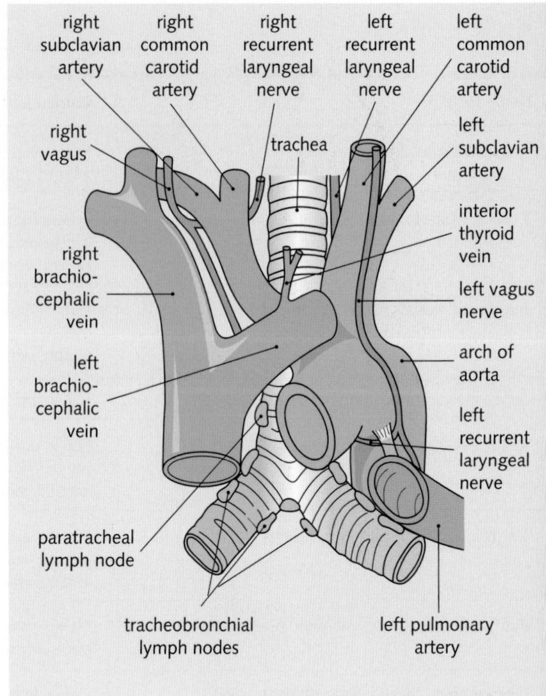

Fig. 3.27 Trachea and its main relations anteriorly and laterally.

disease, can be seen using their accessory muscles of respiration, which include the sternocleidomastoid , pectoralis, scalene, and anterior abdominal muscles. Signs include intercostal recession, suprasternal recession, pursing of the lips, and contraction of the sternocleidomastoid and platysma.

Types of respiration

In adult males diaphragmatic movement is greater than thoracic movement and, therefore, males are said to have an abdominal type of respiration. Adult females have a greater thoracic movement and, therefore, use a thoracic type of respiration. Infants under the age of two depend on diaphragmatic movements because their ribs are horizontal.

Blood and nerve supply, and lymphatic drainage of the oesophagus

	Upper oesophagus	Middle oesophagus	Lower oesophagus
Arterial supply	Inferior thyroid artery	Oesophageal branches of the aorta	Left gastric artery
Venous drainage	Brachiocephalic vein	Azygos vein	Oesophageal tributaries of the left gastric veins which drain finally into the portal vein
Nerve supply	Recurrent laryngeal nerves and sympathetic fibres from cell bodies in the middle cervical ganglion running on the inferior thyroid artery	Fibres from the anterior and posterior oesophageal plexus from the vagus nerves; sympathetic fibres from the sympathetic trunks and greater splanchnic nerves	
Lymphatic drainage	Deep cervical nodes near the origin of the inferior thyroid artery	Tracheobronchial and posterior mediastinal nodes	Preaortic nodes of the coeliac group

Fig. 3.28 Blood and nerve supply, and lymphatic drainage of the oesophagus. (Adapted from *Anatomy as a Basis for Clinical Medicine*, by E C B Hall-Craggs. Williams & Wilkins.)

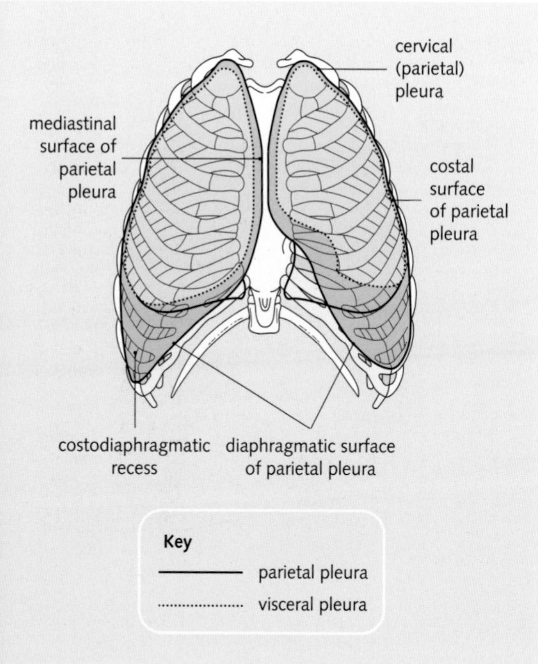

Fig. 3.29 Pleurae.

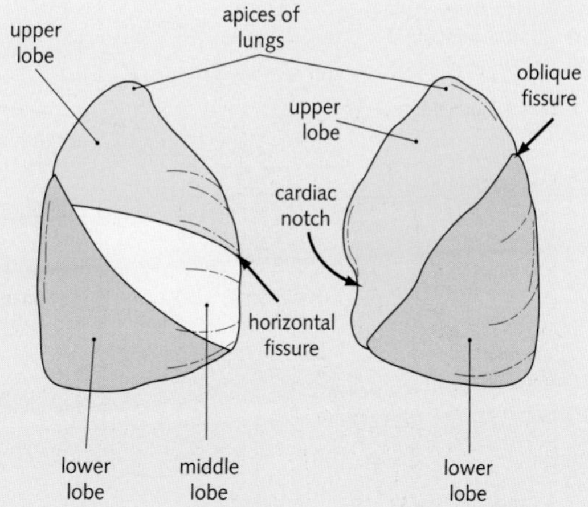

Fig. 3.31 Surface of the lungs.

Nerve supply of the pleura	
Pleura	**Nerve supply**
Costal	Segmentally by the intercostal nerves
Mediastinal	The phrenic nerve
Diaphragmatic	The phrenic nerve centrally, the lower five intercostal nerves peripherally
Visceral	Autonomic nerve supply from the pulmonary plexus

Fig. 3.30 Nerve supply of the pleura.

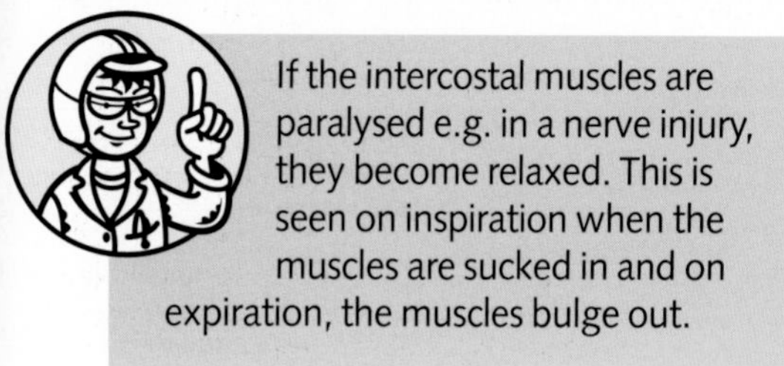

If the intercostal muscles are paralysed e.g. in a nerve injury, they become relaxed. This is seen on inspiration when the muscles are sucked in and on expiration, the muscles bulge out.

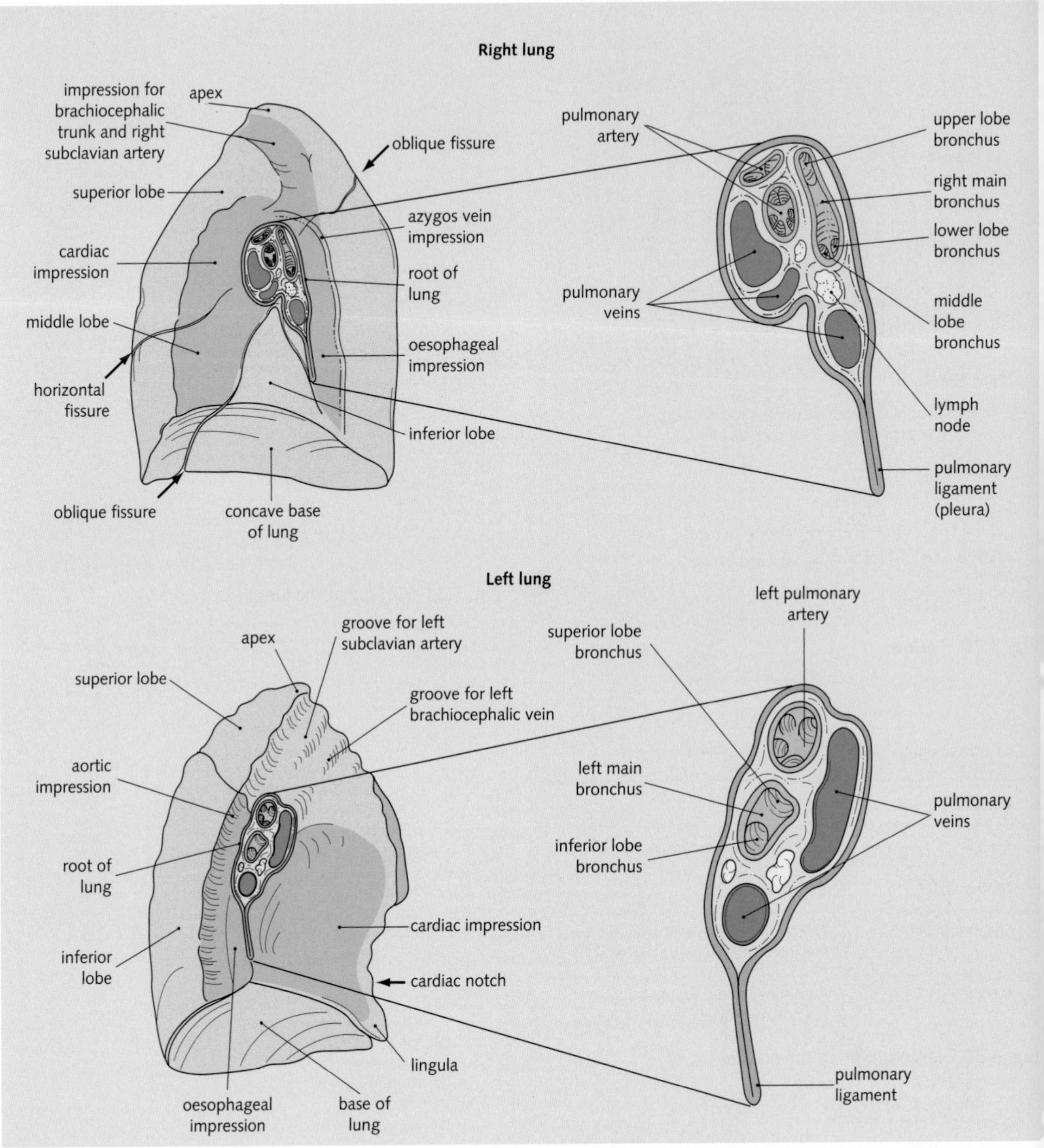

Right lung

impression for brachiocephalic trunk and right subclavian artery
apex
superior lobe
cardiac impression
middle lobe
horizontal fissure
oblique fissure
concave base of lung
oblique fissure
azygos vein impression
root of lung
oesophageal impression
inferior lobe
pulmonary artery
pulmonary veins
upper lobe bronchus
right main bronchus
lower lobe bronchus
middle lobe bronchus
lymph node
pulmonary ligament (pleura)

Left lung

apex
groove for left subclavian artery
superior lobe
groove for left brachiocephalic vein
aortic impression
root of lung
inferior lobe
cardiac impression
cardiac notch
lingula
oesophageal impression
base of lung
superior lobe bronchus
left pulmonary artery
left main bronchus
inferior lobe bronchus
pulmonary veins
pulmonary ligament

Fig. 3.32 Medial aspect of the right and left lungs and contents of their roots. Impressions are only seen in fixed lungs.

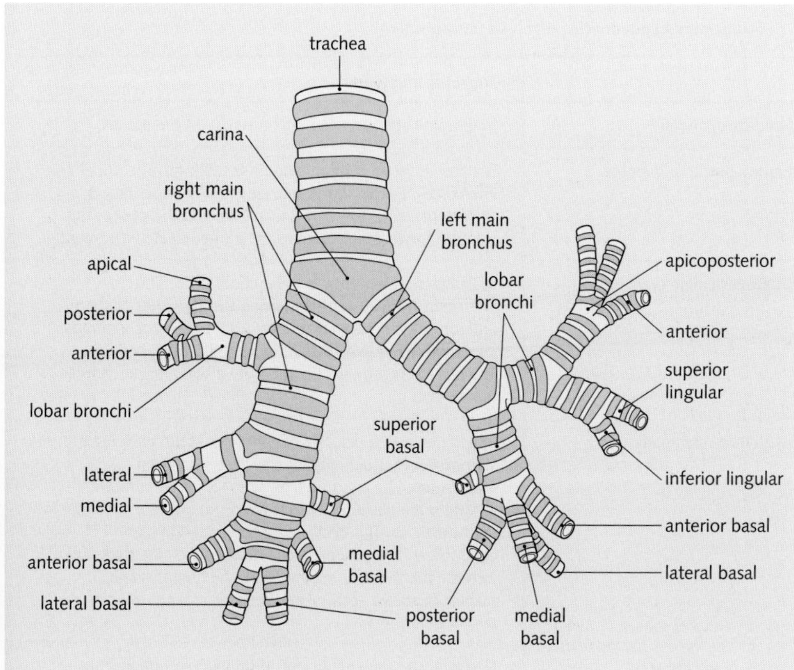

Fig. 3.33 The bronchial tree of the lung.

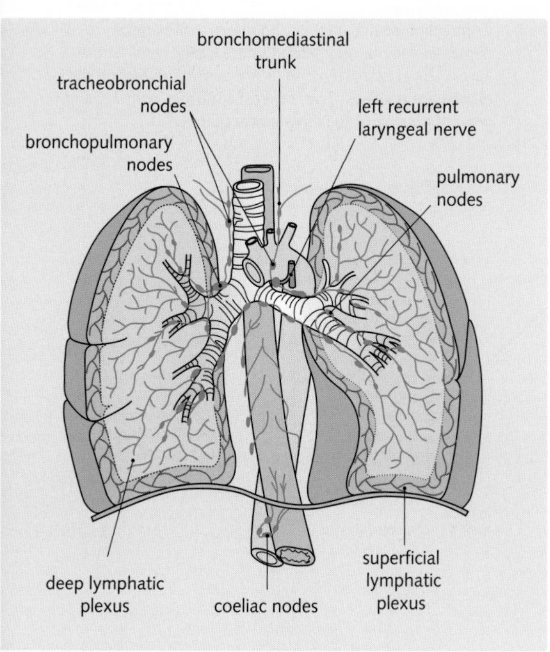

Fig. 3.34 Lymphatic drainage of the lungs.

Movements and muscles involved in respiration		
Movements	**Effect**	**Muscles involved**
Quiet inspiration	Increased vertical diameter	Diaphragm contracts causing flattening of the domes
Quiet inspiration	Increased anteroposterior diameter	Scalenus muscles contract and fix the first rib. The upper intercostal muscles contract to elevate the upper ribs at their sternal ends towards the first rib, and this pushes the sternum forwards—known as the pump handle movement
Quiet inspiration	Increased transverse diameter	Scalenus muscles contract fixing the first rib. The intercostal muscles contract raising the lower ribs along an anteroposterior axis that runs through the costochondral joint (anteriorly) and costovertebral joint (posteriorly). The ribs are raised upward and outward in a bucket handle movement
Forced inspiration	Increase of all three diameters	In addition to the diaphragm contracting, the scalenus and sternocleidomastoid muscles elevate the ribs and manubrium. The intercostal muscles forcefully contract elevating the ribs. Quadratus lumborum lowers and fixes the twelfth rib. This allows a forceful diaphragmatic contraction. The erector spinae muscles arch the back and increase the thoracic volume. With the humerus and scapula fixed the pectoral and serratus anterior muscles raise the ribs
Quiet expiration	Decrease of all three diameters	Elastic recoil of the lungs and controlled relaxation of the intercostal and diaphragmatic muscles causes this passive movement
Forced expiration	Decrease of all three diameters	Contraction of the anterior abdominal wall muscles depresses the ribs and reinforces the elastic recoil of the lungs. The abdominal contents are pushed up forcing the diaphragm upwards. The intercostal muscles contract and prevent bulging of their intercostal spaces

Fig. 3.35 Movements and muscles involved in respiration.

- Outline the surface markings of the lung and pleurae.
- Outline the surface anatomy of the heart and its valves.
- What are the surface markings of the great vessels?
- Describe the lymphatic drainage of the breast and its importance.
- Describe the differences between typical and atypical ribs.
- List the boundaries of the thoracic inlet and outlet.
- Describe a typical intercostal space.
- Describe the anatomy of the diaphragm, including structures passing through and peripheral to it.
- List the divisions of the mediastinum and their contents.
- Describe the pericardium and how the pericardial sinuses are formed.
- Describe the structure of the heart chambers.
- Define the blood supply to the heart, including left/right dominance and venous drainage.
- Describe the thoracic sympathetic trunk.
- Discuss the lymphatic drainage of the thorax.
- Discuss the anatomy of the oesophagus, include its course, innervation and blood supply.
- Outline the blood supply and innervation of the pleurae.
- List the structures and their relative positions in the lung root (hilum).
- Describe the divisions of the bronchial tree.
- Outline the blood supply and innervation of the lungs.
- Discuss the mechanics of respiration.

4. The Abdomen

Regions and components of the abdomen

The abdominal cavity is separated from the thoracic cavity by the diaphragm. Because the domes of the diaphragm arch high above the costal margin, the upper part of the abdomen—including the liver, the spleen, the upper poles of the kidneys, and the suprarenal glands—is protected by the bony thoracic cage. The lower part of the abdominal cavity lies in the bony pelvis.

Posteriorly, the vertebral column protects the abdominal contents, but anteriorly and laterally the abdomen is more vulnerable to injury, with only a muscular wall for protection.

Over the anterior abdominal wall, a part of the superficial fascia condenses into a strong but thin membranous layer under the fat. This allows the fatty layer of the superficial fascia to move freely during thoracic and abdominal movements e.g. respiration. This membranous superficial fascia (Scarpa's fascia) fades over the thoracic wall superiorly, and it fuses with the fascia lata of the thigh inferiorly. In males it continues into the scrotum and penis as the superficial perineal fascia (of Colles) and the superficial fascia of the penis respectively. In the female the superficial perineal fascia lines the labia majora, and it is split centrally by the presence of the vagina.

The anterolateral abdominal wall is made up of a muscular sheet composed of three muscle layers. These are separated laterally, but they fuse anteriorly to surround the rectus abdominis.

The abdominal cavity contains most of the alimentary tract (stomach, duodenum, and small and large intestines) together with its derivatives (liver and pancreas). Parts of the viscera (e.g. small intestines and transverse colon) are attached to double folds of peritoneum, called mesentery, whilst others (e.g. duodenum) are bound down to the posterior abdominal wall. The kidney, suprarenal glands, and ureters lie in the posterior abdominal wall, behind the peritoneum.

Abdominal pain is a very common presentation—knowledge of the embryology and anatomy of the region is crucial in obtaining the correct diagnosis (Fig. 4.1).

Surface anatomy and superficial structures

To facilitate description, the abdomen is divided into regions. The simplest method is to divide the abdomen into four quadrants by vertical and horizontal lines through the umbilicus; however, for more accurate description, it is divided into nine regions by two vertical and two horizontal lines (Fig. 4.2):

- The vertical line on each side corresponds to the midclavicular line, which extends down to the midinguinal point.

Origin and blood supply of the abdominal viscera

Part of fetal gut	Organs	Blood supply	Usual site of presentation of abdominal pain
Foregut	Oesophagus, stomach, first and second part of duodenum, liver, spleen, pancreas	Coeliac artery	Epigastric region
Midgut	Remainder of duodenum, jejunum, ileum, caecum, appendix, ascending colon, right two-thirds of transverse colon	Superior mesenteric artery	Umbilical region
Hindgut	Remainder of transverse colon, descending colon, rectum	Inferior mesenteric artery	Suprapubic region

Fig. 4.1 Origin and blood supply of the abdominal viscera.

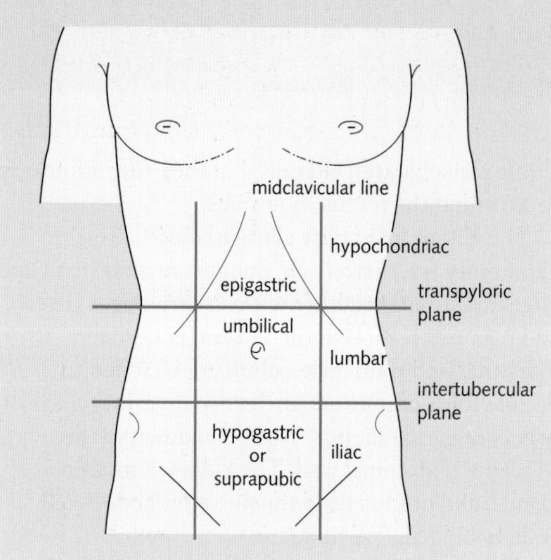

Fig. 4.2 Regions of the abdomen.

- The lower transverse line runs between the two tubercles of the iliac crest (intertubercular plane).
- The upper transverse line lies midway between the pubic symphysis and jugular notch (transpyloric plane).

The linea alba is a midline depression running from the xiphisternum to the pubis. The linea semilunaris is a smooth, curved line, representing the lateral margin of the rectus abdominis.

The inguinal ligament runs from the anterior superior iliac spine to the pubic tubercle. The deep inguinal ring lies at the midinguinal point (halfway between the anterior superior iliac spine and the pubic tubercle).

The body wall has three muscle layers in the abdomen. These layers fuse centrally to form the rectus sheath.

Liver

The inferior border extends from the right 10th costal cartilage in the midclavicular line to the left 5th rib in the midclavicular line. The upper border runs between the left and right 5th ribs; both points are in the midclavicular line. The right border runs from the 5th right rib to the 10th costal cartilage.

A liver biopsy is performed in the 10th intercostal space in the midaxillary line. The patient holds their breath after expiring to reduce the costodiaphragmatic recess and thus the chance of lung damage.

Fundus of the gall bladder

This lies deep to the intersection of the linea semilunaris with the costal margin in the transpyloric plane. At this point the fundus of the gall bladder lies behind the 9th costal cartilage.

Spleen

The spleen lies deep to the 9th, 10th, and 11th ribs on the left behind the midaxillary line. It is not palpable unless it is enlarged, at which point the spleen extends inferiorly and anteriorly along the 10th rib to below the costal margin.

Pancreas

The head of the pancreas lies in the 'C' shaped concavity of the duodenum at the level of the L2 vertebra. The neck of the pancreas lies anterior to the L1 vertebra in the transpyloric plane. The pancreas continues to the left, curving upwards towards the hilum of the spleen.

Kidneys

The hilum lies in the transpyloric plane, 5 cm from the midline. The upper pole of the kidneys lie deep to the 12th rib posteriorly. The right kidney is lower than the left due to the presence of the liver, but they both lie roughly opposite the first three lumbar vertebrae.

Ureters

Each ureter begins at the hilum of the kidney in the transpyloric plane. They run inferiorly over the psoas major muscle anterior to the tips of the transverse processes of the lumber vertebrae (as seen on a urogram) to the sacroiliac joint to enter the pelvis.

The abdominal wall

Skeleton

Fig. 4.3 shows the skeleton of the abdominal and pelvic cavities.

The costal margin and floating ribs have been described previously (see Chapter 3).

The hip bones articulate with the sacrum at the sacroiliac joint and with each other at the pubic symphysis. Each pelvic bone is formed by ilium, ischium, and pubis.

The iliac bones protect the underlying structures, providing a site for muscle attachment. The upper border—the iliac crest—is limited by the anterior superior iliac spine (ASIS) and the posterior superior iliac spine (PSIS), anteriorly and posteriorly, respectively. The iliac tubercle lies behind the ASIS.

The three sheets of muscle of the anterolateral abdominal wall arise from the iliac crest. The latissimus dorsi, quadratus lumborum, and thoracolumbar fascia are also attached to the crest.

The pectineal line lies on the superior surface of the superior ramus of the pubic bone, and medial to it lie the pubic tubercle and pubic crest.

Thoracolumbar fascia

The lumbar part of this fascia arises in three sheets from:

- Tips of the lumbar spines (posterior sheet).
- Tips of the lumbar transverse processes (middle sheet).
- The anterior aspect of the lumbar transverse processes (anterior sheet).

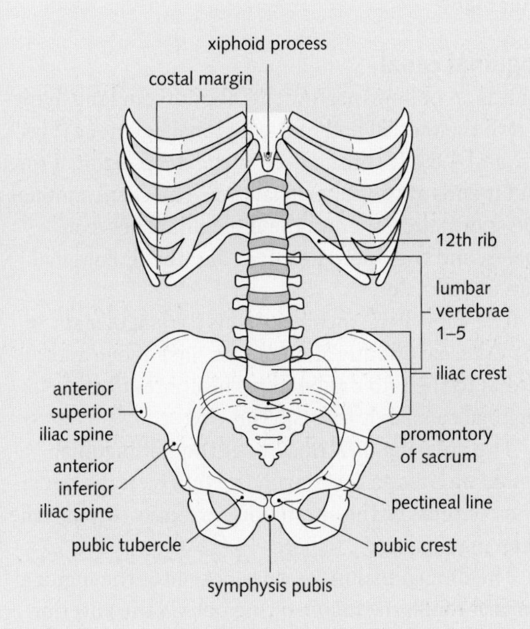

Fig. 4.3 Skeleton of the abdomen and pelvis.

The anterior and middle sheets enclose the quadratus lumborum muscle; the middle and posterior sheets enclose the erector spinae muscle. The three sheets fuse laterally and provide attachment for the internal oblique and transversus abdominis muscles. The thoracic part is formed only by the posterior sheet. This attaches to the thoracic spines and angles of the ribs as far as the first rib.

Muscles of the anterolateral abdominal wall

Fig. 4.4 outlines the muscles of the anterolateral abdominal wall.

Rectus sheath

Each rectus abdominis muscle is enclosed in a fibrous sheath formed by the aponeurotic tendons of the three lateral muscles (Fig. 4.5).

The external oblique muscle contributes to the anterior layer of the sheath over its entire extent. Below the costal margin, the internal oblique aponeurosis splits around the muscle, forming the anterior and posterior layers. The aponeurosis of transversus abdominis contributes to the posterior layer.

Midway between the symphysis pubis and the umbilicus, the posterior wall of the sheath becomes deficient, and all the aponeuroses pass anterior to the rectus muscle. The free posterior margin thus formed is called the arcuate line. The inferior epigastric artery enters the sheath here, and it runs on the deep surface of rectus abdominis. Below the arcuate line, the posterior surface of the rectus abdominis is in contact with the transversalis fascia.

The posterior wall of the sheath is also deficient above the costal margin, where the rectus muscle lies directly on the underlying costal cartilages.

Nerve and blood supply to the anterolateral abdominal wall

The principal nerves and arteries of the anterolateral abdominal wall are shown in Fig. 4.6.

Venous drainage of the anterolateral abdominal wall

The superficial veins include the superficial epigastric and thoracoepigastric veins. These drain ultimately into the femoral vein and axillary veins, respectively.

The superior and inferior epigastric veins and the deep circumflex iliac veins follow the arteries and drain into the internal thoracic and external iliac veins.

Muscles of the anterolateral abdominal wall			
Name of muscle (nerve supply)	**Origin**	**Insertion**	**Action**
External oblique (T6–T12 spinal nerves)	Lower ribs	Becomes aponeurotic and attaches to the xiphoid process, linea alba, pubic crest, pubic tubercle, and iliac crest	Flexes and rotates trunk; pulls down ribs in forced expiration
Internal oblique (spinal nerves T6–T12, iliohypogastric and ilioinguinal nerves)	Lumbar fascia, iliac crest, lateral two-thirds of inguinal ligament	Lower three ribs and costal cartilages, xiphoid process, linea alba, symphysis pubis; forms conjoint tendon with transversus	Assists in flexing and rotating trunk; pulls down ribs in forced expiration
Transversus abdominis (spinal nerves T6–T12, iliohypogastric and ilioinguinal nerves)	Lower six costal cartilages, lumbar fascia, iliac crest, lateral third of inguinal ligament	Xiphoid process, linea alba, symphysis pubis, forms conjoint tendon with internal oblique	Compresses abdominal contents with external and internal oblique
Rectus abdominis (spinal nerves T6–T12)	Symphysis pubis and pubic crest	Costal cartilages 5–7 and xiphoid process	Compresses abdominal contents and flexes vertebral column

Fig. 4.4 Muscles of the anterolateral abdominal wall.

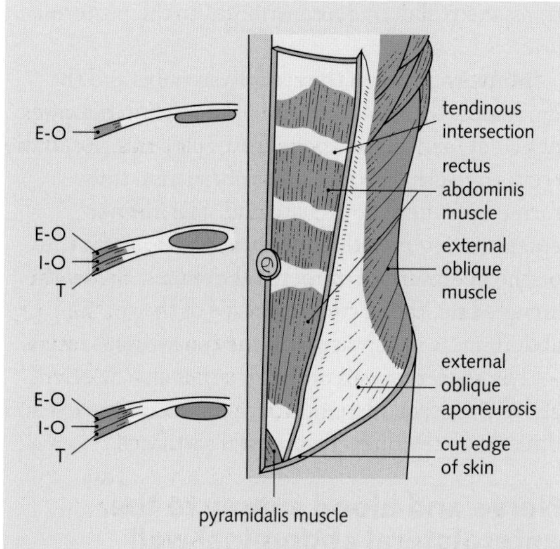

Fig. 4.5 Rectus sheath and rectus abdominis muscle. (E-O, external oblique; I-O, internal oblique; T, transversus abdominis)

The lower two posterior intercostal veins drain into the azygos veins. Of the four lumbar veins, the lower two drain into the inferior vena cava. The upper two join to form the ascending lumbar vein, and, with the subcostal vein, drain into the azygos vein on the right and hemiazygos vein on the left. Blood may return to the heart via the superficial abdominal veins if the inferior vena cava becomes obstructed.

Inguinal region

Inguinal ligament

The inguinal ligament is the lower free edge of the aponeurosis of the external oblique muscle. It extends from the ASIS to the pubic tubercle, and it gives origin to the internal oblique and transverse abdominis muscles and the fascia lata of the thigh.

Inguinal canal

This is an oblique narrow slit, about 6 cm long, lying above the medial half of the inguinal ligament (Figs 4.7 and 4.8). It commences at the deep inguinal ring, and it ends at the superficial ring. The canal contains the spermatic cord and the ilioinguinal nerve in males, and the round ligament and the ilioinguinal nerve in females.

The superficial inguinal ring is a triangular slit in the external oblique aponeurosis, just above and lateral to the pubic tubercle. The contents of the inguinal canal exit through this ring.

The deep inguinal ring lies at the midinguinal point, and it is an opening in the transversalis fascia. The contents of the spermatic cord pass through the deep inguinal ring.

The ilioinguinal nerve does not enter the inguinal canal through the inguinal ring but via the anterior wall of the canal, by running between the external oblique aponeurosis and the internal oblique muscle.

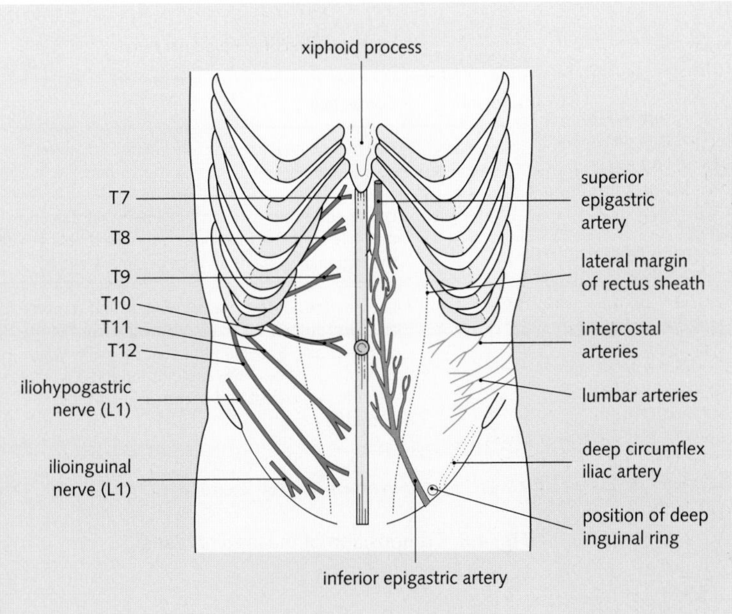

Fig. 4.6 Innervation (left) and arterial supply (right) of the anterolateral abdominal wall.

Spermatic cord

The structures entering the deep inguinal ring pick up three coverings from the layers of the abdominal wall as they pass through the canal to form the spermatic cord (Fig. 4.9). The spermatic cord with all of its coverings is not complete until it emerges from the superficial inguinal ring.

Contents of the spermatic cord comprise:

- The ductus deferens.
- Arteries—the testicular artery (from the abdominal aorta) and the artery to the ductus deferens (from the superior and inferior vesical arteries).
- Veins—the pampiniform plexus of veins.
- Lymphatics—accompany the veins from the testis to the para-aortic nodes.
- Nerves—the genital branch of the genitofemoral nerve supplies the cremaster muscle and sympathetic nerves go to the arteries.
- The processus vaginalis—the obliterated remains of the peritoneal connection with the tunica vaginalis of the testis.

In the male, the genitofemoral nerve's genital branch supplies the cremaster muscle. Its femoral branch supplies a small skin area on the thigh. Stimulation of the femoral branch causes the cremaster muscle to contract, raising the testis. This is called the cremasteric reflex.

Scrotum

This is a sac-like structure lying below the root of the penis. It contains the testis, the epididymis, and the lower end of the spermatic cord.

Scrotal skin is thin and wrinkled. Beneath this is the superficial fascia, which contains the dartos muscle but no fat. This smooth muscle contracts in response to cold, pulling the testes closer to the body and wrinkling the skin. It is supplied by sympathetic nerves. The fascia also forms a median partition that separates the testes. Beneath the dartos muscle is a layer of membranous fascia: the superficial perineal fascia (of Colles). This is continuous with the membranous layer of the superficial fascia of the anterior abdominal wall (Scarpa's fascia).

Blood supply, lymphatic drainage, and nerve supply of the scrotum

The superficial and deep external pudendal arteries (of the femoral artery of the thigh) supply the scrotum. The venous drainage is to the great saphenous vein through the superficial and deep venous tributaries. The lymphatic drainage is the medial superficial inguinal lymph nodes of the thigh.

The nerve supply to the anterior one third of the scrotum is by the ilioinguinal nerve. The posterior

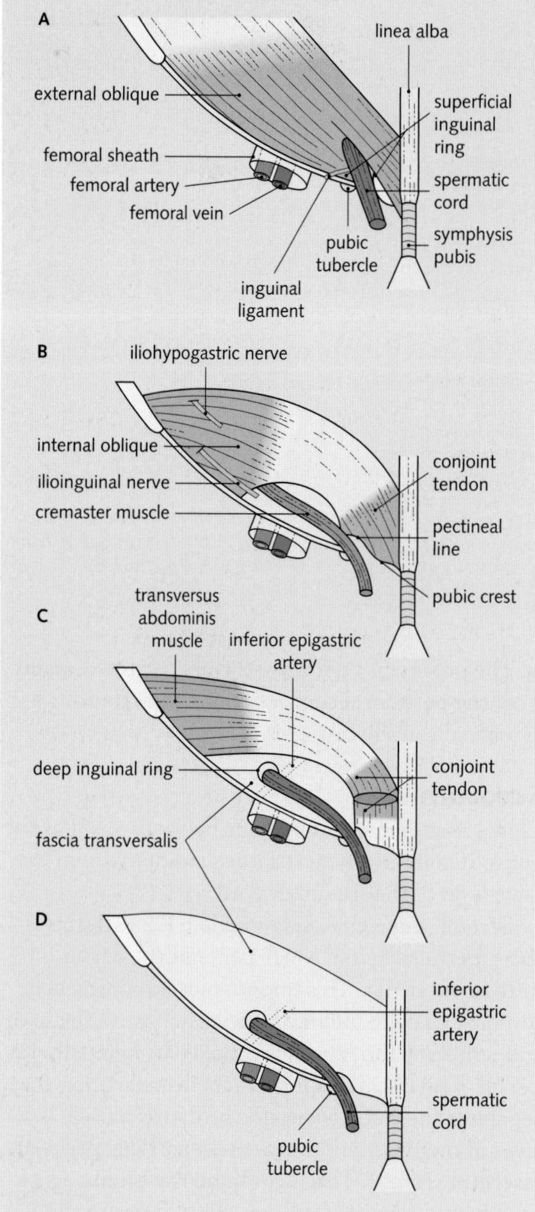

Fig. 4.7 Inguinal canal viewed at different levels. (A) Superficial inguinal ring in external oblique muscle. (B, C) Internal oblique and transversus muscles and the conjoint tendon. (C, D) Deep inguinal ring in fascia transversalis.

two thirds is innervated by the posterior scrotal branch of the perineal nerve (medially) and the perineal branch of the posterior femoral cutaneous nerve (laterally).

Testis

This oval organ lies at the lower end of the spermatic cord (Fig. 4.9). It has the epididymis attached to its posterolateral surface. The anterior and posterior surfaces lie free in a serous space formed by the tunica vaginalis, a remnant of the fetal processus vaginalis.

The testis has a tough fibrous coat, the tunica albuginea. This sends numerous fibrous septules into the gland, dividing it into testicular lobules that contain the seminiferous tubules. Posteriorly the tubules form the rete testis and the efferent ducts, which open into the head of the epididymis.

Composition of the inguinal canal	
Region	**Components**
Anterior wall	External oblique aponeurosis; reinforced laterally by internal oblique
Floor	Lower edge of the inguinal ligament; reinforced medially by the lacunar ligament, which lies between the inguinal ligament and the pectineal line
Roof	Lower edges of the internal oblique and transversus muscles: these muscles arch over the front of the cord laterally, to behind the cord medially, where their joint tendon—the conjoint tendon—is inserted into the pubic crest and pectineal line of the pubic bone
Posterior wall	The strong conjoint tendon medially and the weak transversalis fascia laterally

Fig. 4.8 Composition of the inguinal canal.

Epididymis

This is a long coiled tube attached to the posterior border of the testis. Its head lies at the upper pole of the testis, and it is joined by the efferent ductules. The head gives rise to the body and tail.

The epididymis stores sperm and allows them to mature.

Ductus deferens

This tube transmits sperm from the testis and epididymis to the prostatic urethra. It receives its blood supply from a small artery from the superior vesical artery. This small artery anastomoses with the testicular artery.

Blood supply of the testis

The testicular artery runs in the spermatic cord, and it supplies the testis and the epididymis. The pampiniform plexus provides venous drainage. In the inguinal canal the plexus merges into four veins, which join to form two veins that leave the deep inguinal ring. The left testicular vein drains into the

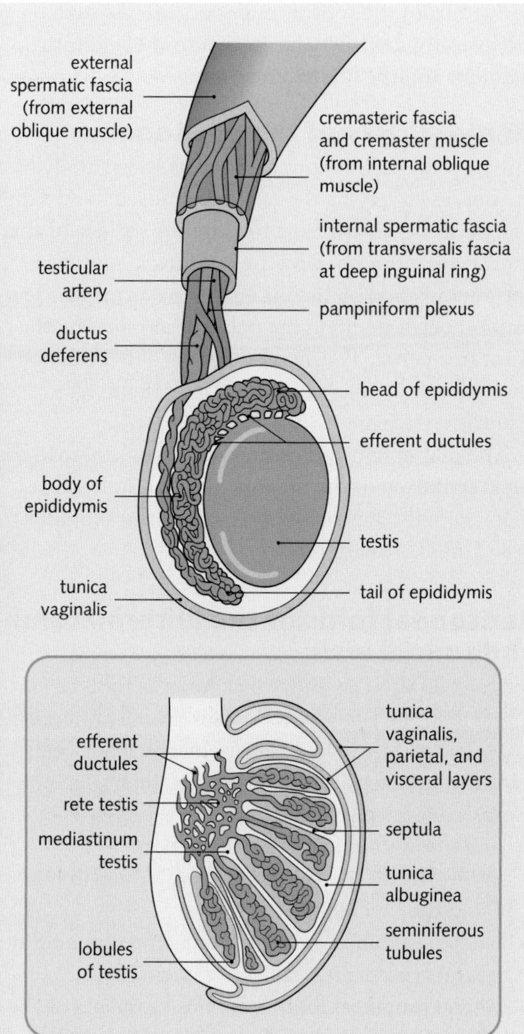

Fig. 4.9 Left testis, epididymis, coverings, and contents of the spermatic cord.

left renal vein, and the right into the inferior vena cava. Renal tumours may obstruct the left renal vein, causing dilatation of the veins in the testis, to form a varicocoele.

The testicular artery and its branches are very closely associated with the pampiniform venous plexus. This acts as a countercurrent heat exchanger. For spermatogenesis to occur, the testicular temperature has to be 2–3 °C below body temperature.

Lymphatics drain into the para-aortic nodes.

Descent of the testis

The testis develops in the posterior abdominal wall of the embryo, but then migrates—through the inguinal canal—into the scrotum: it reaches the deep inguinal ring by 4 months, and it is inside the canal at 7 months, then progresses rapidly through the superficial ring to reach the scrotum at around the time of birth.

A diverticulum of peritoneum—the processus vaginalis—precedes the testis as it passes through the inguinal canal into the scrotum. The processus is normally obliterated except at its lower end, where it becomes the tunica vaginalis.

The mechanism of descent of the testis is not known, but it is preceded by a gubernaculum.

The cremaster muscle elevates the testes towards the inguinal canal as part of the cremasteric reflex. This reflex is very active in children, often leading to a misdiagnosis of undescended testes. Failure of the testis to descend is a serious condition—it may result in impaired fertility, and undescended testes may undergo malignant change.

A hernia is a protrusion of a viscus or part of a viscus through its coverings into an abnormal situation, most frequently into the inguinal canal (inguinal hernia). Direct hernias are common in adults, and they result from a defect in the muscle layers of the abdominal wall. The hernia pushes through the posterior wall of the inguinal canal and its neck lies medial to the inferior epigastric artery. Indirect hernias are seen often in babies, and these result from failure of the processus vaginalis to obliterate. There is no defect in the abdominal wall and the hernia passes through the inguinal canal, and it may enter the scrotum.

The peritoneum

The peritoneum is a serous membrane lined by mesothelium. It has two layers continuous with each other:

- The parietal layer lines the anterior and posterior abdominal wall, the inferior surface of the diaphragm, and the pelvic cavity.
- The visceral layer leaves the abdominal wall and invests the viscera to a greater or lesser degree. This is the serous covering for many of the viscera.

Embryology

During development, the foregut, midgut, and hindgut are suspended from the posterior abdominal wall by a dorsal mesentery (a mesentery is a double layer of peritoneum that encloses an organ and connects it to the body wall—Fig. 4.10). These organs are known as intraperitoneal.

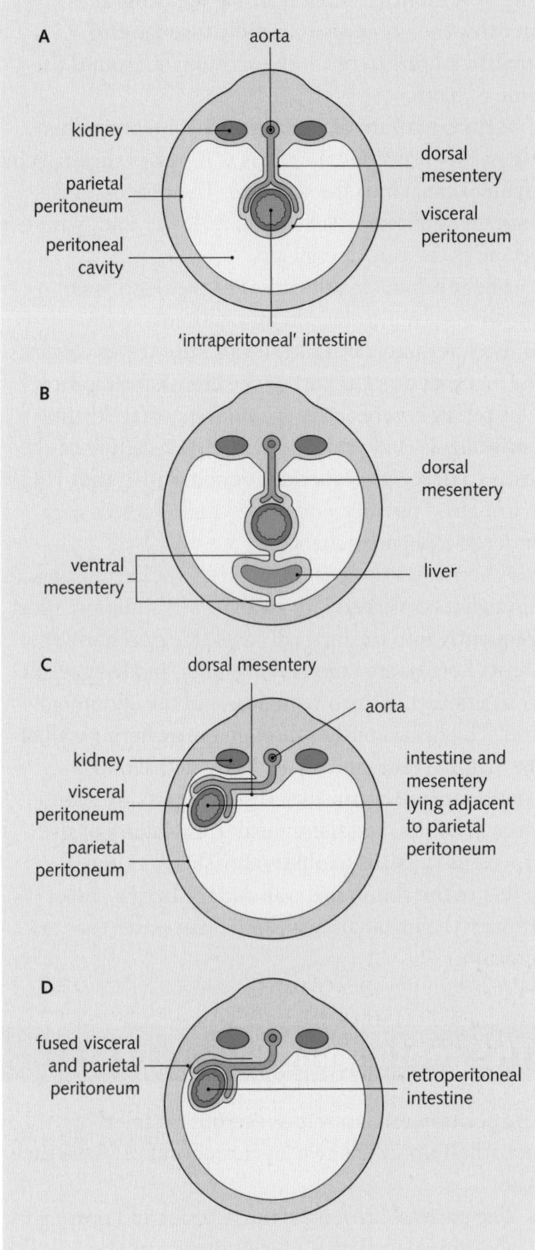

Fig. 4.10 The embryonic dorsal (A) and ventral (B) mesenteries and the formation of the retroperitoneal part of the intestines (C and D). (Adapted from *Anatomy as a Basis for Clinical Medicine*, by ECB Hall-Craggs. Courtesy of Williams & Wilkins.)

Some organs lie against the posterior abdominal wall, and they are covered by peritoneum on their anterior surface only (e.g. the kidneys). These organs are known as retroperitoneal.

A ventral mesentery is present only in the terminal parts of the oesophagus and stomach, and the upper part of the duodenum (foregut). It is derived from the septum transversum. Growth of the liver divides the mesentery into the falciform ligament and the lesser omentum.

Nerve supply of the peritoneum

The parietal peritoneum is supplied segmentally by the nerves supplying the overlying muscles and skin. The peritoneum covering the inferior surface of the diaphragm is supplied by the intercostal nerves peripherally and by the phrenic nerve centrally. The parietal peritoneum in the pelvis is supplied by the obturator nerve. The visceral peritoneum does not have a somatic innervation, and it is, therefore, insensitive to pain.

The diaphragm and its innervation originate in the neck. Irritation of the diaphragm by abdominal or thoracic pathology may cause pain in the shoulder tips. This is termed 'referred pain'.

Peritoneal folds of the anterolateral abdominal wall

Peritoneal folds can either be known as folds or ligaments. The latter is not meant literally but used to describe a fold of peritoneum running between two structures e.g. phrenicocolic ligament is a transverse peritoneal fold between the splenic flexure of the colon and the diaphragm.

Anterolateral peritoneal folds are shown in Fig. 4.11 and they include:

- Median umbilical fold—contains the remnant of urachus (median umbilical ligament).
- Medial umbilical fold—contains remnants of the umbilical arteries (medial umbilical ligaments).
- Lateral umbilical fold—contains the inferior epigastric vessels.

The falciform ligament (an anterior peritoneal fold between the diaphragm and the umbilicus) contains the ligamentum teres hepatis (the remnant of the umbilical vein) in its free margin.

Greater and lesser sacs

The space between the parietal and visceral peritoneum is only potential, and it contains a small quantity of peritoneal fluid. This is the general peritoneal cavity. In males this peritoneal cavity is completely closed, however, in females the cavity communicates externally via the uterine tubes. The peritoneal cavity is divided into greater and lesser sacs.

The lesser sac (omental bursa) is a diverticulum of the peritoneal cavity behind the stomach. It forms because of the change in position of the liver,

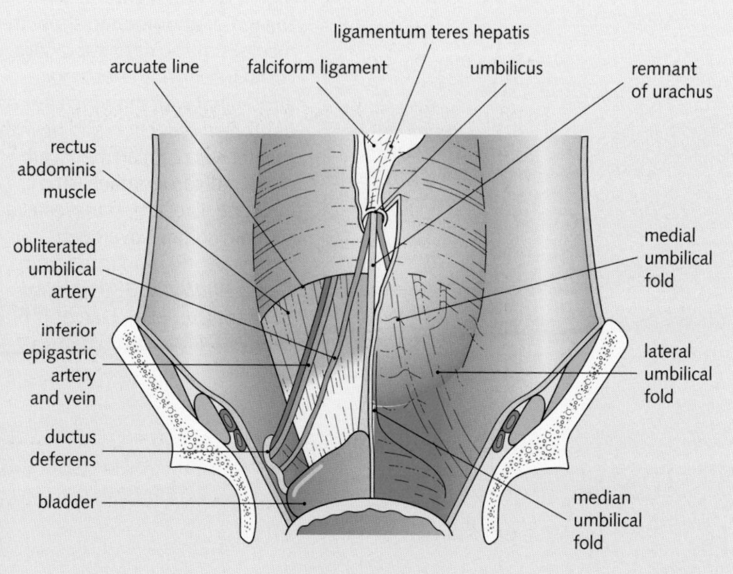

Fig. 4.11 Peritoneal folds of the anterior abdominal wall.

stomach, and spleen during development. The lesser sac communicates with the greater sac through the epiploic (omental) foramen. The epiploic (omental) foramen is bounded by:

- Superiorly—caudate process of liver.
- Anteriorly—portal vein in the free edge of the lesser omentum.
- Inferiorly—first part of the duodenum.
- Posteriorly—inferior vena cava.

The greater sac, which is the remainder of the peritoneal cavity, can be divided descriptively, into compartments by the transverse mesocolon. The supracolic is above, with the infracolic and pelvic compartments below this mesentery (Fig. 4.12).

The infracolic compartment is subdivided further by the root of the mesentery (of the small intestine) into upper right and lower left. The compartment is bounded laterally by the paracolic gutters. The right paracolic gutter communicates with the supracolic compartment.

The supracolic compartment is divided into left and right parts by the falciform ligament. Between the upper surface of the liver and the diaphragm on either side of the falciform ligament are two spaces, each called a subphrenic recess. Under the liver on the right of the falciform ligament, but above the right kidney, is the right subhepatic recess. The left subhepatic recess has similar boundaries, but it is better known as the lesser sac.

In the pelvic compartment the peritoneum lies over and between pelvic viscera. The pouches formed differ between genders. Males have a rectovesical pouch. Females have a vesicouterine and rectouterine pouch (see Chapter 5).

These recesses and pouches are important since they determine the spread of fluids that enter the peritoneal cavity, and they are potential sites where infection or fluid may accumulate depending on whether a patient is prone or supine.

Infection in the subphrenic recess can irritate the diaphragmatic peritoneum, which is innervated by the phrenic nerve. As a result, referred pain is felt in the skin of the shoulder due to a shared C3 and C4 root value between phrenic and cutaneous nerves.

Greater and lesser omenta

The greater omentum is the largest fold of peritoneum. It arises from the greater curvature of the stomach and the superior duodenum. It is filled with fat. The transverse colon and its mesentery are fused to the posterior aspect of the omentum.

The lesser omentum connects the lesser curvature of the stomach and the proximal part of the duodenum to the liver. Although it is a single entity, it can be divided into two parts: a hepatogastric ligament

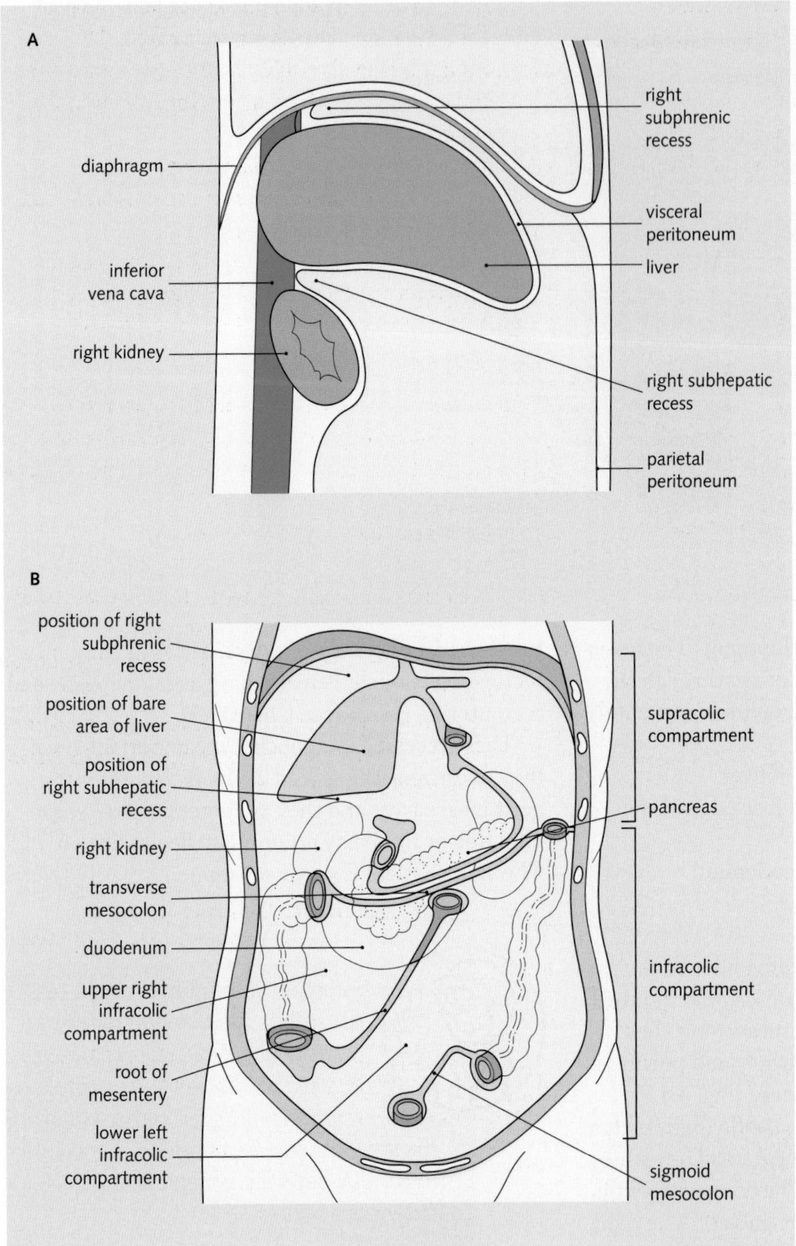

Fig. 4.12 (A) Sagittal section of the upper abdomen to show the recesses of the right supracolic compartment. (B) Posterior abdominal wall showing lines of peritoneal reflection and how the greater sac compartments are divided (liver, stomach, small intestine, caecum, transverse and sigmoid colons have been removed). (B) is adapted from *Gray's Anatomy,* 38th edn. by L H Bannister et al. Harcourt Brace and Co.

between the liver and stomach and a hepatoduodenal ligament between the liver and duodenum.

The greater omentum is the 'policeman' of the abdomen—it can be passively moved to a site of infection and adhere to it, preventing spread.

The abdominal organs

Oesophagus

After passing through the diaphragm, the oesophagus turns forward and to the left to enter the cardiac part of the stomach. Blood and nerve supply are shown in Fig. 3.28.

Gastro-oesophageal reflux is a very common problem in young children and in adults. A number of factors normally prevent reflux of stomach

contents into the oesophagus. These include:

- The sphincteric action of the lower oesophageal muscle.
- The sling of the right crus making an angle between the oesophagus and stomach. This compresses the oesophagus when the diaphragm contracts.
- A mucosal flap.
- Positive intra-abdominal pressure acting on the abdominal oesophagus, and reduced intrathoracic pressure.

Stomach

This is a dilated muscular bag lying between the oesophagus and the duodenum (Fig. 4.13). It is a relatively mobile organ, being fixed at its ends. The gastro-oesophageal junction lies at the level of T10 vertebra, and the pyloric sphincter (gastroduodenal sphincter) lies at the level of L1 vertebra. The stomach is enclosed in peritoneum that passes from the lesser curvature to the greater curvature. At the lesser curvature the lesser omentum splits into its two layers; one layer passes anteriorly and the other layer passes posteriorly over the stomach.

The stomach is capable of considerable dilation and, therefore, has a rugose inner surface. The mucosa is extensively folded, and there is an outer longitudinal, middle circular, and inner incomplete oblique muscle layer.

The relations of the stomach comprise:

- Anterior—the anterolateral abdominal wall, left costal margin, and diaphragm.
- Posterior (stomach bed)—the left suprarenal gland, upper pole of the left kidney, pancreas, spleen, splenic artery, and left colic flexure.

The stomach and oesophagus are foregut derivatives, and, therefore, they get their blood supply from the coeliac trunk: a branch of the abdominal aorta (see Fig. 4.13).

Lymphatic drainage is illustrated in Fig. 4.13.

The nerve supply to the stomach and oesophagus comprises:

- Sympathetic—from the coeliac plexus distributed along arteries.
- Parasympathetic—from the anterior and posterior vagal trunks. Stimulation increases secretions and peristaltic activity.

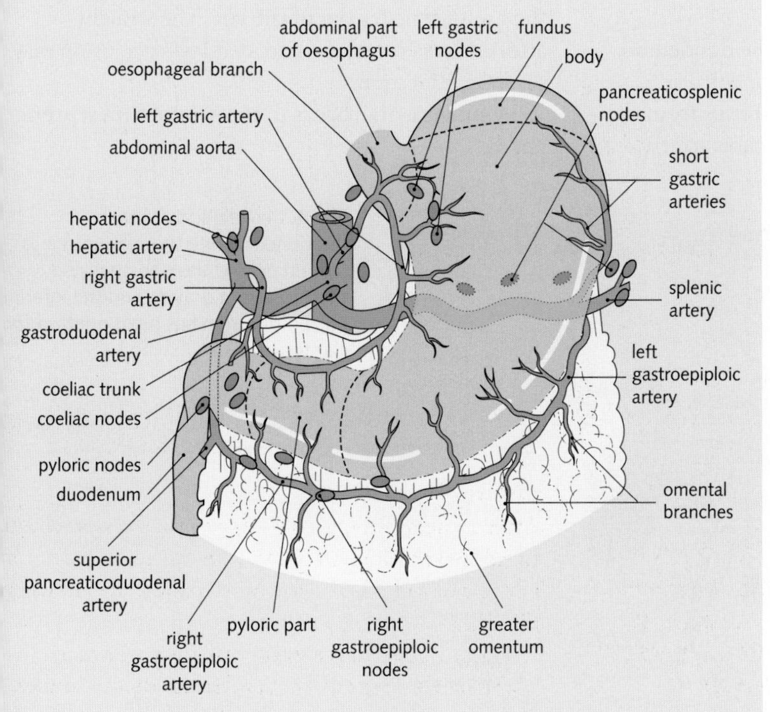

Fig. 4.13 Blood supply and lymphatic drainage of the stomach.

In a sliding hiatus hernia the abdominal oesophagus, cardia, and fundus of the stomach slide superiorly through the oesophageal hiatus. This is because the sling action of the diaphragmatic right crus weakens.

Duodenum

The duodenum is a C-shaped tube. Most of it is retroperitoneal and firmly attached to the posterior abdominal wall. It is divided into four parts:

- The first part passes posteriorly to the right side of the vertebral column, along the transpyloric plane at the L1 vertebral level.
- The second part passes downwards and receives the hepatopancreatic ampulla (of Vater), the opening of the bile duct and main pancreatic duct at the level of the L2 vertebra.
- The third part crosses the vertebral column at the level of L3 vertebra.
- The fourth part ascends to the level of L2 vertebra and opens into the jejunum.

The first part of the duodenum is very susceptible to peptic ulcers. This may be due to infection with *Helicobacter pylori*.

Fig. 4.14 shows the relations of the duodenum.

The duodenum gets its blood supply from:

- Superior pancreaticoduodenal arteries from the gastroduodenal branch of the hepatic artery.
- Inferior pancreaticoduodenal arteries from the superior mesenteric artery.

Duodenal lymph drains into channels that accompany the superior and inferior pancreaticoduodenal vessels to the coeliac and superior mesenteric nodes.

If a duodenal ulcer perforates, gas, digested food, and intestinal bacteria enter the peritoneal cavity. This causes very painful inflammation of the peritoneum (peritonitis).

Jejunum and ileum

The jejunum and ileum lie free in the abdomen. They are attached to the posterior abdominal wall by the mesentery. Their total length is approximately 6–7 m.

Fig. 4.15 outlines the differences between the jejunum and ileum.

The blood supply to the jejunum and ileum is from the jejunal and ileal branches of the superior mesenteric artery. The arteries form a series of anastomotic loops to make arterial arcades. From these arcades, straight arteries (vasa recta) pass to the mesenteric border of the gut. The straight arteries (vasa recta) are end arteries—occlusion may result in infarction.

Lymphatic drainage is to the superior mesenteric nodes.

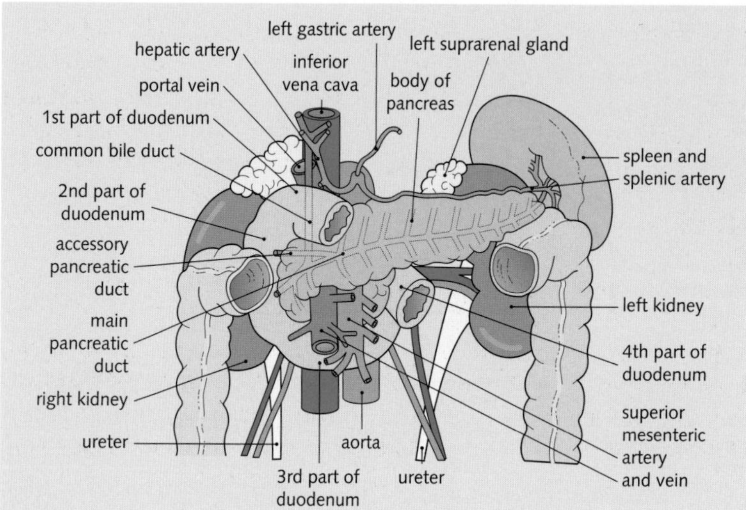

Fig. 4.14 Relations of the duodenum. Note the splenic vein is hidden behind the pancreas and, therefore, not drawn, and the inferior mesenteric vein has been omitted for clarity.

Distinguishing characteristics of the jejunum and ileum		
Characteristic	**Jejunum**	**Ileum**
Colour	Deep red	Paler pink
Wall	Thick and heavy	Thin and light
Vascularity	Greater	Less
Vasa recta	Long	Short
Arcades	A few large loops	Many short loops
Peyer's patches (aggregated lymphoid follicles)	No	Yes—towards the terminal part of the ileum
Plicae circulares (mucosal folds increasing surface area)	More and larger	Less and smaller/absent
Fat	Less—stops at the mesenteric border with the jejunum	More—encroaches onto the ileum

Fig. 4.15 Distinguishing characteristics of the jejunum and ileum.

Parasympathetic fibres from the vagus nerve increase peristalsis and secretion, whereas sympathetic fibres from the lateral horn of T9 and T10 inhibit peristalsis.

Large intestine

This consists of the caecum, appendix, colon, rectum, and upper part of the anal canal.

Caecum and vermiform appendix

The caecum and the vermiform appendix lie in the right iliac fossa.

The caecum lies free in the abdominal cavity, invested by peritoneum. The ileum enters the caecum obliquely, and partially invaginates into it, forming the ileocaecal orifice. The blood supply to the caecum is from the ileocolic artery.

The vermiform appendix is a worm-shaped blind-ending tube with lymphoid tissue in its wall, and it is usually 6–9 cm long. It opens into the posteromedial wall of the caecum, 2 cm below the ileocaecal orifice. The vermiform appendix has its own mesentery, the mesoappendix, and the taeniae coli of the colon merge to form a complete layer of longitudinal muscle.

Blood supply is from the appendicular artery, a branch of the posterior caecal artery. It is an end artery, and any swelling of the vermiform appendix may obstruct the artery, causing necrosis and perforation.

Appendicitis is a common surgical emergency. It usually presents with central abdominal pain, which later spreads to the right iliac fossa. The base of the vermiform appendix is usually identified externally at McBurney's point, one third of the way along a line joining the anterior superior iliac spine and the umbilicus. Internally, the base of the vermiform appendix lies at the point of convergence of the taeniae coli.

Colon

Most of the colon has an incomplete longitudinal muscle layer represented by three bands: the taeniae coli. The longitudinal muscle of the vermiform appendix divides into the taeniae coli and travel along the length of the colon. In the sigmoid (pelvic) colon the taeniae coli are wider, and in the terminal part they coalesce to form a complete longitudinal muscle layer.

Haustrations (sacculations) are pouches along the length of the colon. These occur because the three taeniae coli are 'too short' thus shortening the colonic wall.

Bulbous pouches of peritoneum distended with fat project from the serous coat. These are the appendices epiploicae. They become larger and more developed along the length of the colon. The blood supply reaches these appendices from the colonic mucosa by perforating the muscular wall.

Mucous membrane may herniate through the perforations in the muscle layer of the colon made by the blood vessels supplying the appendices epiploicae. Such a hernia is called a diverticulum.

Ascending colon

This extends from the ileocolic junction to the right colic (hepatic) flexure. On the medial and lateral sides of the ascending colon the peritoneum runs posteriorly forming the paracolic gutters. The colon is fixed by extraperitoneal fibrous tissue to the iliac and lumbar fascia, upon which it lies.

Transverse colon

This extends from the right colic (hepatic) flexure to the left colic (splenic) flexure, the former being lower due to the presence of the right lobe of the liver. It is completely invested in peritoneum and hangs free on the transverse mesocolon.

Descending colon

This extends from the left colic (splenic) flexure to the pelvic brim, and it is retroperitoneal. There are paracolic gutters on its medial and lateral sides formed by the peritoneum running posteriorly. It is connected to the iliac and lumbar fascia by extraperitoneal fibrous tissue.

Sigmoid or pelvic colon

This extends from the descending colon to the pelvic brim. It hangs free from the sigmoid mesocolon. The mesocolon is an inverted V-shape, the base of which lies over the sacroiliac joint. Of the two stems, one runs to the midinguinal point along the external iliac vessels, and the other runs to the level of the third sacral segment where the rectum begins.

The colon is supplied by the superior mesenteric artery (artery of the midgut) up to the proximal two thirds of the transverse colon, and the inferior mesenteric artery thereafter (Fig. 4.16).

The branches anastomose near the medial margin of the entire colon, forming an arterial circle, the marginal artery, from which short vessels pass to the gut wall. The weakest part of the marginal artery supply to the colon is between the middle colic and left colic arteries at the splenic flexure. This site is thus most prone to ischaemia and infarction.

The venous drainage follows the arterial supply to the portal venous system.

The ascending and descending colons are held onto the posterior abdominal wall, i.e. they are retroperitoneal. The transverse and sigmoid colons are suspended by mesenteries and are mobile.

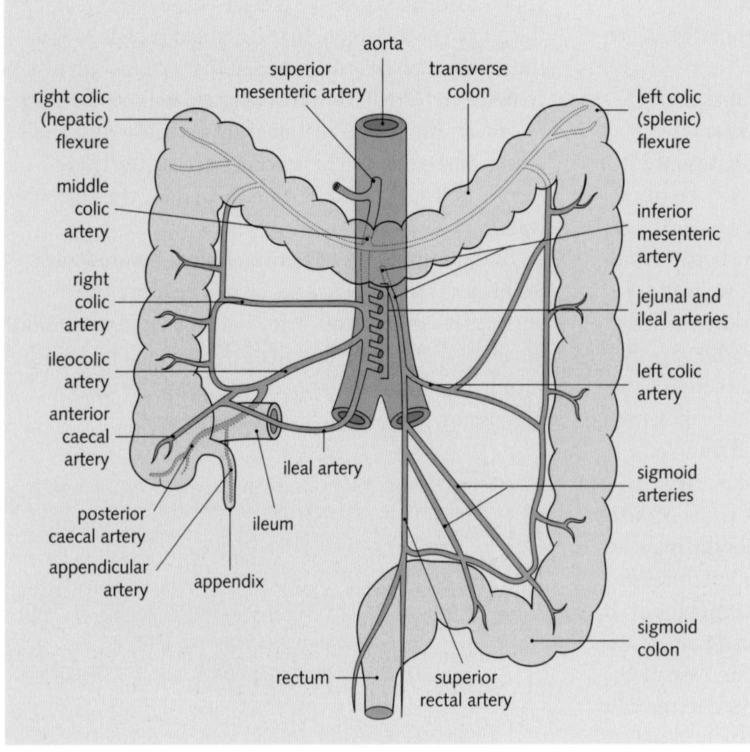

Fig. 4.16 Colon and its blood supply.

Spleen

The spleen is a large lymphoid organ. It removes particulate matter and aged or defective cells from the circulation, and it helps in mounting an immunological response against blood-borne pathogens. The importance of the spleen as an immunological organ is emphasized in people, particularly children, who have their spleens removed. These patients are very susceptible to infections by encapsulated organisms such as the pneumococcus and meningococcus, both of which may cause fatal meningitis. These individuals often need antibiotics life long to prevent infection, together with vaccinations.

The spleen has a convex diaphragmatic surface that fits into the concavity of the diaphragm. The anterior and superior borders are notched and sharp, but the posterior and inferior borders are rounded.

The spleen contacts the posterior wall of the stomach. It is connected to the greater curvature of the stomach by the gastrosplenic ligament and to the posterior abdominal wall at the left kidney by the splenorenal ligament. It is completely enclosed by peritoneum except at the hilum.

The splenic artery from the coeliac trunk is a tortuous vessel that passes along the superior border of the pancreas and anterior to the left kidney. Between the layers of the splenorenal ligament, the splenic artery divides into five or more branches, which enter the hilum.

The splenic vein joins the inferior mesenteric vein, and this runs posterior to the body of the pancreas to unite with the superior mesenteric vein to form the hepatic portal vein.

Lymphatic drainage is to the pancreaticosplenic and coeliac nodes.

The spleen is very susceptible to abdominal trauma, and it may bleed profusely, resulting in collapse of the patient and shock. Emergency splenectomy may be life saving in these cases.

A perforated gastric ulcer in the posterior stomach wall may erode the splenic artery, as it travels along the superior border of the pancreas, causing it to haemorrhage.

Fig. 4.17 Anterior and posterior views of the liver.

Liver

The liver is a wedge-shaped organ lying in the right hypochondrium. It is largely under cover of the costal margin, and it is invested by peritoneum except over its bare area.

The liver has four lobes (Fig. 4.17). The falciform ligament divides the liver into right and left lobes. Posteriorly, there is also a caudate lobe lying between the inferior vena cava and the ligamentum venosum fissure, and a quadrate lobe lying between the gall bladder fossa and the ligamentum teres. Functionally, the quadrate and caudate lobes are part of the left lobe as they are supplied by the left hepatic artery, left branch of the portal vein, and deliver bile to the left bile duct.

The falciform ligament runs up the anterior surface of the liver. At the superior surface of the liver the left leaf passes to the left and returns to form the left triangular ligament. The right leaf forms the upper leaf of the coronary ligament, the right triangular ligament, and the lower leaf of the coronary ligament (Fig. 4.17).

The area between the upper and lower parts of the coronary ligament is the bare area of the liver. This area is devoid of peritoneum and lies in contact with the diaphragm.

The right and left layers of peritoneum meet on the visceral surface of the liver to form the hepatogastric and hepatoduodenal ligaments, both of which are part of the lesser omentum.

Between the caudate and quadrate lobes, the two layers surround the porta hepatis. The porta hepatis, the inferior vena cava, the gall bladder, and the fissures of the ligamentum venosum and ligamentum teres, form an H-shaped pattern. The ligamentum venosum is the remnant of the fetal ductus venosus, which transported blood from the portal and umbilical veins to the hepatic veins.

The porta hepatis contains the right and left branches of the hepatic artery, the hepatic ducts, and the hepatic portal vein (Fig. 4.17):

- The hepatic artery from the coeliac trunk supplies oxygenated blood to the lobes of the liver. The cystic artery arises from it to supply the gall bladder.
- The hepatic portal vein carries the products of digestion from the gut to the liver.
- The right and left hepatic ducts drain bile into the common hepatic duct. The latter joins the cystic duct to form the bile duct.

There are also three hepatic veins that drain the liver. These do not have an extrahepatic course but drain directly into the inferior vena cava.

Lymphatics drain into the hepatic nodes lying around the porta hepatis. They also drain the gall bladder. The hepatic nodes drain into the coeliac nodes. Lymphatics of the bare area drain into the posterior mediastinal nodes.

Nerve supply is from the left vagus nerve and the sympathetic coeliac ganglion.

Gall bladder and biliary tract

The gall bladder lies in a fossa on the visceral surface of the liver. It has a fundus, a body, and a neck (Fig. 4.18).

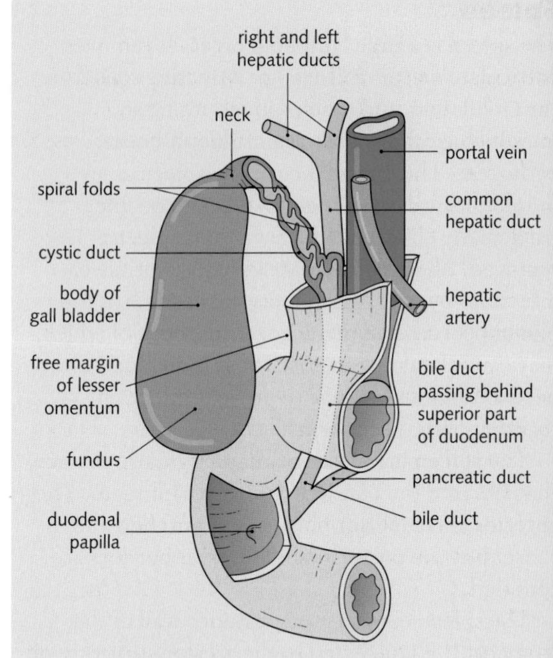

Fig. 4.18 Gall bladder and biliary tract. (Adapted from *Anatomy as a Basis for Clinical Medicine*, by E C B Hall-Craggs. Courtesy of Williams & Wilkins.)

The gall bladder stores and concentrates bile secreted by the liver. The bile is released into the duodenum when the gall bladder is stimulated, e.g. after a fatty meal.

The cystic duct drains the gall bladder and joins the common hepatic duct to form the common bile duct. This passes through the free margin of the lesser omentum (the hepatoduodenal ligament) behind the first part of the duodenum to enter the second part of the duodenum, together with the pancreatic duct, at the hepatopancreatic ampulla (of Vater). The sphincter of Oddi is a layer of circular muscle surrounding the ampulla. It controls the flow of bile and pancreatic secretions into the duodenum. The mucosa lining the neck of the gall bladder and cystic duct is thrown into folds to form a spiral valve.

Obstruction of the biliary system results in the clinical condition of jaundice (yellow skin).

Also, the gall bladder is prone to accumulating gall stones. These may pass into the duct system, causing severe colicky pain (biliary colic).

Blood supply is from the cystic artery, usually a branch of the right hepatic artery. Numerous branches from the hepatic bed may also supply the gall bladder.

Pancreas

The pancreas has both exocrine and endocrine functions. It lies behind the peritoneum on the posterior abdominal wall, roughly at the level of the transpyloric plane (see Fig. 4.14). It has a head, neck, body, and tail:

- The head lies in the concavity of the duodenum, anterior to the inferior vena cava and left renal vein. The bile duct travels through it. A small part of the head, the uncinate process, lies behind the superior mesenteric artery and vein.
- The neck overlies the superior mesenteric vessels and the portal vein.
- The body crosses the left renal vein and the aorta. The splenic vessels run close to this part of the pancreas.
- The tail is accompanied by the splenic vessels and lymphatics in the lienorenal (splenorenal) ligament, to touch the hilum of the spleen.

The main pancreatic duct opens into the duodenum with the bile duct, at the ampulla of Vater. The accessory duct opens into the duodenum 2 cm proximal to the ampulla of Vater.

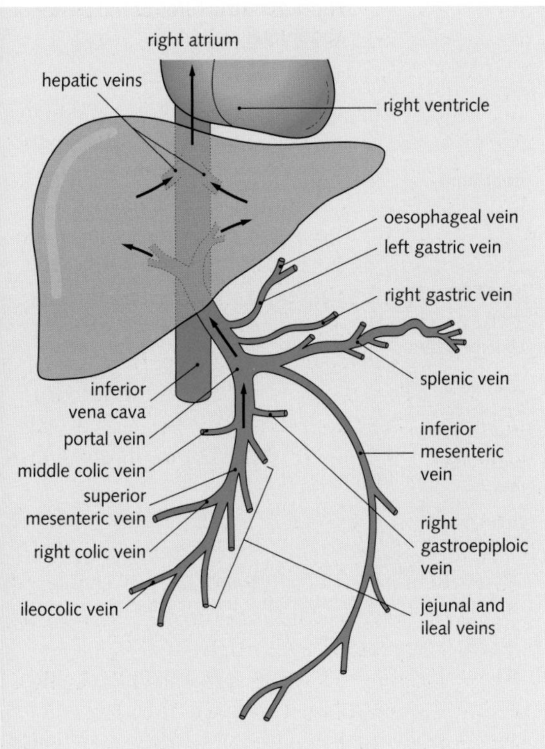

Fig. 4.19 Hepatic portal venous system.

The splenic artery, a branch of the coeliac trunk, supplies the neck, body, and tail of the pancreas. The superior and inferior pancreaticoduodenal arteries supply the head. The splenic vein drains the pancreas. Lymphatics drain into the coeliac and superior mesenteric nodes.

Carcinoma of the pancreas has a very poor prognosis, probably because it is silent and asymptomatic for a long time. If the head of the pancreas is involved, the bile duct running through it may become blocked, leading to jaundice. This may alert the clinician sooner.

Vessels of the gut

The foregut, midgut, and hindgut are supplied by the branches of the coeliac trunk, the superior mesenteric artery, and the inferior mesenteric artery, respectively.

The coeliac trunk arises from the abdominal aorta at the level of T12 vertebra. It gives off the left gastric artery, the common hepatic artery, and the splenic artery (see Fig. 4.13).

The superior mesenteric artery arises from the abdominal aorta at the level of L1 vertebra (transpyloric plane). It gives off the inferior pancreaticoduodenal artery, jejunal and ileal arteries, and the ileocolic, right colic, and middle colic arteries (see Fig. 4.16).

The inferior mesenteric artery arises from the abdominal aorta, opposite L3 vertebra. It gives off the left colic artery, sigmoid arteries, and the superior rectal artery (see Fig. 4.16).

Venous drainage of the gut

Venous blood rich in nutrients from the intestines travels in the hepatic portal system of veins to the liver (Fig. 4.19). The portal vein is formed by the union of the superior mesenteric vein with the splenic vein. The inferior mesenteric vein usually joins the splenic vein.

The portal vein passes posterior to the pancreas and first part of the duodenum to enter the lesser omentum (hepatoduodenal ligament). On reaching the porta hepatis the vein divides into left and right branches supplying the left and right lobes of the liver. From the liver the blood passes to the inferior vena cava via the hepatic veins and then to the heart.

Portosystemic anastomoses

These are areas with both a portal and a systemic venous drainage. They include the oesophagus, the anal canal, the retroperitoneum, and the umbilical region. The most significant of these is the oesophagus.

Portal hypertension caused by liver diseases obstructs portal blood flow and blood is diverted to the systemic veins. This increased volume through the systemic veins causes them to dilate, forming varices. These may rupture if traumatized, causing severe haemorrhage and even death.

Nerve supply of the gastrointestinal tract

All parts of the gut receive sympathetic and parasympathetic nerves that travel with the gut arteries. Sympathetic fibres come from the sympathetic chain and from the coeliac, superior mesenteric, and inferior mesenteric plexuses. Parasympathetic fibres for the foregut and midgut enter the abdomen in the vagus nerves, and they are distributed either directly or via the coeliac and superior mesenteric plexuses. Parasympathetic supply to the hindgut ascends from the pelvis (S2–S4) in the hypogastric plexus.

Sympathetic fibres inhibit peristalsis and secretion; parasympathetic fibres increase them.

The autonomic nervous system supplies the gut.

Lymphatic drainage of the gut

Lymphatics run with the arteries and end ultimately in lymph nodes lying anterior to the aorta (preaortic nodes) at the roots of the three gut arteries.

Lymph from the mucosa of the gut passes through a number of filters including:

- Lymphoid follicles, e.g. Peyer's patches.
- The 'epi' group of nodes, e.g. the epicolic nodes, which lie in the gut margin of the mesentery.
- The 'para' group of nodes, e.g. the paracolic nodes, which lie in the mesentery between the gut margin and the root of the mesentery.

All the lymph eventually enters the coeliac nodes and from here passes into the cisterna chyli—the origin of the thoracic duct.

The posterior abdominal wall

The posterior abdominal wall offers good protection to the abdominal contents. It is composed of the bodies of the five lumbar vertebrae and their intervertebral discs, and the psoas, iliacus, and quadratus lumborum muscles.

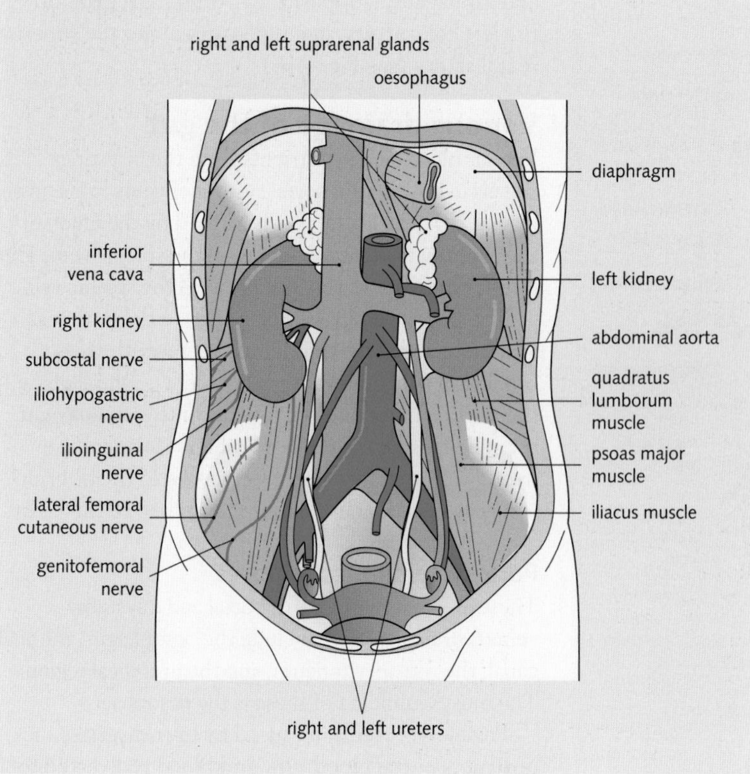

Fig. 4.20 Structures of the posterior abdominal wall.

The lumbar vertebrae project forwards into the abdominal cavity with a forward convexity (lumbar lordosis). The inferior vena cava and the aorta lie in front of the bodies of the vertebrae.

On either side of the vertebral bodies lie the paravertebral gutters. The kidneys and suprarenal glands lie in the superior aspect of these gutters (Fig. 4.20).

Muscles of the posterior abdominal wall

These are outlined in Fig. 4.21.

Fascia of the posterior abdominal wall

The muscles of the posterior abdominal wall are covered by thick, strong fascia, which provides firm support for the peritoneum and retroperitoneal viscera.

Vessels of the posterior abdominal wall

Abdominal aorta

The abdominal aorta passes through the diaphragm at the level of T12 vertebra. It passes inferiorly on the bodies of the lumbar vertebrae. In front of the body of L4 it divides into the common iliac arteries (Fig. 4.22).

Inferior vena cava

This vessel is formed on the right side of the aortic bifurcation, at the level of the L5 vertebra, by the union of the two common iliac veins (see Fig. 3.23). It ascends to the right of the aorta and passes behind the liver to pierce the diaphragm at the level of T8 vertebra and almost immediately enters the heart.

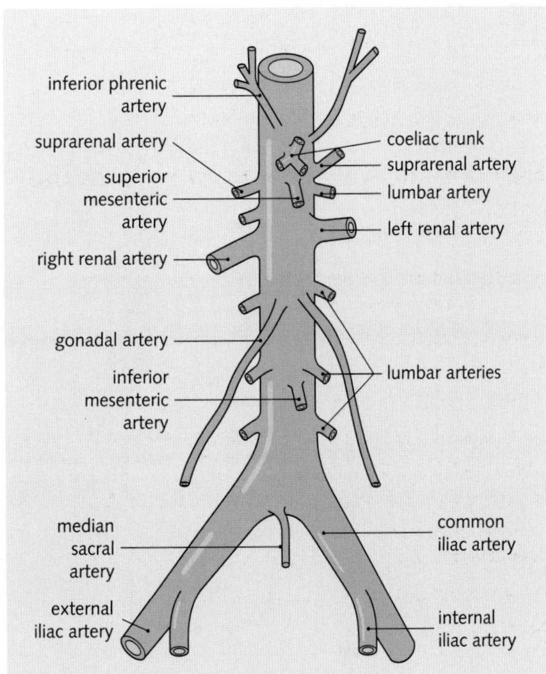

Fig. 4.22 Branches of the abdominal aorta.

Nerves of the posterior abdominal wall

Somatic nerves

The upper four lumbar spinal nerves emerge from their intervertebral foramina into the psoas major muscle, which they supply. Their anterior rami divide and unite to form the lumbar plexus (Fig. 4.23), which is mostly concerned with sensory and motor innervation to the lower limb. However some branches are motor and sensory to the anterior abdominal wall, e.g. iliohypogastric nerve, and sensory to the parietal peritoneum e.g. obturator nerve. The lumbosacral trunk joins the first three sacral nerves to contribute to the sacral plexus.

Muscles of the posterior abdominal wall			
Name of muscle (nerve supply)	**Origin**	**Insertion**	**Action**
Psoas major (LI–L3)	Transverse process, bodies, and intervertebral discs of T12 and L1–L5 vertebrae	Lesser trochanter of femur	Flexes thigh on trunk
Quadratus lumborum (T12–L3)	Iliolumbar ligament, iliac crest, transverse processes of lower lumbar vertebrae	12th rib	Depresses 12th rib during respiration; laterally flexes vertebral column
Iliacus (femoral nerve)	Iliac fossa	Lesser trochanter of femur	Flexes thigh on trunk

Fig. 4.21 Muscles of the posterior abdominal wall. (Adapted from *Clinical Anatomy, An Illustrated Review with Questions and Explanations*, 2nd edn, by R S Snell. Little Brown & Co.)

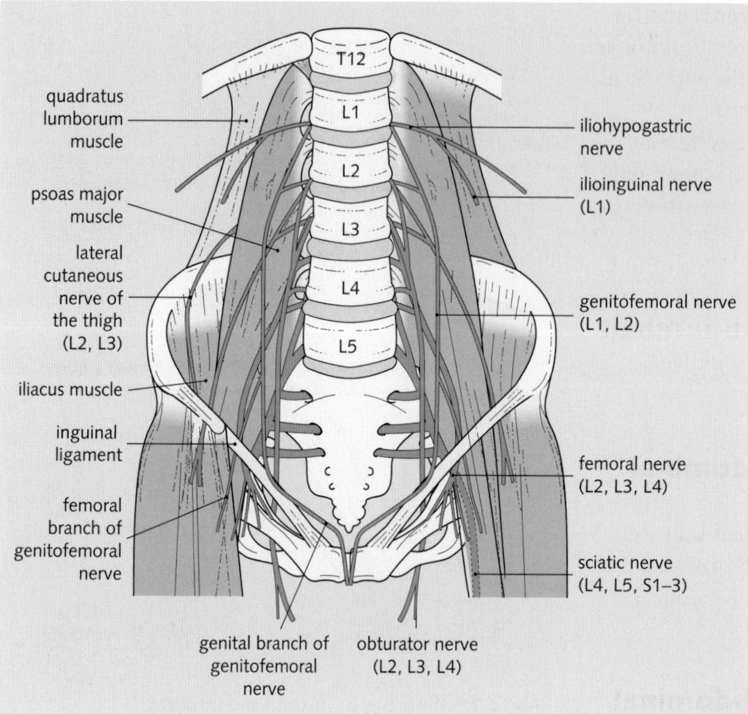

Fig. 4.23 Lumbar plexus and the relationship of the branches to the psoas muscle. The spinal root values of the branches are shown in parentheses. Note the sciatic nerve is not part of the lumbar plexus and is only shown for completeness.

Autonomic nerves

The autonomic nervous system of the abdomen is composed of the following:

- The vagus nerves and pelvic splanchnic nerves (parasympathetic).
- The lumbar sympathetic trunks, the thoracic splanchnic nerves, and the lumbar splanchnic nerves (sympathetic).

Sympathetic nerves

The lumbar sympathetic trunk comprises preganglionic fibres from the lower thoracic trunk and from L1 and L2 nerves (via white rami). This trunk enters the abdomen posterior to the medial arcuate ligament of the diaphragm. It runs down on the medial border of psoas major.

There are usually four lumbar ganglia. These give somatic branches (grey rami communicantes) to all five lumbar nerves, supplying the body wall and lower limb, and visceral branches (lumbar splanchnic nerves) that join the coeliac, aortic, and superior hypogastric plexuses. Fibres from the third and fourth ganglia join with fibres from the aortic plexus in front of L5 vertebra to form the superior hypogastric plexus. The superior hypogastric plexus divides into the right and left hypogastric nerves. These run into the pelvis to join the inferior hypogastric plexus. There are no branches to the abdominal viscera.

The greater and lesser splanchnic nerves pierce the crura of the diaphragm to enter the coeliac ganglion. These splanchnic nerves are almost totally preganglionic, and they relay in the coeliac ganglia. The least splanchnic nerves relay in a small renal ganglion close to the renal artery.

From the coeliac ganglion, postganglionic fibres form a rich network around the aorta—the coeliac plexus. This is situated around the origin of the coeliac trunk, and it supplies all the abdominal viscera via the visceral branches of the aorta. Fibres passing to the kidneys pick up branches from the renal ganglion to form the renal plexus around the renal artery.

The suprarenal gland has a second supply in addition to the coeliac plexus. Preganglionic fibres from the lesser splanchnic nerve pass without relay to the cells of the suprarenal medulla. Stimulation causes the release of adrenaline.

Functions of the sympathetic nerves include vasomotor, motor to the sphincters, inhibition of peristalsis, and transport of sensory fibres from all of the abdominal viscera.

Parasympathetic nerves

The vagal trunks enter the abdomen on the surface of the oesophagus. Branches to the coeliac plexus supply the gut up to the transverse colon. Branches to the renal plexus pass to the kidneys. The distal

part of the transverse colon and the descending and sigmoid colons receive parasympathetic innervation from the pelvic splanchnic nerves.

Functions of the parasympathetic nerves comprise motor and secretomotor to the gut and glands.

Kidneys

The kidneys are retroperitoneal organs that lie largely under cover of the costal margin in the paravertebral gutters of the posterior abdominal wall. The position of the kidneys varies with respiration, but they lie approximately opposite the first three lumbar vertebrae. The right kidney is lower than the left kidney, as it lies below the liver (Fig. 4.24).

Each kidney is surrounded by perinephric fat. The renal fascia encloses these two structures and separates the kidney from the suprarenal gland. The fascia is firmly attached to the renal vessels and the ureter at the hilum of the kidney.

A renal artery supplies each kidney (Fig. 4.25). At the hilum of the kidney the main artery divides into anterior and posterior branches. These are further subdivided into segmental arteries and then into interlobular arteries. Venous drainage is via the segmental veins, which join together to form a renal vein. The renal veins join the inferior vena cava at the L2 vertebral level. The left renal vein is longer since it has to pass in front of the aorta to reach the inferior vena cava on its right side.

A sympathetic nerve supply arises from the coeliac, renal, and superior hypogastric plexuses. These carry pain afferent fibres. A parasympathetic nerve supply is from the vagus but this has an unknown function.

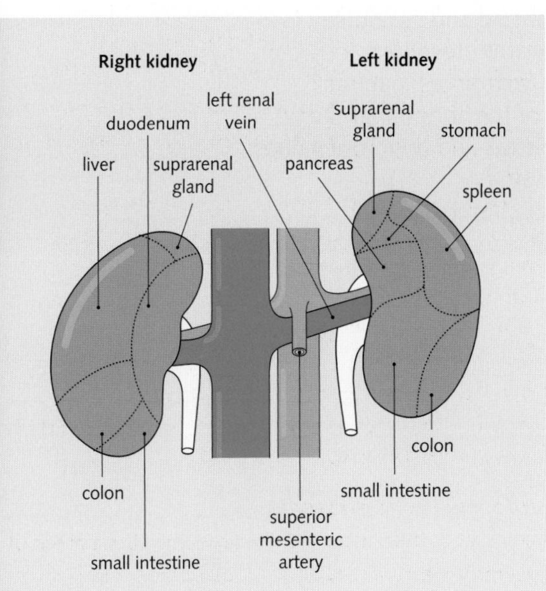

Fig. 4.24 Kidneys and their main anterior relations.

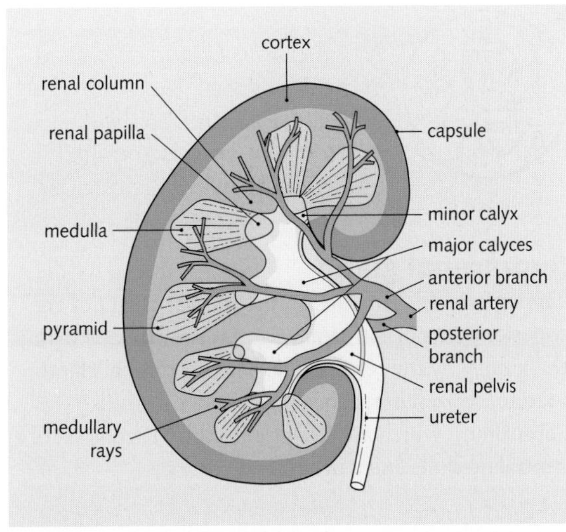

Fig. 4.25 Macroscopic structure and arterial supply of the kidney.

Ureters

The ureters are formed at the renal pelvis. They descend on the psoas muscle behind the peritoneum, and they cross the common iliac artery at its bifurcation at the pelvic brim. They turn towards the bladder at the level of the ischial spine.

The ureter narrows in three places:

- At the pelvoureteric junction.
- Where it crosses the pelvic brim.
- At its termination in the bladder.

Blood supply is from the renal artery, the abdominal aorta, the gonadal and vesical arteries, the common and internal iliac arteries, and the middle rectal artery.

The ureter has a sympathetic nerve supply from the coeliac and hypogastric plexuses. Parasympathetic fibres come from the pelvic splanchnic nerves. The pain afferents accompany the sympathetic nerves.

Carcinoma of the cervix is a very common tumour in middle-aged women. Advanced cases may cause renal failure by obstructing the ureters, and they are often fatal.

In the hilum of the kidney, the structures from anterior to posterior are vein, artery, ureter.

A transplanted kidney is placed in the iliac fossa. The renal vessels are sutured to the external iliac vessels and the ureter is sutured to the bladder.

Suprarenal gland

The suprarenal gland lies on the medial aspect of the superior pole of each kidney, and it is separated from the kidney by the renal fascia. The suprarenal gland secretes several hormones (e.g. corticosteroids, adrenaline), which are essential for life. It consists of a central medulla and a peripheral cortex.

The suprarenal gland is supplied by three main vessels:

- The suprarenal branch of the inferior phrenic artery.
- The suprarenal branch of the renal artery.
- The suprarenal artery from the aorta.

The venous drainage of the gland differs slightly on either side. On the right a single suprarenal vein drains into the inferior vena cava directly. On the left a single suprarenal vein drains into the left renal vein.

Lymph from the kidneys and the suprarenal glands drains into the para-aortic lymph nodes.

- Outline the boundaries of the abdominal cavity, both skeletal and muscular.
- Discuss two methods for dividing the abdomen into regions.
- Outline the surface markings of four abdominal organs.
- What is the rectus sheath?
- What are the boundaries and contents of the inguinal canal?
- List the contents of the spermatic cord.
- Describe the structure and nerve supply of the peritoneum.
- Describe how the greater sac is divided into recesses/pouches. Why are these important?
- What four factors prevent gastric reflux into the oesophagus?
- List the relations of the stomach.
- Outline the differences between the jejunum and ileum.
- Describe the anatomy of the colon.
- Describe the ligaments on the liver formed by peritoneal reflections.
- Describe how the biliary tree is formed and its course.
- Discuss the blood supply of the gastrointestinal tract.
- Describe the nerve supply of the gastrointestinal tract.
- Describe the lymphatic drainage of the abdominal cavity.
- List the branches of the lumbar plexus and their root values.
- Describe the anatomy of the kidney.
- Describe the blood supply to the suprarenal glands.

5. The Pelvis and Perineum

Regions and components of the pelvis

The pelvis lies below and behind the abdomen, and it is where the trunk communicates with the lower limbs. It is enclosed by bony, muscular, and ligamentous walls.

The bony pelvis is formed by the two hip bones, the sacrum, and coccyx. It has an upper part, the greater pelvis, flanked by the iliac bones, and a lower part, the lesser pelvis. The greater and lesser pelves meet at the pelvic brim (Fig. 5.1).

The pelvic cavity is continuous with the abdominal cavity, and it is, therefore, lined by the peritoneum of the greater peritoneal sac. The peritoneum passes down into the pelvis to cover partially the terminal portions of the alimentary tract, the bladder, and the internal reproductive organs of the female.

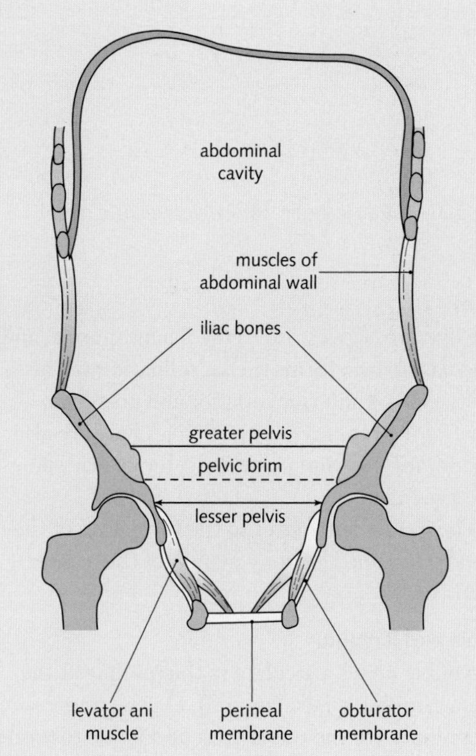

Fig. 5.1 Outline of the pelvic cavity.

The contents of the pelvis include:

- The coils of the small intestine.
- The rectum and sigmoid colon.
- The ureters and bladder.
- The ovaries, uterine tubes, uterus, and vagina in females.
- The ductus deferens, seminal vesicles, and prostate in males.
- The lumbosacral trunk, obturator nerve, sympathetic trunks, and sacral plexus.
- The common iliac arteries, gonadal arteries, and superior rectal arteries.

Surface anatomy and superficial structures

Bony landmarks

The iliac crest can be felt along its entire length. The anterior superior iliac spine is at the anterior border of the iliac crest and lies in the fold of the groin superiorly. The posterior superior iliac spine is at the posterior end of the iliac crest. It lies under a skin dimple at the level of S2 vertebra.

The pubic tubercle can be felt on the upper border of the pubis. The symphysis pubis joins the two pubic bones, and it may also be palpated. The pubic crest is a ridge of bone on the superior surface of the pubic bone, medial to the pubic tubercle. The pubic arch can be felt from the pubic symphysis to the ischial tuberosity.

The spinous processes of the sacrum fuse to form the median sacral crest. The crest can be felt beneath the skin in the buttock cleft. The sacral hiatus is found at the lower end of the sacrum, about 5 cm above the coccyx. The coccyx may be palpated about 2.5 cm behind the anus.

Viscera

The bladder is a pelvic organ in the adult, but when full it may be palpated through the anterior abdominal wall.

The non-pregnant uterus is not usually palpable. In pregnancy the fundus of the uterus may be palpated from about week 12. At term, the fundus is usually at the level of the xiphisternum.

In a rectal examination of a male the following structures can be palpated:

- The bulb of the penis (anteriorly).
- The membranous urethra (anteriorly).
- The prostate (anteriorly).
- The sacrum (posteriorly).
- The coccyx (posteriorly).
- The ischial spines (posteriolaterally).

In the female the structures palpated posteriorly are the same as in the male. Anterior structures palpated are:

- The body of the uterus.
- The cervix.

In a vaginal examination the following are palpated:

- The lips of the external os.
- The base of bladder and urethra (anteriorly).
- The uterine tubes (laterally).
- The ovaries (laterally).
- The rectouterine pouch (posteriorly).

The bony pelvis and pelvic wall

Bony pelvis

The bony pelvis is formed by the two hip bones, the sacrum, and the coccyx (Fig. 5.2). The hip bones meet anteriorly at the pubic symphysis; posteriorly they articulate with the sacrum at the sacroiliac joints. The bony pelvis thus forms a ring that protects the pelvic contents.

The pelvis is divided into the greater pelvis (false pelvis), which lies above the pelvic brim (pelvic inlet), and the lesser pelvis (true pelvis), which lies between the pelvic inlet and pelvic outlet (Fig. 5.3). The pelvic inlet lies at about 45 degrees to the pelvic outlet.

Sacrum

The sacrum consists of the fused five sacral vertebrae (Fig. 5.4). There are anterior and posterior sacral foramina for passage of the anterior and posterior rami of the sacral spinal nerves. The median sacral crest represents the fused spinal processes of the sacral vertebrae.

The sacrum articulates with the hip bone via its articular surface at the sacroiliac joints.

Hip bone

This is formed by the fusion of the ilium, the ischium, and the pubic bone shortly after puberty (Fig. 5.5).

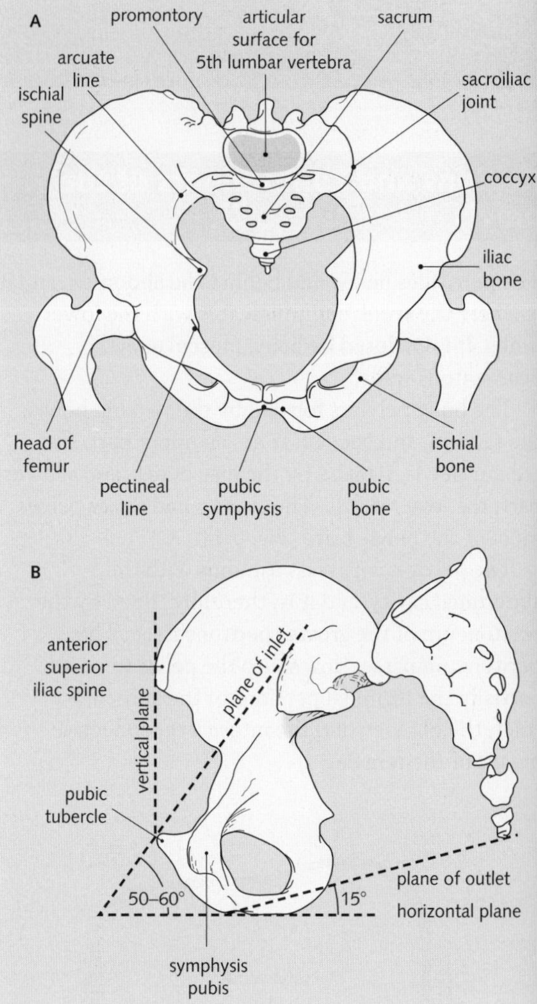

Fig. 5.2 (A) Pelvic girdle. (B) Pelvic inlet and outlet.

Ilium

The iliac fossa gives rise to the iliacus muscle, and the articular surface forms the sacroiliac joint. The iliac crest, and the anterior superior and posterior superior iliac spines lie superiorly. There are also anterior inferior and posterior inferior iliac spines (see Fig. 5.5).

The ilium contributes to the formation of the acetabulum and the bony margin of the greater sciatic notch.

Pubis and ischium

The pubic bones articulate in the midline at the pubic symphysis (see Fig. 5.5). On the upper surface of the body are the pubic crest and pubic tubercle. Each pubic bone has a superior and inferior ramus. The superior ramus forms the superior border of the

Boundaries of pelvic apertures	
Pelvic inlet	**Pelvic outlet**
Superior border of the pubic symphysis	Inferior margin of the pubic symphysis
Posterior border of the pubic crest	Inferior ramus of the pubis and the ischial tuberosity
Pectineal line	Sacrotuberous ligaments
Arcuate line of the ilium	Tip of the coccyx
Anterior border of the ala of the sacrum	
Sacral promontory	

Fig. 5.3 Boundaries of pelvic apertures.

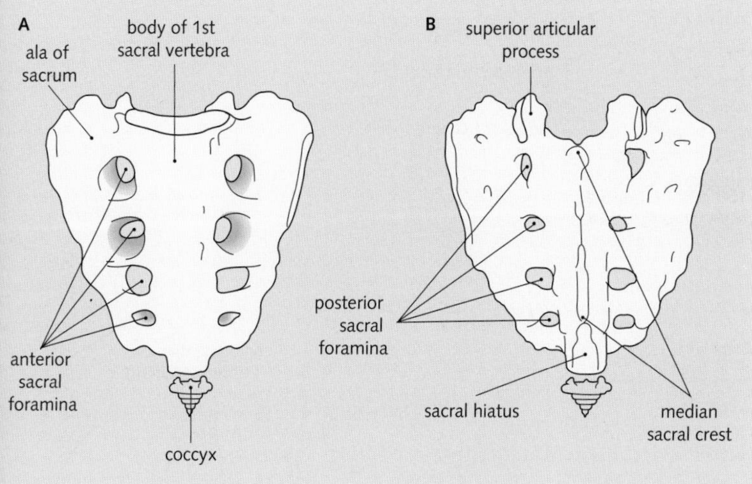

Fig. 5.4 (A) Anterior and (B) posterior views of the sacrum.

obturator foramen. The inferior ramus unites the pubis with the ischial bone to form the ischiopubic ramus. This leads to the body of the ischium and the ischial tuberosity.

The posterior border of the ischium contributes to the formation of the greater and lesser sciatic notches. The two notches are separated by the ischial spine. The sacrotuberous and sacrospinous ligaments transform the notches into the greater and lesser sciatic foramina.

The three bones of the hip all contribute to the formation of the acetabulum.

The position of the pelvis

The pelvis in a standing individual is tilted, such that the anterior superior iliac spine and the superior border of the pubic symphysis lie in the same vertical plane. A horizontal plane runs through the superior border of the pubic symphysis, ischial spine, and coccyx.

Male and female pelves

The male and female pelves may show a great deal of sexual dimorphism (Fig. 5.6).

The largest diameter of the pelvic inlet is the transverse diameter (Fig. 5.7), while the largest diameter of the pelvic outlet is the anteroposterior diameter. As the fetal head enters the pelvic inlet its maximum diameter lies across the pelvis, but as it descends through the birth canal the head rotates through 90 degrees, so that its maximum diameter

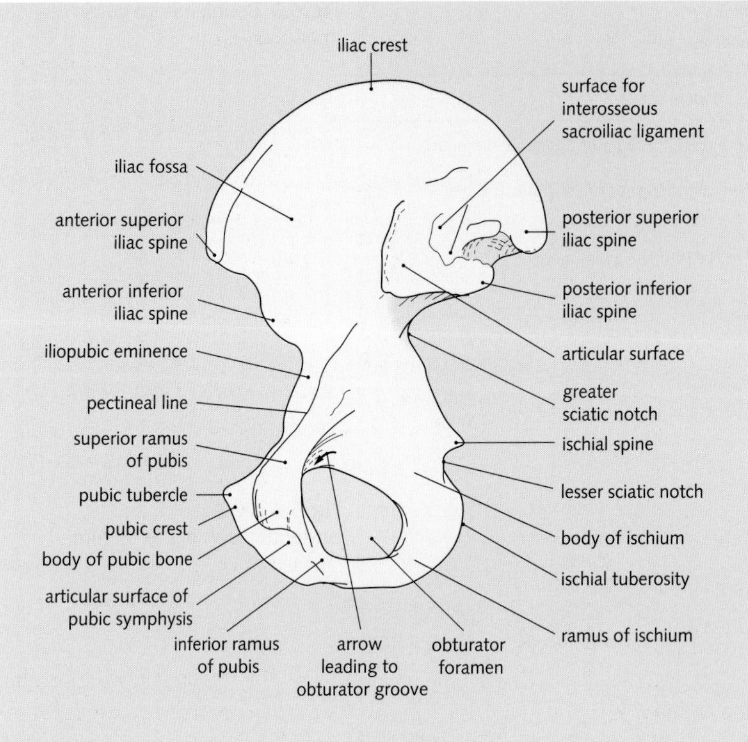

Fig. 5.5 Medial view of the hip bone.

Differences between the male and female pelves		
	Male	**Female**
Acetabulum	Large	Small
Build	Robust	Light
Inferior pelvic aperture	Relatively small	Relatively large
Obturator foramen	Round	Oval
Pubic arch	Narrow	Wide
Superior pelvic aperture	Usually heart-shaped	Usually oval or rounded

Fig. 5.6 Differences between the male and female pelves.

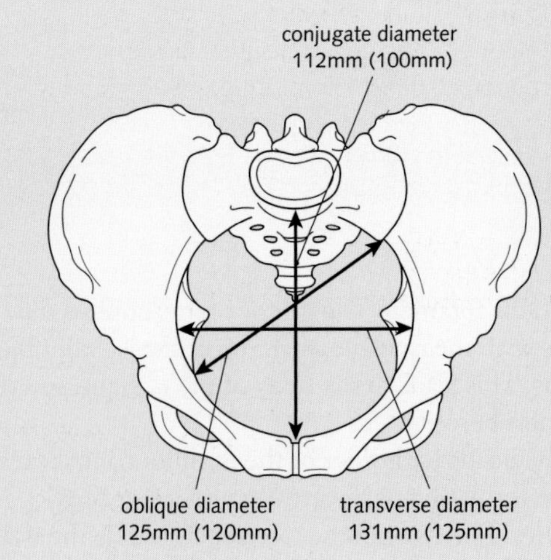

Fig. 5.7 Female pelvic inlet and its average diameters (male average diameters).

lies anteroposteriorly at the pelvic outlet. Failure of this rotation leads to arrest in the delivery, and instrumental assistance (e.g. forceps) or a caesarean section may be required.

Pelvic joints

Pubic symphysis

This is a secondary cartilaginous joint between the two pubic bones (see Fig. 5.5). It is usually immobile, and it is reinforced by the superior pubic ligament and the arcuate pubic ligament.

Sacroiliac joint

This is a synovial joint, but it allows only minimal movement. It is strengthened by strong interosseous ligaments. Weaker anterior and posterior ligaments

also stabilize the joint. The joint transmits the weight of the upper body to the hip bones.

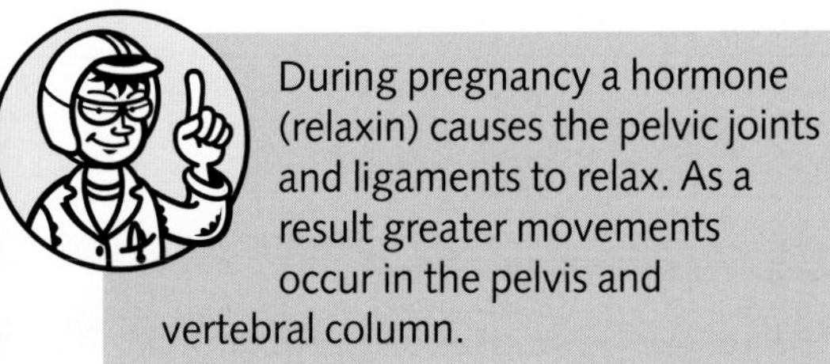

During pregnancy a hormone (relaxin) causes the pelvic joints and ligaments to relax. As a result greater movements occur in the pelvis and vertebral column.

Pelvic wall and floor

The side wall of the pelvis is formed by the hip bone with the obturator internus muscle (Fig. 5.8). The posterior wall is formed by the sacrum and the piriformis muscle as it passes into the greater sciatic foramen.

The pelvic floor forms a gutter of muscle around the terminal parts of the rectum, prostate, and urethra in the male, and the vagina and urethra in the female (Fig. 5.9).

Fig. 5.10 outlines the muscles of the pelvic wall and floor. The levator ani muscle takes origin from the arcus tendineus. This is a thickening of the obturator internus fascia and runs from the body of the pubis to the ischial spine.

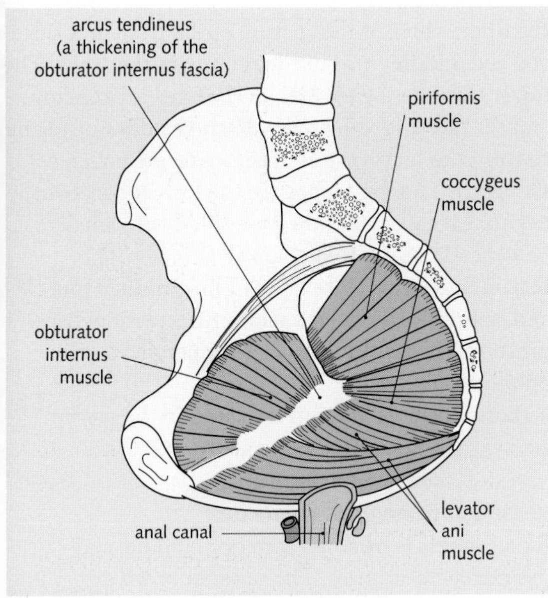

Fig. 5.8 Pelvic wall and its muscles.

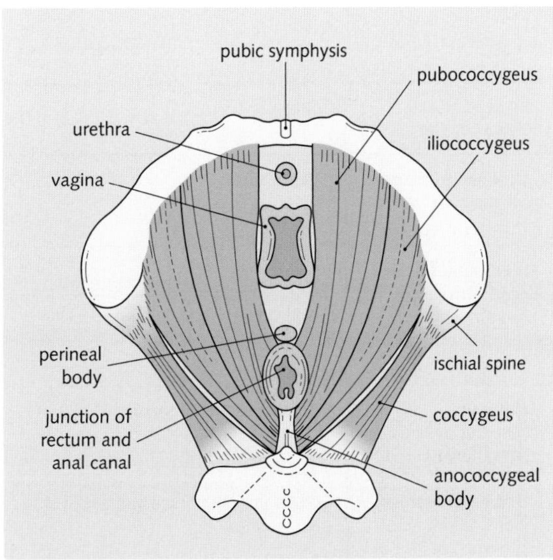

Fig. 5.9 Pelvic floor viewed from below.

Perineal body

The perineal body is a midline knot of fibromuscular tissue lying posterior to the prostate or vagina. Parts of the levator ani, the bulbospongiosus muscle, the external sphincter of the anal canal, and the superficial and deep transverse perineal muscles are fused with it. The position and insertion of muscles into the perineal body provides an essential supporting role for pelvic and perineal structures.

Damage to the perineal body during childbirth may result in prolapse of the bladder, vagina, and uterus.

Anococcygeal body

The anococcygeal body is a midline raphe running from the anorectal junction to the tip of the coccyx, and into which the levator ani muscle inserts. The raphe also separates the two ischiorectal fossae behind the anal canal.

Pelvic fascia

Over the pelvic wall the fascia is a strong membrane covering the obturator internus and piriformis muscles. The spinal nerves lie external to the fascia and the vessels lie internal to it. The sacral plexus lies between the fascia and the piriformis muscles.

Over the pelvic floor, the fascia consists of loose areolar tissue. The fascia condenses around the neurovascular bundles to form ligaments, and it also gives rise to the puboprostatic and pubovesical ligaments in the male and female, respectively. These fibromuscular bands, on either side of the median plane, run from the pubic bone to the bladder neck.

Muscles of the pelvic wall and floor			
Name of muscle (nerve supply)	**Origin**	**Insertion**	**Action**
Coccygeus (4th and 5th sacral nerves)	Ischial spine	Inferior aspect of sacrum and coccyx	Supports pelvic viscera, flexes coccyx
Levator ani (pudendal nerve, 4th sacral nerve)	Ischial spine Body of pubis Fascia of obturator internus	Perineal body Anococcygeal body Walls of prostate, vagina, rectum, and anal canal	Supports pelvic viscera; sphincter to anorectal junction and vagina Counteracts increased abdominal pressure, e.g. defaecation, parturition
Piriformis (first and second sacral nerves)	Anterior aspect of sacrum	Greater trochanter of femur	Rotates femur laterally at hip and stabilizes hip joint
Obturator internus (nerve to obturator internus; L5, S1, S2)	Obturator membrane and adjacent hip bone	Greater trochanter of femur	Rotates femur laterally at hip and stabilizes hip joint

Fig. 5.10 Muscles of the pelvic wall and floor.

They immobilize the bladder neck and support the bladder. Between the ligaments the deep dorsal vein of the penis (or clitoris) passes.

The fascia varies in thickness over the pelvic viscera.

Pelvic peritoneum

In the male, the peritoneum from the pelvic brim lines the pelvic wall and cavity inferiorly. From the anterior abdominal wall the peritoneum is reflected onto and attaches to the superior surface of the bladder. This attachment means that as the bladder fills and enlarges it peels the peritoneum away from the anterior abdominal wall. Behind the bladder, the peritoneum descends before ascending onto the rectum then sacrum, and it forms the rectovesical pouch (Fig. 5.11).

In the female the peritoneum turns superiorly to adhere to the uterus from the bladder, and it forms the vesicouterine pouch. From the back of the uterus and upper vagina the peritoneum ascends to cover the rectum then sacrum. This reflection forms the rectouterine pouch (Fig. 5.16).

The pelvic contents

Pelvic organs

Rectum

The rectum commences as a continuation of the sigmoid colon (where the sigmoid mesocolon ends) at the level of the third piece of the sacrum. It ends at the anorectal junction by piercing the pelvic floor at the border of the puborectalis muscle to become the anal canal.

The rectum has three lateral curves and its lowest part dilates as the rectal ampulla. There are also three transverse folds containing both mucous membrane and circular muscle. The folds correspond to the positions of the curves, and they may provide support for faecal material.

The rectum has no mesentery. Peritoneum covers the upper third of the rectum at the front and sides, and the middle third of the rectum at the front. The lower third lies below the level of the peritoneum, and the latter is reflected onto the bladder or vagina to form the rectovesical or rectouterine pouch (of Douglas). These pouches are the lowest parts of the peritoneal cavity, and they are filled with small bowel and the sigmoid colon (Fig. 5.11). The rectum has a complete layer of longitudinal muscle and, therefore, no haustrations (sacculations) are present. The rectum also lacks appendices epiplociae.

Posteriorly the rectum is related to the sacrum, coccyx, and pelvic floor.

Vessels and nerves of the rectum

Blood supply is from the superior, middle, and inferior rectal arteries. The superior rectal artery is a continuation of the inferior mesenteric artery. The

others are discussed in the section describing the blood supply to the pelvis.

The rectal plexus of veins drains into the inferior mesenteric vein (portal system of veins). The rectal plexus is also drained by the middle and inferior rectal veins. These are systemic veins, i.e. there is portosystemic anastomosis in the rectum. The longitudinal venous channels of the rectum may dilate to form haemorrhoids. Unlike oesophageal varices, however, this is rarely due to portal obstruction.

The nerve supply of the rectum consists of:

- Sympathetic—hypogastric plexus.
- Parasympathetic—pelvic splanchnic nerves, which are motor to the rectal muscles.

The rectum is distinguished from the sigmoid colon by the lack of a mesentery.

In Hirschsprung's disease the autonomic plexuses are absent from the wall of the rectum. The rectum is collapsed, and this leads to bowel obstruction, constipation, and vomiting.

Lymphatics accompany branches of the superior and middle rectal arteries and eventually drain to the preaortic nodes at the origin of the inferior mesenteric vessels.

Ureters in the pelvis

The ureters cross the pelvic brim at the bifurcation of the common iliac vessels (Fig. 5.12). They continue into the lesser pelvis towards the ischial spines and cross the obturator nerve and vessels. At the pelvic floor they run forward to enter the base of the bladder near its superior angle. Here, in males, the ductus deferens crosses the ureter superiorly; in females, the uterine artery crosses the ureter.

The ureter may be damaged in hysterectomy (removal of the uterus) when it may be tied while attempting to tie off the uterine artery.

To recall that the female ureter is crossed by the uterine artery superiorly remember 'bridge over troubled water'.

Bladder

The undistended bladder is a pyramid-shaped organ (Fig. 5.13). The apex points towards the pubic symphysis and the median umbilical ligament is attached to it.

The base is triangular. In males, the base lies largely below the rectovesical pouch and it is not

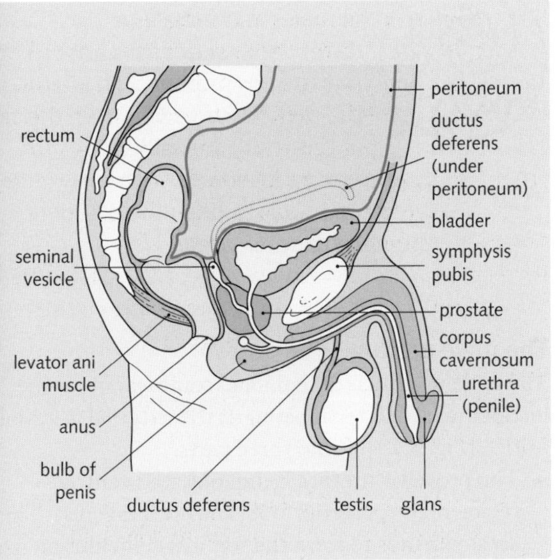

Fig. 5.11 A section through the male pelvis, illustrating the rectum and rectovesical pouch.

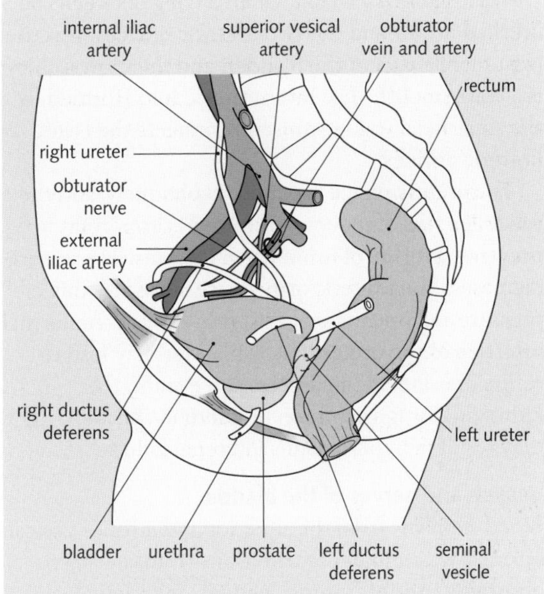

Fig. 5.12 Course of the ureter in the pelvis of the male.

covered by peritoneum. The ductus deferens and the seminal vesicles are attached to the surface, and the ureter enters the bladder at its superolateral surface. In females, the base is firmly attached to the vaginal wall and the upper part of the cervix by connective tissue.

Two inferolateral surfaces become continuous with each other at the retropubic space.

The bladder wall is composed of three layers. An outer serous (peritoneal) layer lines the superior surface of the bladder only. A smooth muscle layer, the detrusor muscle, is an interlacing network of fibres, and this gives the bladder a trabeculated appearance. The detrusor muscle has a parasympathetic nerve supply. However a second superficial layer of smooth muscle exists in the trigone and extends into the proximal urethra. This differs from the detrusor muscle histologically, and it has a sympathetic nerve supply. An inner mucosal layer of urothelium (transitional epithelium) lines the bladder.

The urethra leaves the neck of the bladder. In the neck, in males, the muscle is arranged in a circular fashion. This constitutes the internal urethral sphincter, which prevents retrograde ejaculation (sperm entering the bladder). In females the muscle is arranged longitudinally and this sphincter is lacking.

Internal surface of the bladder

When the bladder is empty, the mucosa is thick and folded. As the bladder fills, the mucosa becomes thinner and smoother.

The trigone is a triangular area lying between the urethral orifice and the two ureteric orifices. It is the least mobile part of the bladder, and the mucosa here is always smooth. The interureteric fold (formed by the superficial trigone muscle) connects the two ureteric orifices.

The ureters pierce the mucosa obliquely, and the valve-like flap of mucosa produced is important in preventing reflux of urine when intravesical pressure increases. The ureteric orifices are closed by this pressure and opened by peristaltic activity. Abnormal insertion of the ureters in the bladder may lead to reflux of urine up the ureters and even to the kidneys. This is a common problem in children and it may result in hypertension and renal failure.

Vessels and nerves of the bladder

Blood supply is from the superior and inferior vesical arteries, with minor contributions from the obturator, uterine, inferior gluteal, and vaginal arteries.

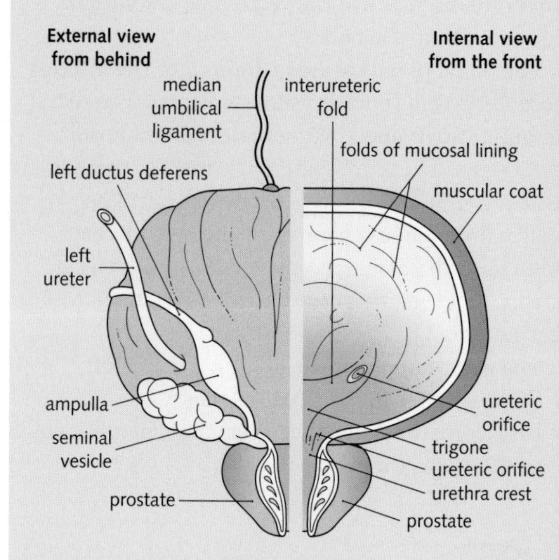

Fig. 5.13 Base of the bladder and related structures in the male.

In males, veins form the vesicoprostatic plexus, which drains into the internal iliac veins. A similar plexus is formed in females.

The nerve supply comprises:

- Parasympathetic (motor)—pelvic splanchnic nerves.
- Sympathetic—the superior hypogastric and pelvic plexuses.

In males and females a distended bladder rises above the symphysis pubis and lifts the peritoneum away from the anterior abdominal wall. A needle can be inserted above the symphysis pubis to drain the bladder (suprapubic cystostomy).

The male urethra

The male urethra is 20 cm long, commencing at the bladder neck and terminating at the external urethral orifice (Fig. 5.14). It has three parts:

- The prostatic urethra has an elevated central region on its posterior wall, the urethral crest. The crest expands to form the seminal colliculus on which lies the prostatic utricle. The orifices of the ejaculatory ducts open on either side of this.

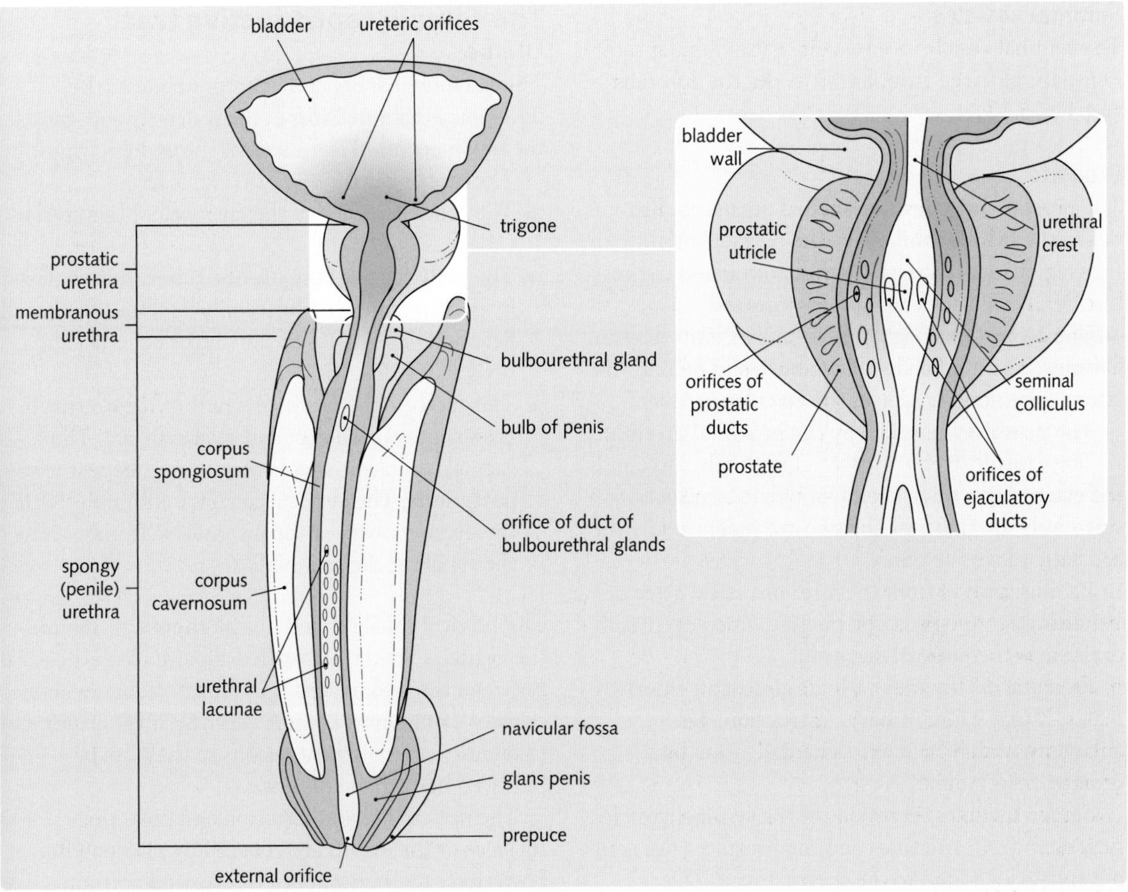

Fig. 5.14 Male urethra (showing details of the prostatic urethra).

- The membranous urethra lies between the apex of the prostate and the bulb of the penis. It is surrounded by the sphincter urethrae and the perineal membrane. The bulbourethral glands lie on either side of it.
- The spongy urethra passes through the bulb, corpus spongiosum, and glans of the penis. Immediately before the external urethral orifice, the urethra expands to form the navicular fossa.

Numerous urethral glands open throughout the course of the urethra.

The female urethra

The female urethra is only 4 cm long. It runs from the bladder neck, through the pelvic floor and the perineal membrane, to open into the vestibule, anterior to the vaginal opening.

The female urethra, being shorter than that of the male, is more prone to urinary tract infections by ascending organisms.

Male reproductive organs in the pelvis

Ductus deferens

The ductus deferens passes from the epididymis to the pelvic cavity via the inguinal canal. At the deep inguinal ring it hooks around the inferior epigastric artery, crossing the external iliac vessels to enter the pelvic cavity. It crosses the obturator neurovascular bundle and the ureter to reach the base of the bladder (see Fig. 5.13). The terminal part dilates, forming the ampulla, and it joins the duct of the seminal vesicle to form the ejaculatory duct, which opens into the prostatic urethra on the colliculus.

Seminal vesicles

The seminal vesicles are two elongated lobular sacs lying lateral to the ampulla of the ductus deferens (see Fig. 5.13).

Prostate

The prostate is a chestnut-shaped organ that lies below the bladder and above the perineal membrane (see Fig. 5.14). It has a base that is attached to the bladder, and an apex that points inferiorly.

The prostate has right and left lobes united by an isthmus. The median lobe lies above and behind the lateral lobes and it receives the ejaculatory ducts.

The prostate is pierced by the proximal urethra.

The capsule of the prostate completely surrounds the gland, and a thick sheath of pelvic fascia surrounds the capsule. The two are separated by the prostatic plexus of veins.

Blood supply is from the inferior vesical artery. Veins drain to the prostatic plexus, which eventually drains into the internal iliac veins.

Prostatic enlargement affects almost all elderly males. It may cause urinary obstruction. The enlargement may be due to a benign lesion or a prostatic carcinoma.

Semen includes secretions of the seminal vesicles (e.g. vitamin C, fructose) and the prostate (e.g. acid phosphatase) added to the spermatozoa. This is delivered to the prostatic part of the urethra. A single ejaculate is about 3.0 mL.

The female reproductive tract

Uterus

This is a muscular organ that accommodates the developing embryo. Most of the wall is smooth muscle, the myometrium. The mucosa is the endometrium.

The uterus has three parts (Fig. 5.15):

- The fundus lies above the entrance of the uterine tube.
- The body receives the uterine tubes. It is enclosed by peritoneum, which laterally becomes the broad ligament. The cavity of the uterus occupies the body.
- The cervix is the narrowest part of the uterus. It has a supravaginal part and a vaginal part. The vaginal fornix surrounds the cervix, deepest posteriorly. The cervical canal is continuous with the uterine cavity at the internal os. It opens into the vagina at the external os.

The cervical canal is lined by columnar epithelium; the vagina is lined by stratified squamous epithelium. Between the two regions is a transition zone where cervical carcinomas arise. A cervical smear attempts to identify premalignant lesions so they can be removed before cancers arise.

The upper anterior, superior, and posterior surfaces of the uterus are covered by peritoneum. Posteriorly the peritoneum is reflected onto the rectum to form the rectouterine pouch (Fig. 5.16). The peritoneum continues from the lateral surface of

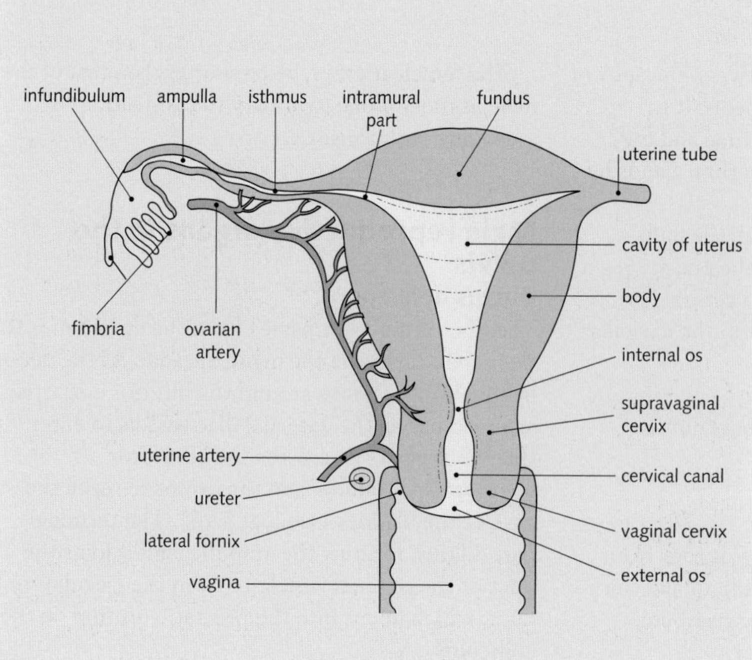

Fig. 5.15 Uterus and its blood supply.

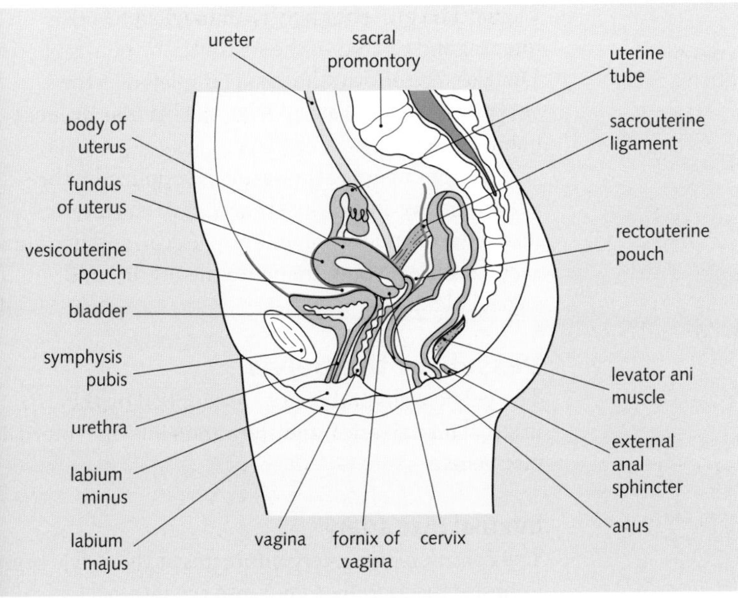

Fig. 5.16 Sagittal section through the female pelvis.

the uterus to the pelvic side wall as the broad ligament.

Uterine (fallopian) tubes

The uterine tubes extend from the junction of the body and fundus of the uterus. They run in the upper edge of the broad ligament (see Fig. 5.15). The peritoneum investing each tube is the mesosalpinx.

The uterine tube is composed of the isthmus, the ampulla, and the infundibulum. The outer end of the infundibulum is fimbriated. One of these fimbria, the ovarian fimbria, is attached to the ovary.

When the ovum is shed into the peritoneal cavity, it is taken up by the infundibulum and passed to the uterus. Fertilization usually occurs in the tubes, but occasionally the fertilized ovum may implant in the tube (an ectopic pregnancy).

Ligaments of the uterus

The broad ligament is a double fold of peritoneum that is attached to the uterus and pelvic side wall (Fig. 5.17). It forms the mesosalpinx for the uterine tubes, mesometrium for the uterus, and the mesovarium for the ovary.

The suspensory ligament of the ovary contains the ovarian vessels and lymphatics. The anterior layer of the broad ligament is pushed forward by the round ligament of the uterus. The posterior layer bulges backwards as the mesovarium, suspending the ovary.

The round ligament extends from the body of the uterus to the pelvic brim. It passes through the inguinal canal to the labium majus. It is the remnant of the gubernaculum (see Testis, p. 70).

Transverse cervical ligaments (cardinal ligaments) are thickened connective tissue at the base of each broad ligament, extending from the cervix and vaginal fornix to the side wall of the pelvis. They stabilize the cervix laterally.

The uterosacral ligaments extend from the cervix to the fascia over piriformis, passing on either side of the rectum. They keep the cervix pulled back against the forward pull from the round ligament.

The pubocervical ligaments extend from the cervix and upper vagina to the posterior aspect of the pubic bones.

The parametrium is the tissue lying between the two peritoneal layers of the broad ligament. It contains the uterine and ovarian vessels and lymphatics, the round ligament, and the suspensory ligament of the ovary.

Vessels of the uterus and uterine tubes

The uterine artery, a branch of the internal iliac artery, runs in the broad ligament. It anastomoses with branches of the ovarian artery. Each artery gives rise to tubal branches, which anastomose and supply the uterine tubes.

The venous drainage is into a uterine plexus. Two uterine veins arise from the plexus, and these drain into the internal iliac vein.

Ovary

The ovary is an ovoid organ that lies in the ovarian fossa in the angle between the internal and external iliac vessels and closely related to the obturator nerve. It produces the ova and sex hormones.

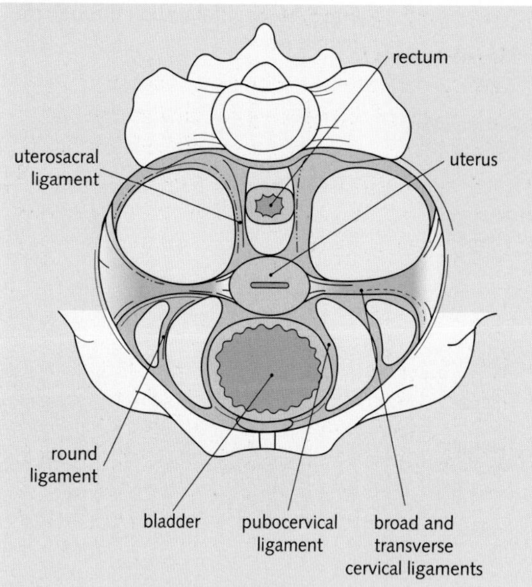

Fig. 5.17 Ligaments of the uterus.

The ovary is attached to the posterior leaf of the broad ligament by a double fold of peritoneum: the mesovarium. It is attached to the uterus by the ligament of the ovary, which runs between the two layers of the broad ligament and is continuous with the round ligament, both being remnants of the gubernaculum.

Blood supply is from the ovarian artery, a branch of the abdominal aorta. It runs in the broad ligament and it gives branches to the uterus and uterine tubes. It anastomoses with the uterine artery. An ovarian venous plexus communicates with the uterine venous plexus. Two ovarian veins follow the artery: the right ovarian vein drains into the inferior vena cava; the left vein drains into the left renal vein.

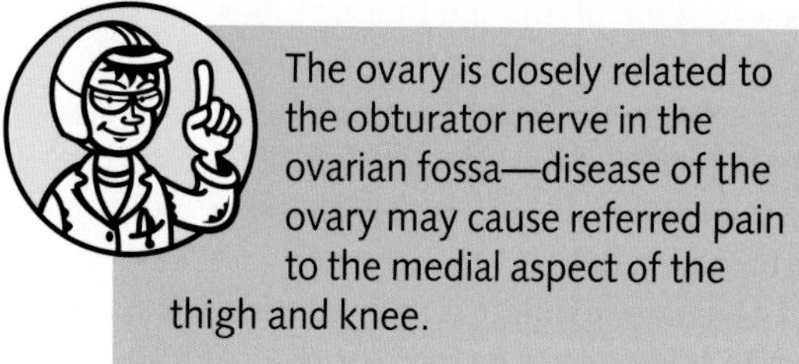

The ovary is closely related to the obturator nerve in the ovarian fossa—disease of the ovary may cause referred pain to the medial aspect of the thigh and knee.

Vagina

The vagina is continuous with the cervix at the external os, and it opens into the perineum at the vaginal orifice (see Fig. 5.16). The vaginal fornix surrounds the part of the cervix that projects into the vagina. The posterior part of the vaginal fornix is deepest and related to the rectouterine pouch of Douglas. Anteriorly, the vagina is related to the cervix, and it is separated from the bladder by loose connective tissue.

The upper part of the vagina is supplied by the uterine artery and vaginal branches of the internal iliac, internal pudendal, inferior vesical, and middle rectal arteries. Veins drain into the uterine and vaginal plexuses.

Vessels of the pelvis

The pelvic walls and cavity are supplied by the internal iliac arteries, and they drain into the internal iliac veins.

Internal iliac artery

The common iliac artery bifurcates at the pelvic brim opposite the sacroiliac joint into the internal and external iliac arteries (Fig. 5.18). The internal iliac artery passes inferiorly and branches into anterior and posterior divisions. The external iliac artery is concerned mainly with the blood supply to the lower limb.

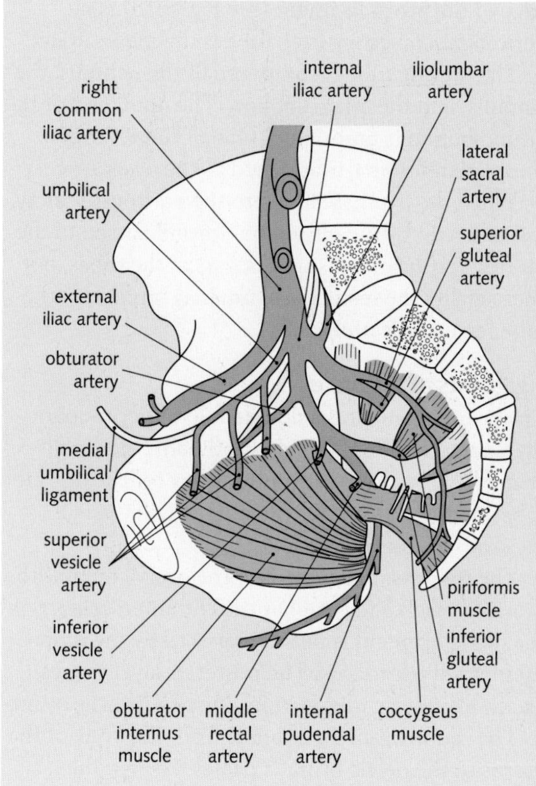

Fig. 5.18 Branches of the internal iliac artery.

Internal iliac vein

The internal iliac vein commences at the greater sciatic notch by the confluence of the gluteal veins and others that accompany branches of the internal iliac arteries. It passes superiorly and out of the pelvis, lying posterior to the artery on the medial surface of psoas major. Here, it joins the external iliac vein to form the common iliac vein.

Tributaries include:

- Veins corresponding to the arteries.
- The uterine and vesicoprostatic venous plexuses.
- The rectal venous plexuses.
- The lateral sacral veins (it communicates with the vertebral venous plexus via these veins).
- The obturator vein.

If venous drainage of the lower limb becomes obstructed. The pelvic veins enlarge and provide an alternative route for venous return.

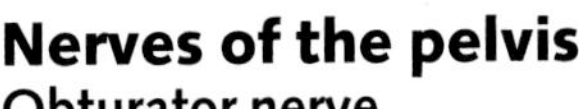

Nerves of the pelvis

Obturator nerve

The obturator nerve supplies the adductor compartment of the thigh by piercing the medial border of psoas and passing along the side of the pelvis to the obturator foramen. It passes through the obturator foramen into the thigh with the obturator artery and vein below it.

The obturator nerve may be damaged during ovary removal (oophrectomy). This results in the spasm of adductor muscles of the thigh and cutaneous sensory loss over the medial thigh and knee.

Sacral plexus

The sacral plexus is formed by the anterior rami of L4 and L5 spinal nerves (the lumbrosacral trunk) and the anterior rami of S1–S4 spinal nerves (Fig. 5.19). The plexus lies on the piriformis muscle, and it is covered by the pelvic fascia. The lateral sacral

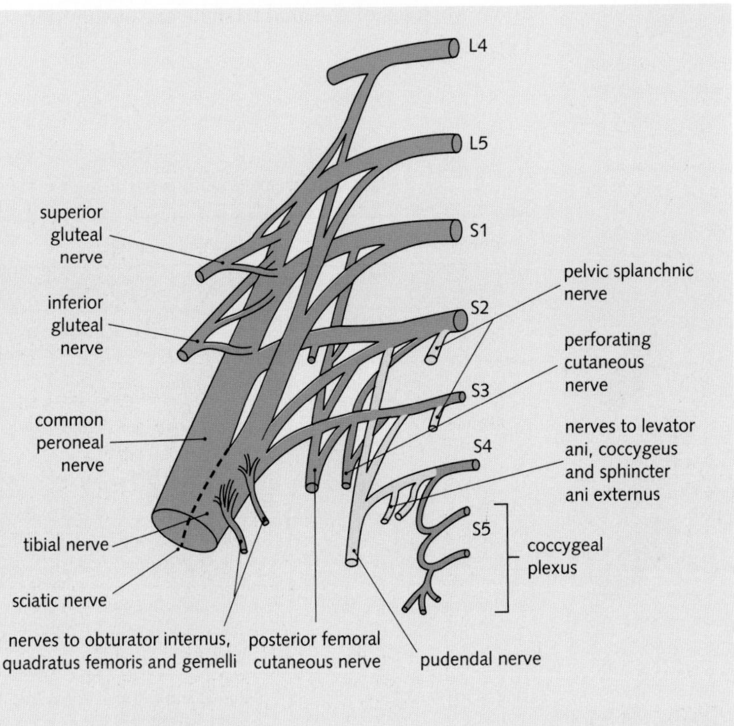

Fig. 5.19 Sacral plexus.

arteries and veins lie anterior to the plexus. The sacral nerves and the lumbosacral trunk give off branches and then divide into the anterior and posterior divisions.

Sacral sympathetic trunk

The sacral sympathetic trunk crosses the pelvic brim behind the common iliac vessels. There are four ganglia along the trunk. The trunks of the two sides unite in front of the coccyx at a small swelling—the ganglion impar. The sacral sympathetic trunk gives somatic branches to all the sacral nerves and visceral branches to the inferior hypogastric plexus.

Inferior hypogastric plexus

The right and left hypogastric plexuses comprise the pelvic plexuses (Fig. 5.20). They lie on the side wall of the pelvis, lateral to the rectum. They receive the right and left hypogastric nerves from the superior hypogastric plexus.

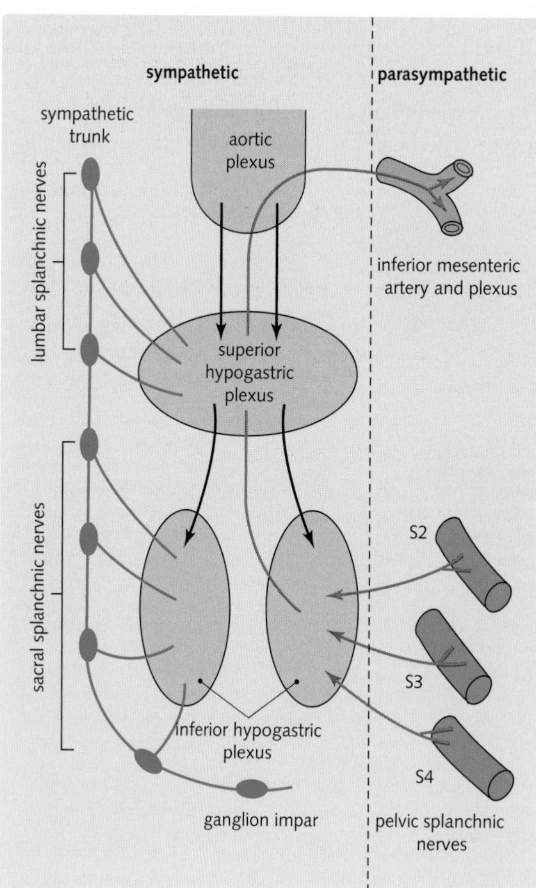

Fig. 5.20 Autonomic plexuses in the pelvis. (Adapted from *Anatomy as a Basis for Clinical Medicine*, by E C B Hall-Craggs. Courtesy of Williams & Wilkins.)

Branches

These are all visceral. Functions include the control of micturition, defaecation, erection, ejaculation, and orgasm.

Lymphatic drainage of the pelvis

This is summarized in Fig. 5.21.

The perineum

The perineum is the region of the trunk lying below the pelvic diaphragm. It is bounded by the pelvic outlet. Fig. 5.22 illustrates the boundaries of the perineum.

The perineum may be divided into an anterior urogenital triangle and a posterior anal triangle.

Anal triangle

The anal triangle has an anterior boundary, which is an imaginary transverse line between the ischial tuberosities. The lateral boundaries are the sacrotuberous ligaments and the apex is the tip of the coccyx (Fig. 5.22). It contains the two ischiorectal fossae separated by the anal canal, the anococcygeal ligament, and the perineal body. The triangle slopes anteriorly and inferiorly from its apex. Muscles of the anal triangle are outlined in Fig. 5.23.

Lymphatic drainage of the pelvis

Structure	Lymphatic drainage
Anal canal	Superficial inguinal nodes
Bladder	Internal and external iliac nodes
Ovary	Aortic nodes
Rectum	Inferior mesenteric nodes Internal iliac nodes Pararectal nodes Preaortic nodes
Urethra	Internal iliac nodes Superficial inguinal nodes
Uterus and uterine tubes	External iliac nodes Internal iliac nodes Sacral nodes
Vagina	Internal and external iliac nodes Superficial inguinal nodes

Fig. 5.21 Lymphatic drainage of the pelvis.

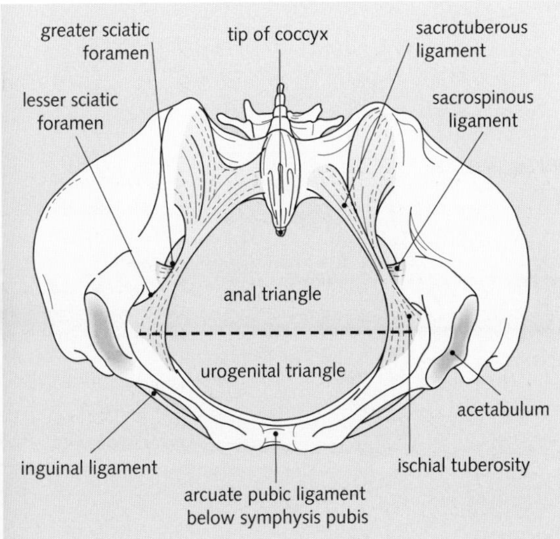

Fig. 5.22 Boundaries of the perineum.

Anal canal

The anal canal commences at the anorectal junction where the rectum passes through the puborectalis muscle. It extends for about 4 cm and ends at the anus.

Lining of the anal canal

The upper part of the anal canal is lined by columnar epithelium with goblet cells that is thrown into folds—the anal columns. Inferiorly, these columns are linked by horizontal folds, forming the anal valves. The recesses between the columns and valves are the anal sinuses where the anal glands open. The lower margins of the anal valves form the pectinate line. Below this there is a transitional zone of stratified squamous epithelium, limited inferiorly by the anocutaneous junction (white line), where the lining of the canal becomes true skin.

Muscular wall of the anal canal

This surrounds the mucous membrane and it is composed of the internal and external anal sphincters (Fig. 5.24).

The internal anal sphincter is a continuation of the circular smooth muscle of the rectum. It surrounds the upper three-quarters of the anal canal and it ends at the white line.

At the anorectal junction the longitudinal muscle joins with puborectalis muscle fibres to form a descending fibroelastic sheet. This is called the conjoint longitudinal coat. It runs between the internal and external anal sphincters piercing them to insert into the perianal skin and ischiorectal fossa fat. It is suggested that the coat might cause puckering of the perianal skin.

The external anal sphincter has subcutaneous, superficial, and deep parts but these fuse together, forming a single muscular mass. The muscle is striated and under voluntary control. The external sphincter is supplied by the perineal branch of S4 spinal nerve and the inferior rectal nerve (from the pudendal nerve). Voluntary contraction of the sphincter delays defaecation.

Fig. 5.25 shows the blood supply and lymphatic drainage of the anal canal.

Ischiorectal fossa and pudendal canal

The ischiorectal fossa lies lateral to the anal canal. It is filled with adipose tissue, and it contains the pudendal canal as well as the inferior rectal arteries and nerves (see Fig. 5.24). The boundaries of the ischiorectal fossa are detailed in Fig. 5.26.

Muscles of the anal triangle

Name of muscle (nerve supply)	Origin	Insertion	Action
External anal sphincter—subcutaneous part (inferior rectal nerve and perineal branch of fourth sacral nerve)	Encircles anal canal, no bony attachments		Voluntary sphincter of anal canal
External anal sphincter—superficial part (inferior rectal nerve and perineal branch of fourth sacral nerve)	Perineal body	Coccyx	
External anal sphincter—deep part (inferior rectal nerve and perineal branch of fourth sacral nerve)	Encircles anal canal	Coccyx	

Fig. 5.23 Muscles of the anal triangle. (Adapted from *Essential Clinical Anatomy*, 1996, by K L Moore. Williams & Wilkins.)

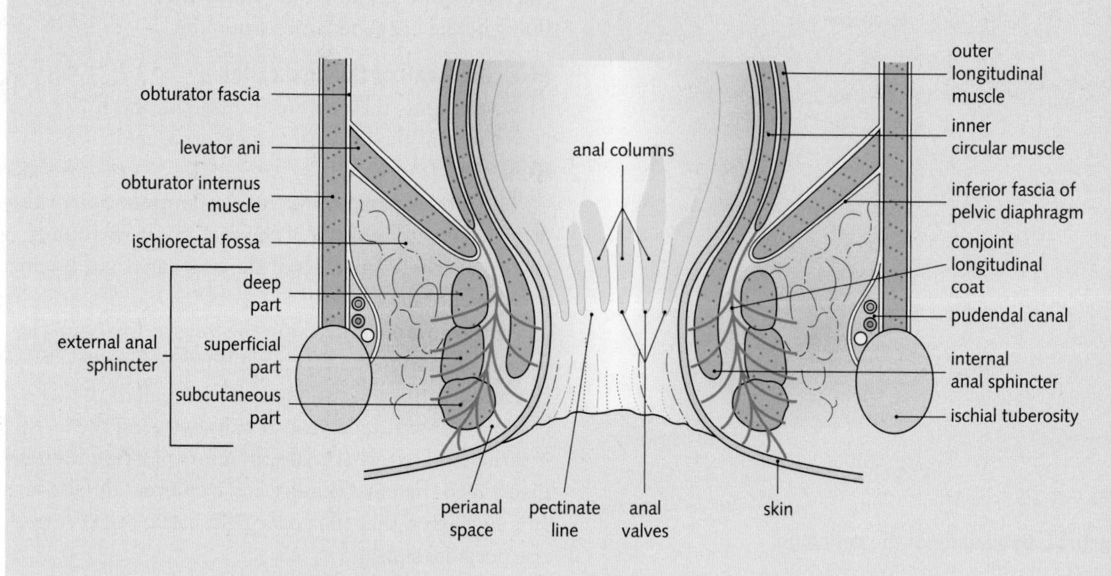

Fig. 5.24 Anal canal and ischiorectal fossa.

Fig. 5.25 Blood supply and lymphatic drainage of the anal canal.

Blood supply and lymphatic drainage of the anal canal	
Arterial supply	**Origin**
Superior rectal artery Middle rectal artery Inferior rectal artery	Inferior mesenteric artery Internal iliac artery Internal pudendal artery
Venous drainage	**Destination**
Internal venous plexus (in submucosa) External venous plexus (outside the muscular coat) Muscular coat of the upper part of the canal	Superior rectal veins and to inferior mesenteric veins Inferior rectal veins drain into internal pudendal veins Middle rectal veins
Lymphatic drainage	**Destination**
Upper part of anal canal Lower part of anal canal (above the anocutaneous junction) Skin	Nodes alongside the rectum and then to preaortic nodes Internal and common iliac nodes Superficial inguinal nodes

The pudendal canal lies on the lateral wall of the ischiorectal fossa—the ischial tuberosity. The canal is roofed by fascia that is continuous with the obturator internus fascia above, and that fuses with the ischial tuberosity below. It contains the internal pudendal vessels and the pudendal nerve, which are conducted from the lesser sciatic foramen to the deep perineal pouch.

The ischiorectal fossa's fat has a poor blood supply. Therefore, it is vulnerable to infection and abscess formation.

Boundaries of the ischiorectal fossa	
Boundary	**Components**
Base	Skin over anal region of perineum
Medial wall	Anal canal and levator ani
Lateral wall	Ischial tuberosity and obturator internus
Apex	Where levator ani is attached to its tendinous origins over obturator fascia
Anterior extension	Superior to deep perineal pouch

Fig. 5.26 Boundaries of the ischiorectal fossa.

The male urogenital triangle

The male urogenital triangle is bounded posteriorly by an imaginary transverse line between the two ischial tuberosities. Its lateral boundaries are the ischiopubic rami, which meet at the apex—the pubic symphysis (Fig. 5.22).

Within the urogenital triangle is the deep perineal space. The space lies below the anterior part of the levator ani muscle. The deep perineal space contains striated muscle that is 'sandwiched' between superior and inferior layers of fascia (Fig. 5.27).

Loose pelvic fascia forms the superior boundary to the deep space, which contains the deep transverse perineal muscles and the sphincter urethrae.

The inferior fascia forms the inferior boundary of the deep perineal space. This tough fascial layer is called the perineal membrane and it is attached to the ischiopubic rami, from just behind the pubic symphysis to the ischial tuberosities.

A superficial perineal fascia layer that is continuous with the membranous superficial fascia (Scarpa's fascia) of the anterior abdominal wall descends and lines the scrotum. Between this superficial perineal fascia (of Colles) and perineal membrane is the superficial perineal space.

The urethra penetrates the deep perineal space and perineal membrane (Fig. 5.27).

Fig. 5.28 outlines the muscles of the urogenital triangle. These muscles are involved in micturition, copulation, and support of the pelvic viscera.

Deep perineal space

The deep perineal space is the region above the perineal membrane. It contains the deep transverse perineal muscles and sphincter urethrae and the following structures:

- The membranous urethra.
- The bulbourethral glands.

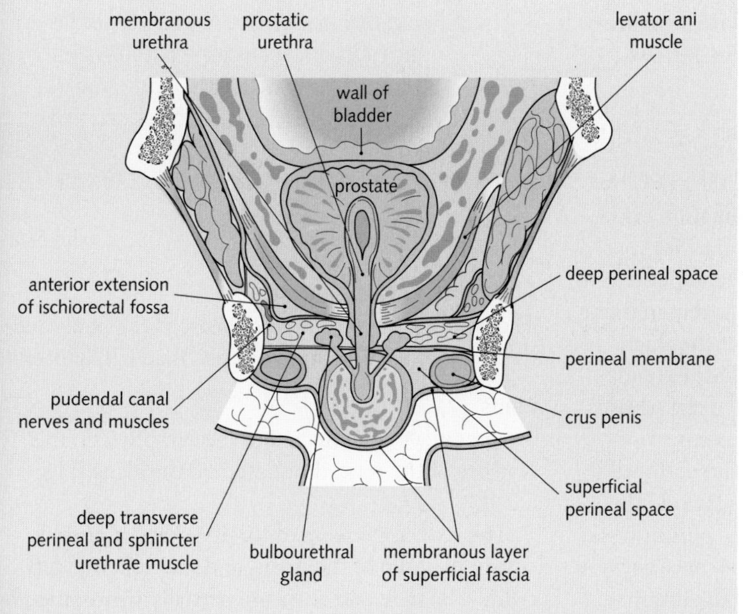

Fig. 5.27 Coronal section of the male perineum.

Muscles of the urogenital triangle			
Name of muscle (nerve supply)	**Origin**	**Insertion**	**Action**
Superficial transverse perineal muscle (perineal branch of pudendal nerve)	Ischial tuberosity	Perineal body	Fixes perineal body
Bulbospongiosus (perineal branch of pudendal nerve)	Perineal body and median raphe in male, perineal body in female	Fascia of bulb of penis and corpora spongiosum and cavernosum in male, fascia of bulbs of vestibule in female	In male, empties urethra after micturition and ejaculation, and assists in erection of penis; in female, sphincter of vagina and assists in erection of clitoris
Ischiocavernosus (perineal branch of pudendal nerve)	Ischial tuberosity and ischial ramus in male and female	Fascia covering corpus cavernosum	Erection of penis or clitoris
Deep transverse perineal muscle (perineal branch of pudendal nerve)	Ramus of ischium	Perineal body	Fixes perineal body
Sphincter urethrae (perineal branch of pudendal nerve)	Pubic arch	Surrounds urethra	Voluntary sphincter of urethra

Fig. 5.28 Muscles of the urogenital triangle. (Adapted from *Essential Clinical Anatomy*, 1996, by K L Moore. Williams & Wilkins.)

- The pudendal vessels (continuing forward from the pudendal canal).
- The dorsal nerve of the penis.

The bulbourethral glands are two small glands lying on either side of the membranous urethra (see Fig. 5.26). Their ducts pierce the perineal membrane to enter the spongy part of the urethra. Secretions contribute to the seminal fluid.

Superficial perineal space

The superficial perineal space is lined inferiorly by the superficial perineal fascia, a continuation of the membranous superficial fascia of the anterior abdominal wall. It is attached to the ischiopubic rami, the posterior border of the perineal membrane, the perineal body, and the fascia lata of the thigh. The superficial perineal fascia (of Colles) extends into the penis as the superficial fascia of the penis (Buck's fascia), attaching to the glans penis.

Superiorly the space is limited by the perineal membrane. The contents of the space include the root of the penis, superficial transverse perineal, bulbospongiosus, and ischiocavernosus muscles. It also contains the perineal branches of the internal pudendal artery and pudendal nerve.

Rupture of the penile urethra causes urine to leak into the superficial perineal space. Due to the superficial perineal fascia's continuity with the anterior abdominal wall and penile superficial fascia, urine can pass into the scrotum and penis, and pass upward into the anterior abdominal wall.

The male external genitalia

The male external genitalia consist of the penis and the scrotum. The scrotum is described in Chapter 4.

Penis

The penis is suspended from the symphysis pubis. It consists of the root, the shaft, and the glans (Fig. 5.29).

The root is made up of three masses of erectile tissue: the bulb of the penis, and the right and left crura. The bulb and crura are partially surrounded by the bulbospongiosus and ischiocavernosus muscles.

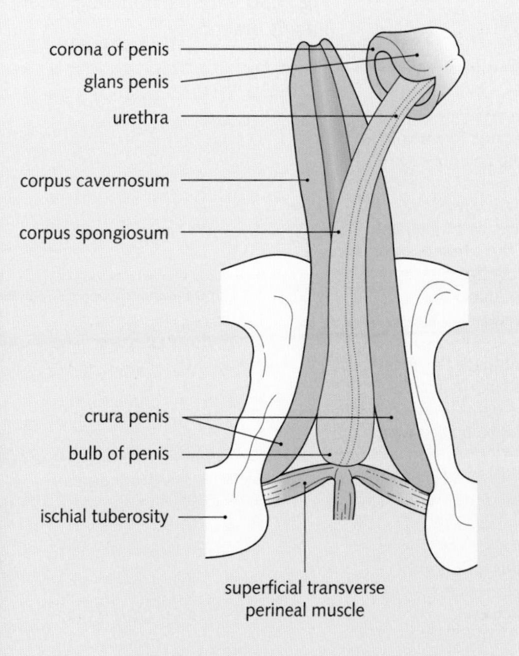

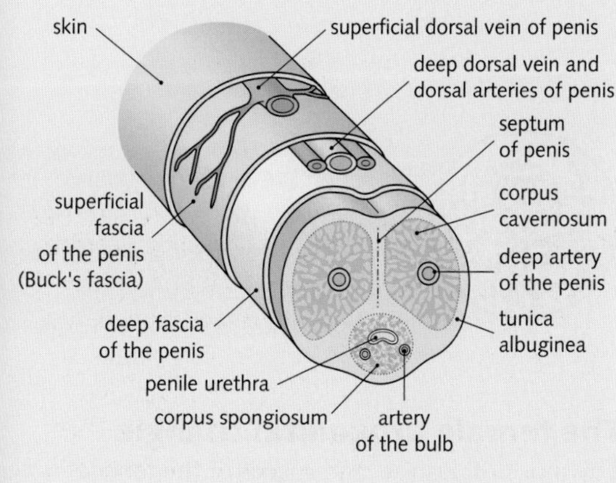

Fig. 5.29 Composition and structure of the penis.

The superficial transverse perineal muscle is also closely related to these.

The crura become the corpora cavernosa; the bulb becomes the corpus spongiosum. The urethra passes into the erectile tissue of the bulb, and it continues in the corpus spongiosum to the external urethral orifice.

The distal end of the corpus spongiosum expands to form the glans penis. The shaft is surrounded by thin loose skin. At the proximal part of the glans penis the skin is reflected upon itself to form the prepuce or foreskin, which covers the glans. The prepuce is attached to the ventral surface of the glans by a fold of skin, the frenulum of the prepuce, which contains a small artery.

A tough tunica albuginea surrounds and passes trabeculae into the three corpora, forming numerous spaces into which arteries empty. The deep fascia of the penis encircles the tunica albuginea. Around this deep fascia runs the superficial fascia of the penis (Buck's fascia).

The crura and corpora cavernosa receive blood from the deep arteries of the penis. The bulb and corpus spongiosum are supplied by the artery to the bulb and the dorsal artery. The dorsal artery supplies the skin and superficial layers as well.

These vessels allow rapid distension of the cavernous spaces to produce an erection.

Venous drainage is to the deep and superficial dorsal veins.

Parasympathetic vasodilator and sensory fibres enter the penis via the pudendal nerves and their terminal branches, the dorsal nerves of the penis.

Vessels of the male urogenital triangle

The internal pudendal artery is a branch of the internal iliac artery. Its course, distribution, and branches are outlined in Fig. 5.30.

The internal pudendal veins are the venae comitantes of the arteries. The deep dorsal vein of the penis drains into the prostatic plexus. The superficial dorsal vein of the penis drains to a superficial pudendal vein and then to the femoral vein.

The blood supply to the scrotum is described in Chapter 4.

Lymphatics of the male urogenital triangle

The penis and scrotum drain into the superficial and deep inguinal nodes.

Nerves of the male urogenital triangle

The pudendal nerve (S2–S4) passes with the internal pudendal artery through the lesser sciatic foramen and the pudendal canal, where it gives rise to the inferior rectal nerve. The inferior rectal nerve supplies the external anal sphincter and the skin around the anus. At the posterior border of the perineal membrane the nerve divides into the perineal nerve and the dorsal nerve of the penis. The perineal nerve passes superficial to the perineal membrane. It supplies the scrotum posteriorly and all the remaining striated muscles of the perineum.

The dorsal nerve of the penis runs with the dorsal artery of the penis. It passes over the dorsum of the

Internal pudendal artery and its branches		
Vessel (in female)	**Source**	**Course and distribution**
Internal pudendal artery	Anterior division of internal iliac artery	Exits the pelvis via the greater sciatic foramen to enter the pudendal canal. At the anterior end of the canal it enters the deep perineal pouch and continues forward on the deep surface of the perineal membrane. It terminates by dividing into the dorsal and deep arteries of the penis (or clitoris)
Inferior rectal artery	Internal pudendal artery	Crosses the ischiorectal fossa to supply the muscles and skin of the anal canal
Perineal branch	Internal pudendal artery	Passes to the superficial perineal space to supply its muscles and the scrotum (or labia in female)
Artery to bulb of penis (or clitoris)	Internal pudendal artery	Supplies the erectile tissue of the bulb and corpus spongiosum
Deep artery of penis (or clitoris)	Internal pudendal artery	Supplies the corpus cavernosum
Dorsal artery of penis (or clitoris)	Internal pudendal artery	Passes to the dorsum of the penis (or clitoris). It supplies the erectile tissue of the corpus cavernosum and superficial structures
Urethral artery	Internal pudendal artery	Supplies the urethra

Fig. 5.30 Internal pudendal artery and its branches.

penis lateral to the artery and terminates in the glans.

The nerve supply to the scrotum is described in Chapter 4.

To recall the pudendal nerve supplies levator ani remember S2, 3, 4 keeps your guts off the floor.

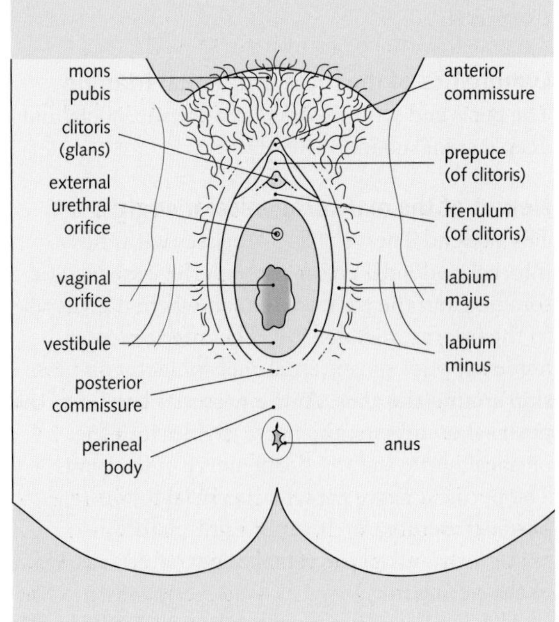

Fig. 5.31 Female external genitalia.

The female urogenital triangle

The muscles, fasciae, and spaces of the female urogenital triangle are similar to those of the male urogenital triangle. However, certain features differ because of the presence of the vagina and external female genitalia.

The deep perineal space differs from the male only in that it is deficient where the vagina pierces it, and it lacks the female equivalent of the bulbourethral gland.

The superficial perineal space again is similar to that of the male, and it is lined by a less-defined superficial perineal fascia. However, it differs because the vagina passes through the space; it is full of fat; it is smaller; and it is found within the labia majora.

Perineal membrane

The perineal membrane is wider, in the female since the pelvis is wider, and it is also weaker than in the

male because of the presence of the vagina piercing it. As the urethra and vagina pierce the membrane their outer fascial covering fuses with it. The crura of the clitoris are attached to the ischiopubic rami.

Perineal body

In the female, the perineal body lies between the vagina and the anal canal. The perineal body lacks support from the perineal membrane because of the presence of the vagina. As a consequence the perineal body in the female has greater mobility. The superficial and deep transverse perineal muscles, the pubovaginalis, bulbospongiosus, and the superficial part of the external anal sphincter are attached to the perineal body.

The external female genitalia

The external female genitalia consist of the mons pubis, clitoris, vagina, vestibule, bulb of the vestibule, greater vestibular glands, external urethral orifice, labia majora, and labia minora. Together these parts form the vulva (Fig. 5.31).

Mons pubis

In the adult (i.e. after puberty), this is the mound of coarse-haired skin and fat anterior to the pubic symphysis. The mons pubis extends posteriorly as the labia majora.

Labia majora

These two fatty folds of skin are joined together anteriorly to form the anterior commissure, and they are continuous with the mons pubis. As the folds pass posteriorly they fade into the skin near the anus. This area is the posterior commissure. The round ligament of the uterus ends in the labium majus.

Labia minora

These are fat free skin folds that lie within the labia majora and surround the vestibule of the vagina. They enclose the clitoris by dividing into the prepuce in front and the frenulum behind.

Clitoris

This homologue of the penis consists of two erectile crura attached to the perineal membrane and ischiopubic rami. Anteriorly, the crura become the corpora cavernosa. These are bound together by fascia to form the body of the clitoris. The glans surmounts the body. It is connected to the bulbs of the vestibule by erectile tissue. These lie on either side of the vaginal orifice.

Vestibule

This contains the openings of the greater vestibular glands, the vagina, and the external urethral orifice.

Bulb of the vestibule

This is the homologue of the penile bulb and corpus spongiosum. Posteriorly these two erectile masses, either side of the vagina, are attached to the perineal membrane. Anteriorly they join the glans of the clitoris. Each is covered by the bulbospongiosus muscle.

Greater vestibular glands

The homologues of the bulbourethral glands, they are found in the superficial perineal pouch in contact with the posterior end of the bulb of the vestibule. Their ducts open into the vestibule and lubricate the vagina in sexual arousal.

Vagina

Superiorly, the vagina passes through the pelvic floor surrounded by part of the puborectalis; anteriorly, it is closely related to the urethra; posteriorly, the perineal body separates it from the anal canal; inferiorly, it opens into the vestibule at the introitus. This opening is partially occluded by a thin membrane—the hymen—which is usually destroyed during sexual intercourse.

External urethral orifice

This lies in the vestibule inferior to the glans of the clitoris but superior to the vagina.

Lymphatics of the female urogenital triangle

The lymph drains to the superficial inguinal lymph nodes.

Vessels and nerves of the female urogenital triangle

The internal pudendal artery has a similar course and distribution in the female as in the male, except:

- Posterior labial branches replace scrotal branches.
- The artery to the bulb and vestibule and the dorsal arteries of the clitoris replace the arteries to the penis.

The blood supply to the external genitalia (e.g. the labia) is the same as the scrotum. It receives the superficial and deep external pudendal arteries (from the femoral artery). It drains via corresponding veins into the great saphenous vein. The pudendal nerve has a similar distribution in both genders. The only

difference is in the naming of the nerves, in which labial replaces scrotal.

The nerve supply to the labia is the same as for the scrotum:

- The anterior third is supplied by the ilioinguinal nerve.
- The posterior two thirds are supplied by the posterior labial branch of the pudendal nerve (medially) and by the perineal branch of the posterior femoral cutaneous nerve (laterally).

A pudendal nerve block relieves pain caused by perineal stretching in child birth. The anaesthetic is injected near the nerve as it crosses the sacrospinous ligament near the ischial spine. The latter is palpable on vaginal examination.

- Describe the surface landmarks of the pelvic bones.
- List the structures palpable on a rectal examination in males and females.
- What are the main features of the pelvic bones and joints, including the position of the pelvis?
- Discuss the differences between male and female pelves.
- Discuss the anatomy of the pelvic floor and wall muscles.
- What is the perineal body and its function?
- Describe the peritoneum in the pelvis.
- Describe the anatomy of the rectum.
- What is the course of the ureters in the pelvis?
- Describe the anatomy of the bladder.
- Describe the anatomy of the male reproductive tract.
- Describe the anatomy of the female reproductive tract.
- Outline the vessels of the pelvis.
- Outline the nerves of the pelvis.
- List the lymphatic drainage of the pelvic viscera.
- Describe how the perineum is divided into triangles. What are their contents?
- Discuss the anatomy of the anal canal.
- Describe how the superficial perineal space is formed and its contents in male and female.
- Describe the male external genitalia.
- Describe the female external genitalia.

6. The Lower Limb

Regions and components of the lower limb

The lower limb is built for support, locomotion, and the maintenance of equilibrium. Weight is transferred from the rigid bony pelvis, through the acetabulum, to the lower limb. Propulsive movements are transmitted in a similar way but in the opposite direction.

The hip joint is formed by the acetabulum and the head of the femur. It is a very stable joint, with a good range of movement.

The femur articulates with the tibia at the knee joint. Only flexion and extension are possible at this joint. The superior part of the fibula serves for muscle attachment only, and it does not take part in the formation of the knee joint or in weightbearing.

Both the tibia and the fibula articulate with the talus to form the ankle joint, where only flexion and extension movements may occur.

The blood supply to the lower limb is from the external iliac artery. This becomes the femoral artery beneath the inguinal ligament, and it supplies the entire thigh region. Behind the knee the femoral artery becomes the popliteal artery, which supplies the leg and the foot.

The gluteal region is supplied by the superior and inferior gluteal arteries, branches of the internal iliac artery.

The lower limb is innervated by the lumbar and sacral plexuses via the femoral, obturator, and sciatic nerves. The gluteal region is also supplied by the superior and inferior gluteal nerves.

Surface anatomy and superficial structures

Hip and thigh region

The iliac crest and iliac spines should be familiar (Chapter 5). The greater trochanter of the femur is the widest palpable structure in the hip region.

The large quadriceps muscle makes up the anterior surface of the thigh. It inserts into the patella and, via the patellar ligament, into the tibial tuberosity. The patella is a sesamoid bone into which the quadriceps tendon inserts. It may be easily palpated.

The iliotibial tract lies on the lateral surface of the thigh. The gluteus maximus and tensor fasciae latae insert into it. It may be demonstrated with the subject standing on tiptoe.

The hamstrings lie in the posterior compartment of the thigh. They can be demonstrated by attempting to flex the knees against resistance.

Knee region

On the lateral side of the knee the head of the fibula, the lateral part of the tibia, and the lateral condyle of the femur are palpable. Posteriorly, the diamond-shaped popliteal fossa is seen when the knees are flexed against resistance.

Leg region

The subcutaneous anteromedial surface of the tibia (shin bone) is easily felt.

Just below the head of the fibula the common peroneal nerve becomes superficial as it runs from the popliteal fossa into the lateral compartment of the leg. It is vulnerable to injury at this point. The peroneal muscles lie laterally on the leg.

The large gastrocnemius and soleus muscles are seen at the back of the leg. These join to form the tendocalcaneus, which inserts into the calcaneum.

Ankle

The malleoli are prominences medially and laterally.

Superficial veins

The foot drains into the dorsal venous arch, which drains laterally into the small saphenous vein and medially into the great saphenous vein.

Great saphenous vein

The great saphenous vein passes anterior to the medial malleolus, and it ascends in the subcutaneous tissue on the medial side of the leg and thigh until it reaches the saphenous opening in the fascia lata (Fig. 6.1A and inset). Here, it perforates the cribriform fascia covering the opening and joins the femoral

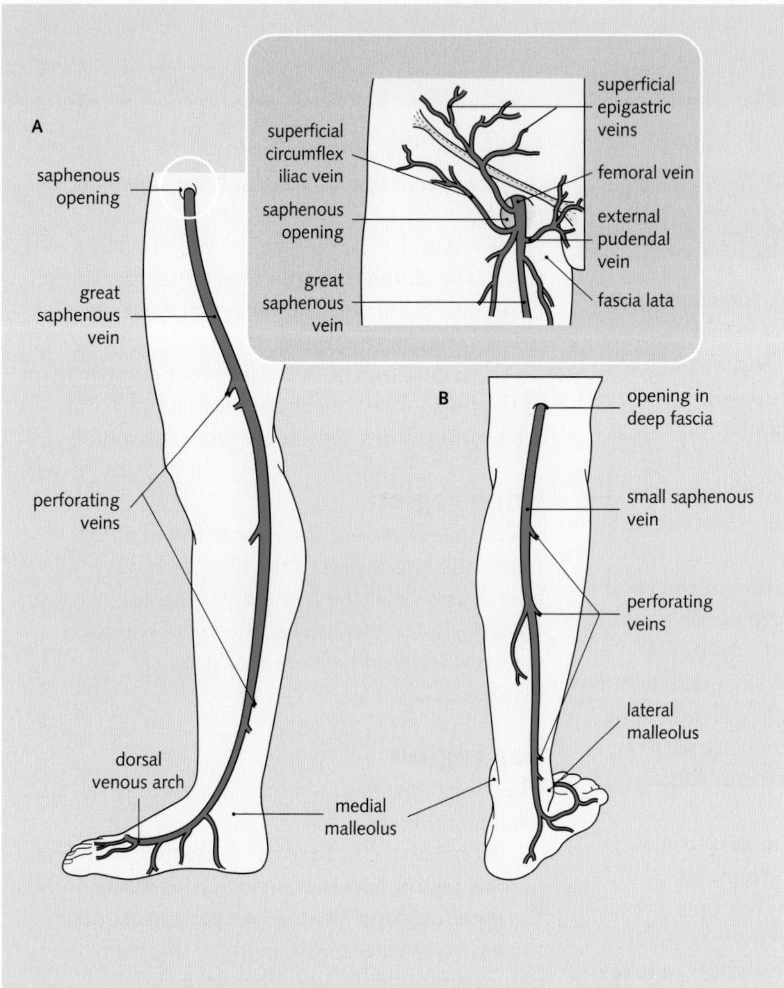

Fig. 6.1 Great (A) and inset and small (B) saphenous veins and their tributaries.

vein. It communicates with the deep veins via perforating veins.

The great saphenous vein is valved, allowing flow of blood in one direction only. Damage to the valves may result in varicose veins.

Tributaries of the great saphenous vein arise from the anterior and medial aspects of the thigh and from the anterior abdominal wall. These include:

- The superficial circumflex iliac vein.
- The superficial epigastric vein.
- The external pudendal vein.

The great saphenous vein is commonly used as a site for cannulation and it may also be used as a graft for a coronary artery bypass operation.

Small saphenous vein

The small saphenous vein passes posteriorly to the lateral malleolus and runs superiorly and posteriorly to pierce the deep fascia of the popliteal fossa, where it joins the popliteal vein (Fig. 6.1B). It drains the lateral part of the leg, and it communicates with the deep veins of the leg via perforating veins.

Lymphatic drainage of the lower limb

Lymphatics from the superficial tissues of the lower limb drain into the superficial inguinal lymph nodes (Fig. 6.2). These lie superficial to the deep fascia around the termination of the saphenous vein. They also receive lymph from the lower part of the perineum, the abdominal wall, and the buttock.

The deep nodes lie alongside the femoral vessels, and they drain the deep tissues of the lower limb. They communicate with the superficial nodes and

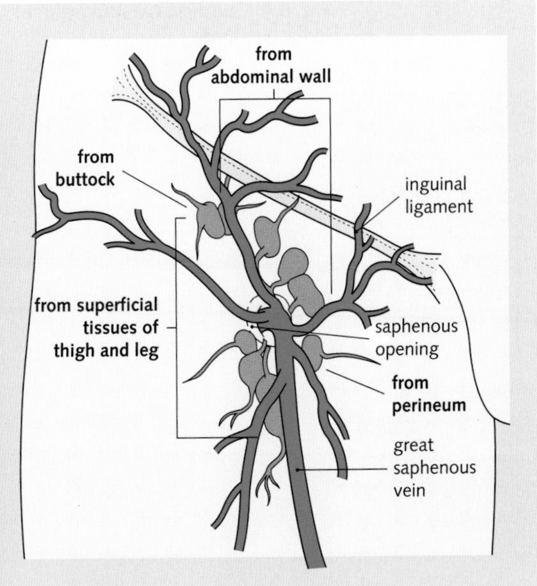

Fig. 6.2 Superficial inguinal lymph nodes.

their efferent vessels drain into nodes lying alongside the external iliac vessels.

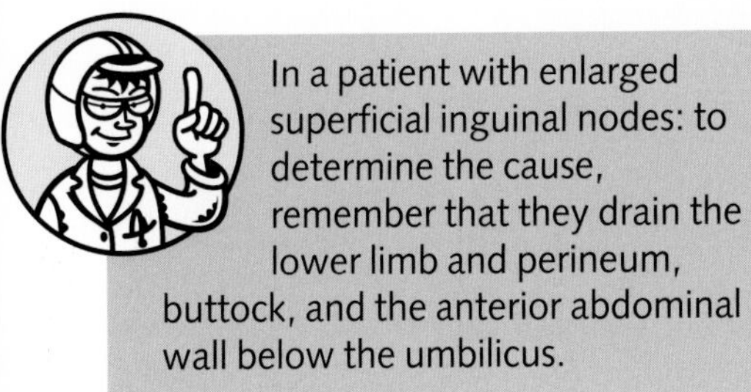

In a patient with enlarged superficial inguinal nodes: to determine the cause, remember that they drain the lower limb and perineum, buttock, and the anterior abdominal wall below the umbilicus.

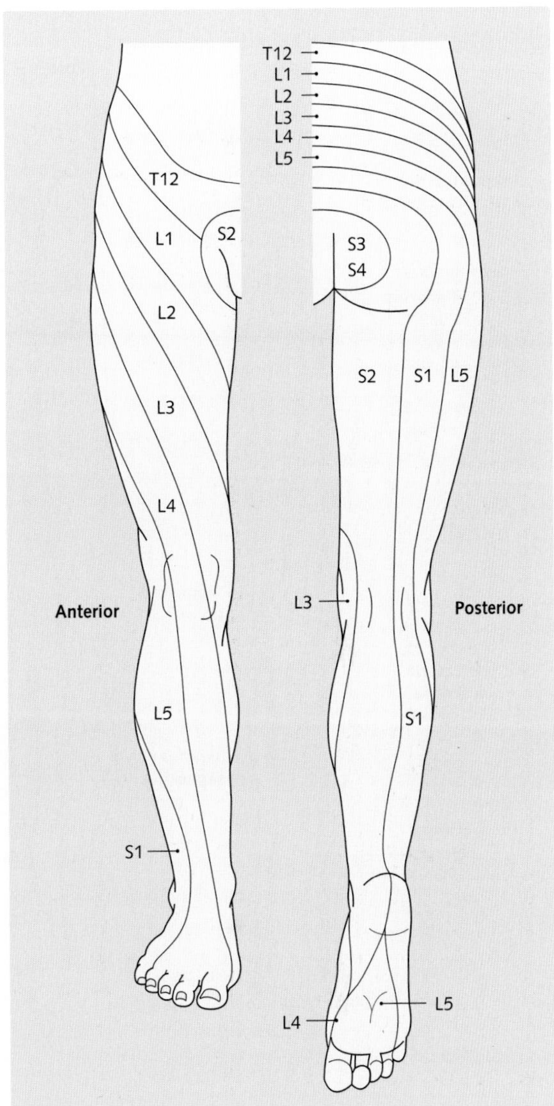

Fig. 6.3 Dermatomes of the lower limb.

Superficial fascia of the lower limb

The membranous superficial fascia (Scarpa's fascia) becomes fused with the fascia lata of the thigh as it extends inferiorly from the anterior abdominal wall. The position at which the fusion occurs is at a skin crease just below the inguinal ligament. Remember, this fusion prevents extravasated fluid (e.g. urine) in the superficial perineal pouch from tracking inferiorly in the lower limb.

Cutaneous innervation of the lower limb

Figs 6.3 and 6.4 illustrate the dermatomes of the lower limb and its cutaneous innervation, respectively.

The gluteal region, hip, and thigh

Skeleton of the hip and thigh

Pelvic girdle

The pelvic girdle protects the pelvic cavity and supports the body weight. It transmits load to the lower limbs via the sacrum, hip bone, and hip joint (see Chapter 5).

Femur

The femur is the long bone of the thigh. The bone and its muscle attachments are illustrated in Fig. 6.5.

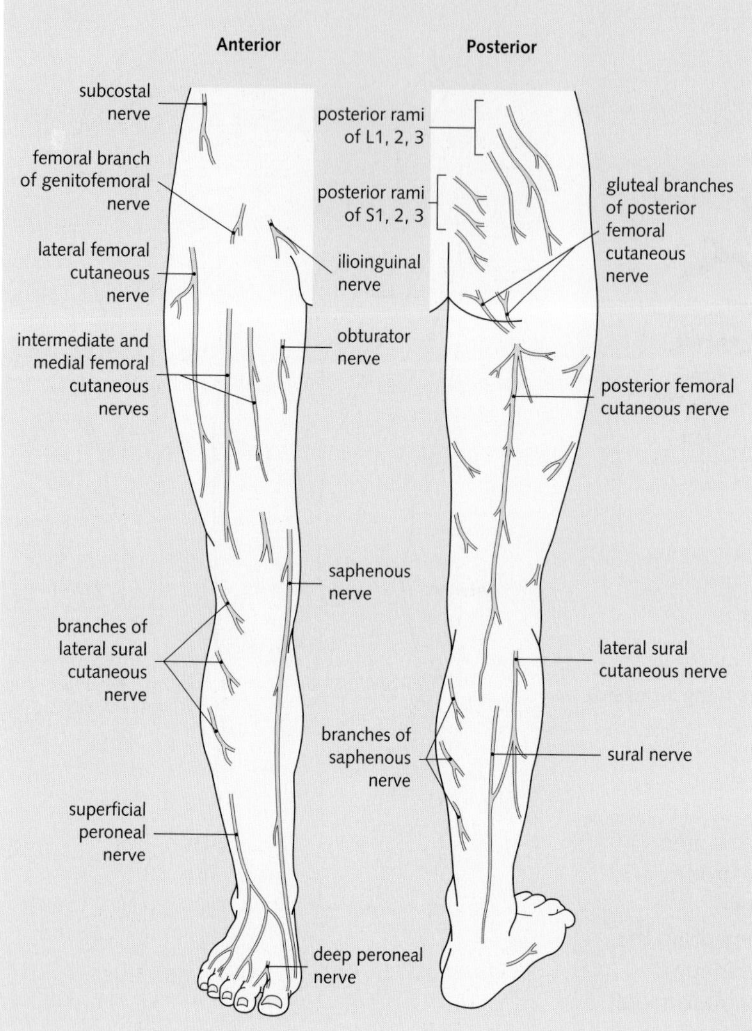

Fig. 6.4 Cutaneous innervation of the lower limb.

Fascia lata

The fascia lata is the deep fascia of the thigh. It lies below the skin and superficial fascia, and it encloses the compartments of the thigh. Its attachments can be traced along the pelvis:

- Ilium, sacrum, and sacrotuberous ligament (posteriorly)
- Ischial tuberosity, ischiopubic ramus and body of the pubic bone (inferiorly)
- Pubic tubercle and pectinate line (anteriorly)
- Iliac crest (superiorly).

Inferiorly the fascia lata is attached to the tibia and fibula and continues below into the deep fascia of the calf.

The iliotibial tract is a thickening of the fascia lata. It begins at the level of the greater trochanter and it extends to the lateral condyle of the tibia. The tract keeps the knee hyperextended when standing stationary.

The thigh is divided into anterior, posterior, and adductor compartments by lateral, medial, and intermediate intermuscular septa (Fig. 6.6).

Gluteal region

The gluteal region, or buttock, lies behind the pelvis and extends from the iliac crest to the fold of the buttock. The greater and lesser sciatic foramina in this region are formed anteriorly by the greater and lesser sciatic notches, respectively. Posteriorly the foramina are completed by the sacrotuberous and sacrospinous ligaments. The muscles, the vessels, and the nerves that exit or enter the sciatic foramina and

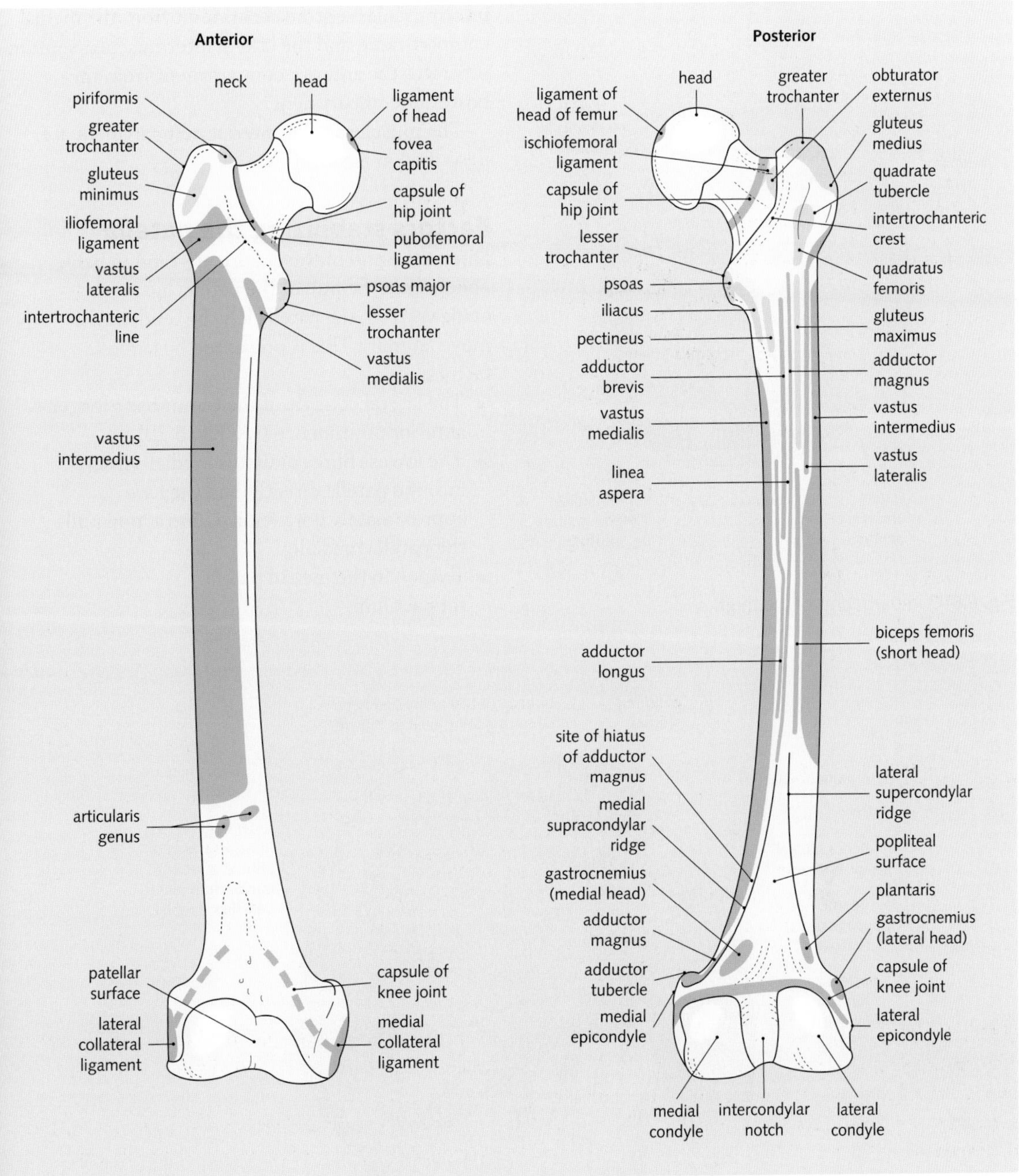

Anterior

neck
head
piriformis
greater trochanter
gluteus minimus
iliofemoral ligament
vastus lateralis
intertrochanteric line
ligament of head
fovea capitis
capsule of hip joint
pubofemoral ligament
psoas major
lesser trochanter
vastus medialis
vastus intermedius
articularis genus
patellar surface
lateral collateral ligament
capsule of knee joint
medial collateral ligament

Posterior

head
greater trochanter
ligament of head of femur
ischiofemoral ligament
capsule of hip joint
lesser trochanter
psoas
iliacus
pectineus
adductor brevis
vastus medialis
linea aspera
adductor longus
site of hiatus of adductor magnus
medial supracondylar ridge
gastrocnemius (medial head)
adductor magnus
adductor tubercle
medial epicondyle
obturator externus
gluteus medius
quadrate tubercle
intertrochanteric crest
quadratus femoris
gluteus maximus
adductor magnus
vastus intermedius
vastus lateralis
biceps femoris (short head)
lateral supercondylar ridge
popliteal surface
plantaris
gastrocnemius (lateral head)
capsule of knee joint
lateral epicondyle
medial condyle
intercondylar notch
lateral condyle

Fig. 6.5 Muscles and ligaments of the anterior and posterior surfaces of the right femur.

supply the gluteal region, are shown in Figs 6.7–6.10. The piriformis muscle is the key to understanding this region, since vessels and nerves either pass superior or inferior to it.

Anterior compartment of the thigh

The anterior compartment of the thigh is bound anterolaterally by the fascia lata. The medial

The buttock is a common site for intramuscular injections. The sciatic nerve is at risk of damage unless injections are given in the upper outer quadrant.

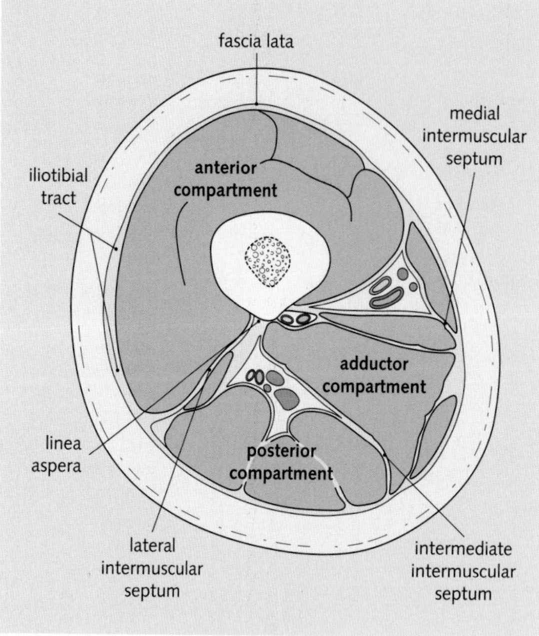

Fig. 6.6 Compartments of the thigh.

intermuscular septum separates it from the medial compartment, and the lateral intermuscular septum separates the anterior compartment from the posterior compartment.

The muscles of the anterior compartment are described in Fig. 6.11.

Factors stabilizing the patella

Due to the pull of the quadriceps tendon being oblique and the pull of the patella ligament being vertical, the patella has the tendency to move laterally. This is prevented by three factors:

- The lateral condyle of the femur has a longer anterior prominence (see Fig. 6.20)
- The lowest fibres of vastus medialis insert into the patella directly, and they are approximately horizontal. Contraction pulls the patella medially.
- Tension in the medial patella retinaculum.

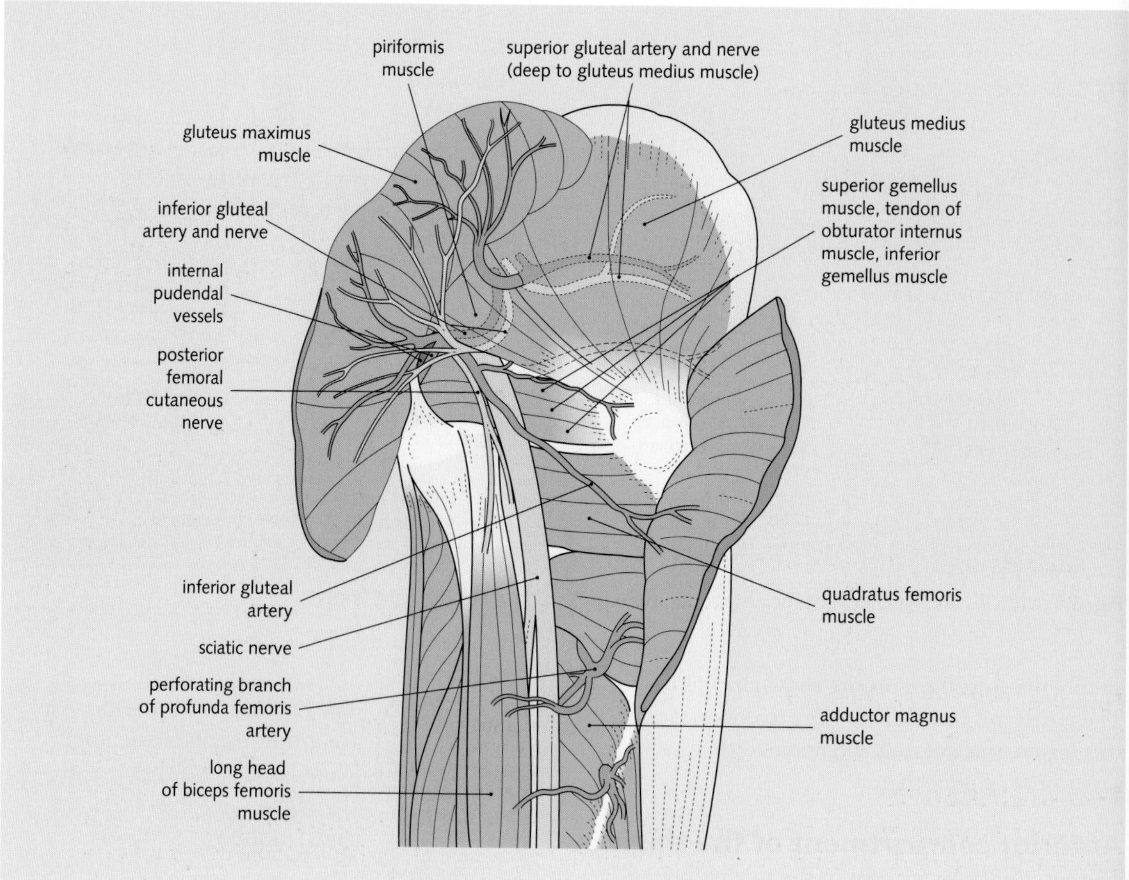

Fig. 6.7 Gluteal region showing the main nerves and vessels.

Muscles of the gluteal region			
Name of muscle (nerve supply)	**Origin**	**Insertion**	**Action**
Tensor fasciae latae (superior gluteal nerve)	Iliac crest	Iliotibial tract	Extends knee joint Tenses iliotibial tract
Gluteus maximus (inferior gluteal nerve)	Ilium, sacrum, coccyx, and sacrotuberous ligament	Iliotibial tract and gluteal tuberosity of femur	Extends and laterally rotates thigh at hip joint; extends knee joint
Gluteus medius (superior gluteal nerve)	Ilium	Greater trochanter of femur	Abducts thigh at hip joint and tilts pelvis when walking
Gluteus minimus (superior gluteal nerve)	Ilium	Greater trochanter of femur	As gluteus medius and medially rotates thigh
Piriformis (S1, S2 nerves)	Anterior surface of sacrum	Greater trochanter of femur	All these muscles rotate thigh laterally at hip joint and stabilize the hip joint
Obturator internus (sacral plexus)	Inner surface of obturator membrane	Greater trochanter of femur	
Gemellus inferior (sacral plexus)	Ischial tuberosity	Greater trochanter of femur	
Gemellus superior (sacral plexus)	Ischial spine	Greater trochanter of femur	
Quadratus femoris (sacral plexus)	Ischial tuberosity	Quadrate tubercle on femur	

Fig. 6.8 Muscles of the gluteal region.

Arteries of the gluteal region	
Artery	**Course and distribution**
Internal pudendal	Passes through the greater sciatic foramen to enter the gluteal region, then passes into the perineum via the lesser sciatic foramen; supplies the muscles of the pelvic region and the external genitalia
Superior gluteal	Enters the gluteal region through the greater sciatic foramen; supplies gluteus maximus, medius and minimus, and tensor fasciae latae
Inferior gluteal	Passes through the greater sciatic foramen to enter the gluteal region; supplies gluteus maximus, obturator internus, and quadratus femoris

Fig. 6.9 Arteries of the gluteal region.

Femoral triangle

The femoral triangle contains the femoral artery, vein, and nerve (Fig. 6.12). These all lie superficially just beneath the skin, superficial fascia, and fascia lata.

The femoral artery and vein and the femoral canal are enclosed by the femoral sheath, which is a continuation of the transversalis and iliac fasciae into the thigh. The femoral nerve is separated from the femoral sheath by the iliopsoas fascia.

The arrangement of the femoral vessels and nerve as they pass under the inguinal ligament can be remembered as 'NAVY' (nerve, artery, vein, Y-fronts!).

Femoral canal

The femoral canal contains efferent lymphatics passing from the deep inguinal nodes to the abdomen (see Fig. 6.12). It provides space for expansion of the femoral vein during times of increased venous return from the lower limb. Its boundaries are the medial part of the inguinal

Nerves of the gluteal region	
Nerve (origin)	**Course and distribution**
Inferior gluteal (anterior rami of L5–S2)	Leaves pelvis through greater sciatic foramen below piriformis, and supplies gluteus maximus
Superior gluteal (anterior rami of L4–S1)	Leaves pelvis through greater sciatic foramen above piriformis and passes between gluteus medius and minimus to supply these muscles and tensor fasciae latae
Nerve to quadratus femoris (anterior rami of L4, L5, and S1)	Leaves pelvis through greater sciatic foramen below piriformis to supply the hip joint, inferior gemellus, and quadratus femoris
Nerve to obturator internus (anterior rami of L5, S1, and S2)	Enters gluteal region through greater sciatic foramen below piriformis, descends posterior to ischial spine, enters lesser sciatic foramen, and passes to obturator internus to supply it and superior gemellus
Posterior femoral cutaneous (sacral plexus—S1–S3)	Leaves pelvis through greater sciatic foramen, below piriformis, runs deep to gluteus maximus, and emerges from its inferior border to supply skin of buttock and then surface skin over posterior of thigh and calf
Pudendal (anterior rami of S2–S4)	Enters gluteal region through greater sciatic foramen below piriformis, descends posterior to sacrospinous ligament, enters perineum through lesser sciatic foramen, and supplies the latter
Sciatic (sacral plexus—L4–S3)	Leaves pelvis through greater sciatic foramen below piriformis, to enter gluteal region—it has no motor branches in the gluteal region

Fig. 6.10 Nerves of the gluteal region.

Muscles of the anterior compartment of the thigh			
Name of muscle (nerve supply)	**Origin**	**Insertion**	**Action**
Quadriceps femoris— rectus femoris; vastus lateralis, medialis, and intermedius (femoral nerve)	Ilium and upper part of femur	Quadriceps tendon into patella then patellar ligament onto tibia	Extends leg at knee joint; flexes hip joint
Sartorius (femoral nerve)	Anterior superior iliac spine	Shaft of tibia	Flexes, abducts, and laterally rotates thigh at hip joint; flexes and medially rotates leg at knee joint
Psoas major (lumbar plexus, L1–L3 nerves)	T12 body, transverse processes, bodies, and intervertebral discs L1–L5	Lesser tronchanter of femur (together with iliacus muscle)	Flexes thigh on trunk
Iliacus (femoral nerve)	Iliac fossa of hip bone	Lesser trochanter of femur	Flexes thigh on trunk
Pectineus (femoral nerve)	Superior ramus of pubis	Upper shaft of femur	Flexes and adducts thigh at hip joint

Fig. 6.11 Muscles of the anterior compartment of the thigh.

ligament, the lacunar ligament, the femoral vein, and pectineus.

To distinguish a femoral hernia from an indirect inguinal hernia, locate the pubic tubercle. If the hernial sac is medial to the tubercle it is an indirect inguinal hernia; if lateral to the tubercle it is a femoral hernia.

Adductor canal

At the apex of the femoral triangle, the femoral vessels disappear beneath sartorius and follow the muscle to the medial aspect of the thigh in a channel—the adductor canal.

It is bounded laterally by the vastus medialis and posteromedially by the adductor longus and adductor magnus.

Vessels of the thigh

Fig. 6.13 outlines the arterial supply of the thigh.

The femoral pulse can be palpated at the midinguinal point, half way between the anterior superior iliac spine and the pubic symphysis.

Cruciate anastomosis

The cruciate anastomosis provides an alternative circulation should the femoral artery be obstructed. It is made up of:

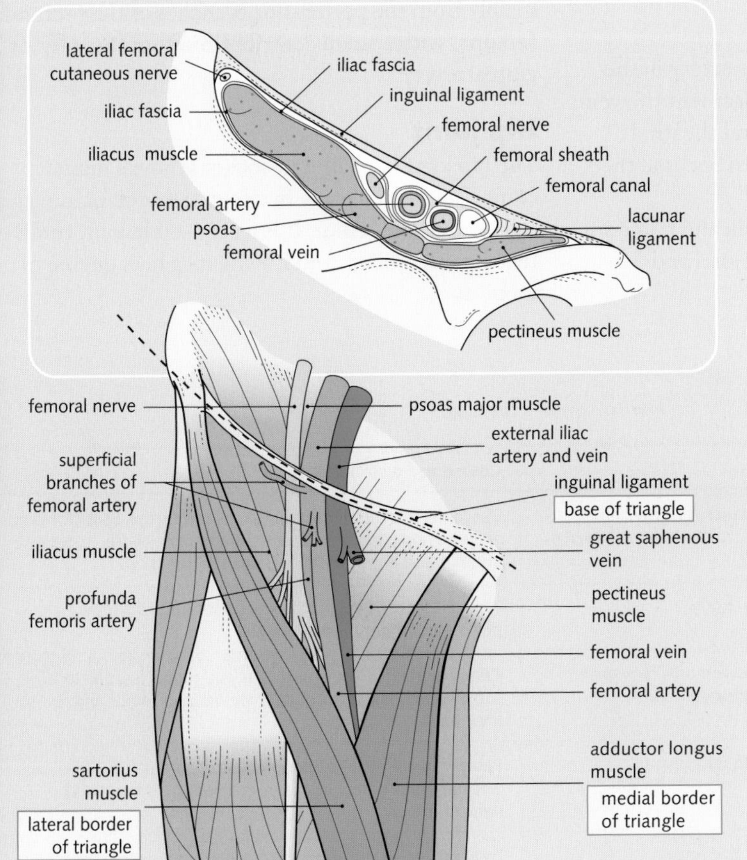

Fig. 6.12 Femoral triangle and its contents.

An accessory obturator artery arises from the inferior epigastric artery in 20% of individuals. Surgeons need to be aware of this as it is next to the lacunar ligament, which is cut to release a femoral hernia.

- The inferior gluteal branch of the internal iliac artery.
- The medial and lateral circumflex femoral arteries.
- The first perforating artery.

All the perforating arteries anastomose with each other and with the muscular branches of the popliteal artery. In this way the internal iliac artery and the popliteal artery are linked.

Femoral vein

The femoral vein lies posterior to the artery in the adductor canal. Below the inguinal ligament the vein lies medial to the artery in the femoral sheath. It passes behind the inguinal ligament to become the external iliac vein.

It receives the great saphenous vein and tributaries corresponding to branches of the femoral and profunda femoris arteries.

Nerves of the thigh

The nerves of the thigh region are outlined in Fig. 6.14.

Adductor compartment of the thigh

The muscles of the adductor compartment are outlined in Fig. 6.15. They are all supplied by the obturator nerve except for the part of the adductor magnus muscle that belongs to the posterior compartment, which is supplied by the sciatic nerve.

The perforating branches of the profunda femoris artery and the muscular branches of the femoral artery provide the majority of the blood supply to the adductor compartment. The obturator artery also contributes proximally to the blood supply of this region.

Posterior compartment of the thigh

The muscles of the posterior compartment are outlined in Fig. 6.16 and illustrated in Fig. 6.17. They are all supplied by the sciatic nerve (see Fig. 6.14).

Blood supply to the posterior compartment is mainly from the perforating branches of the profunda femoris, with a small contribution from the inferior gluteal artery.

Hip joint

The hip joint is a ball-and-socket synovial joint comprising articulation between the acetabulum and the head of the femur. It is a very stable joint (unlike the shoulder joint), and it exhibits a high degree of mobility.

Arterial supply to the thigh

Artery	Origin	Course and distribution
Femoral	Continuation of external iliac artery distal to inguinal ligament	Descends through femoral triangle, enters the adductor canal and ends by passing through the adductor hiatus; supplies anterior and anteromedial surfaces of the thigh
Profunda femoris	Femoral artery	Passes inferiorly, deep to adductor longus, to supply posterior compartment of thigh
Lateral circumflex femoral	Profunda femoris; may arise from femoral artery	Passes laterally deep to sartorius and rectus femoris to supply anterior part of gluteal region, and femur and knee joint
Medial circumflex femoral	Profunda femoris	Passes medially and posteriorly between pectineus and iliopsoas, and enters gluteal region; supplies head and neck of femur
Obturator	Internal iliac artery	Passes through obturator foramen and enters medial compartment of thigh; supplies obturator externus, pectineus, adductors of thigh, and gracilis—muscles attached to ischial tuberosity and head of femur

Fig. 6.13 Arterial supply to the thigh.

Nerves of the thigh	
Nerve (origin)	**Course and distribution**
Ilioinguinal (lumbar plexus—L1)	Supplies skin over femoral triangle
Genitofemoral (lumbar plexus—L1–L2)	Descends on anterior surface of psoas major and divides into genital and femoral branches: femoral branch supplies skin over femoral triangle; genital branch supplies scrotum or labia majora
Lateral femoral cutaneous (lumbar plexus—L2–L3)	Passes deep to inguinal ligament, 2–3 cm medial to anterior superior iliac spine; supplies skin on anterior and lateral aspects of thigh
Medial and intermediate femoral cutaneous (femoral nerve)	Arise in femoral triangle and pierce fascia lata of thigh; supply skin on medial and anterior aspect of thigh
Posterior femoral cutaneous (sacral plexus—S2–S3)	Passes through greater sciatic foramen below piriformis; supplies skin over posterior aspect of thigh, buttock, and proximal leg
Femoral (lumbar plexus—L2–L4)	Passes deep to inguinal ligament; supplies anterior thigh muscles, hip and knee joints, and skin on anteromedial side of thigh
Obturator (lumbar plexus—L2–L4)	Enters thigh through obturator foramen and divides: anterior branch supplies adductor longus, adductor brevis, gracilis, and pectineus; posterior branch supplies obturator externus and adductor magnus
Sciatic (sacral plexus—L4–S3)	Enters gluteal region through greater sciatic foramen below or through piriformis, descends along posterior aspect of thigh, and divides proximal to the knee into tibial and common peroneal nerves; innervates hamstrings by its tibial division (except for short head of biceps femoris – innervated by common peroneal division) and has articular branches to hip and knee joints

Fig. 6.14 Nerves of the thigh.

Muscles of the adductor compartment of the thigh			
Name of muscle (nerve supply)	**Origin**	**Insertion**	**Action**
Adductor brevis (obturator nerve)	Inferior ramus of pubis	Posterior surface of femur	Adducts thigh at hip joint
Adductor longus (obturator nerve)	Body of pubis	Posterior surface of femur	Adducts thigh at hip joint
Adductor magnus (adductor part—obturator nerve; hamstring part—sciatic nerve)	Ischiopubic ramus	Posterior surface of femur, adductor tubercle of femur	Adducts thigh at hip joint; hamstring part extends thigh at hip joint
Gracilis (obturator nerve)	Ischiopubic ramus	Upper part of tibia	Adducts thigh at hip joint and flexes leg at knee joint
Obturator externus (obturator nerve)	Outer surface of obturator membrane	Greater trochanter of femur	Lateral rotation of thigh at hip joint

ig. 6.15 Muscles of the adductor compartment of the thigh.

Stability is mainly achieved from the close fit etween the femoral head and the acetabulum. Great nobility is achieved because the femoral neck is much narrower than the diameter of the head so that considerable movement may occur in all directions before the neck impinges on the acetabular labrum.

Muscles of the posterior compartment of the thigh			
Name of muscle (nerve supply)	**Origin**	**Insertion**	**Action**
Biceps femoris: Long head (common peroneal division of sciatic) Short head (tibial division of sciatic)	Ischial tuberosity Linea aspera	Head of fibula (both heads)	Flex leg at knee joint and extend thigh at hip joint
Semitendinosus (tibial division of the sciatic)	Ischial tuberosity	Upper part of tibial shaft	Flex leg at knee joint and extend thigh at hip joint
Semimembranosus (tibial division of sciatic)	Ischial tuberosity	Medial condyle of the tibia, forms the oblique popliteal ligament	Flex leg at knee joint and extend thigh at hip joint

Fig. 6.16 Muscles of the posterior compartment of the thigh.

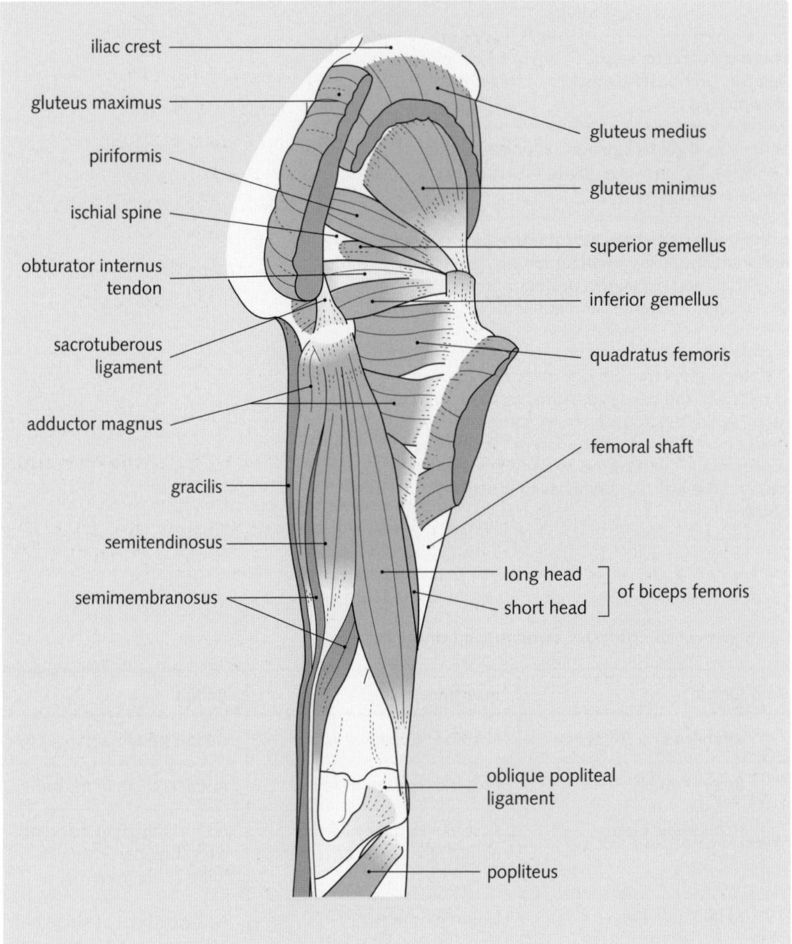

Fig. 6.17 Structure of the posterior aspect of the right thigh

The articular surface of the acetabulum is covered by hyaline cartilage. The peripheral edge of this surface is deepened by a rim of fibrocartilage—the acetabular labrum. The labrum thus contributes further to joint stability. The labrum continues across the acetabular notch as the transverse ligament.

The articular surface of the head of the femur is also covered by hyaline cartilage. The non-articular convexity of the head is excavated into a pit (fovea) for attachment of the ligament of the head of the femur.

The capsule is attached around the labrum and transverse ligament and the neck of the femur. Over the femoral neck the capsule is thrown into folds

called retinacula. It is loose and strong. A synovial membrane lines its internal surface.

The capsule is reinforced by three strong ligaments that extend from the pelvic bone to the femur:

- The pubofemoral ligament—prevents excessive abduction and hyperextension.
- The ischiofemoral ligament—prevents excessive medial rotation and hyperextension.
- The iliofemoral ligament—prevents hyperextension and lateral rotation.

Blood supply is from branches of the following:

- The medial circumflex femoral artery.
- The superior gluteal artery.
- The inferior gluteal artery.
- The obturator artery.

Nerve supply comprises the following:

- The femoral nerve.
- The sciatic nerve.
- The obturator nerve.

Movements of the hip joint

Any movement of the shaft of the femur is accompanied by a different movement of the neck and head (Fig. 6.18).

A femoral neck fracture causes the axis of rotation to lie along the femoral shaft. This shift of axis causes medial rotators (e.g. iliopsoas) to become lateral rotators, and the lower limb rotates laterally in this fracture.

The knee and popliteal fossa

Popliteal fossa

The diamond-shaped popliteal fossa is bordered by the biceps femoris, semitendinosus, and semimembranosus muscles superiorly and by the gastrocnemius muscle inferiorly (Fig. 6.19). It is roofed by the deep fascia, which is pierced by the small saphenous vein and lymphatics. The floor is formed by the body of the femur, the oblique popliteal ligament, and the popliteus muscle.

Movements of the hip joint and the muscles responsible for these movements

Movement of the thigh on the trunk	Movement at the hip joint	Muscles involved	Factors limiting movement
Flexion	Head of the femur moves about a transverse axis passing through both acetabula and causes the shaft to swing anteriorly	Psoas major, iliacus, tensor fasciae latae, pectineus, sartorius	Thigh touching abdomen, hamstring muscle tension if leg is extended
Extension	As flexion but opposite direction	Gluteus maximus, hamstrings	Iliofemoral ligament, pubofemoral ligament
Abduction	Head of the femur moves in the acetabulum about an anteroposterior axis and causes the femoral neck and shaft to swing laterally	Gluteus medius, gluteus minimus	Adductor muscle tension, pubofemoral ligament
Adduction	As abduction but opposite direction	Adductors, gracilis	Gluteus medius, gluteus minimus, other leg
Medial rotation	Rotation of the femoral head in the acetabulum about a vertical axis that passes through the femoral head and medial condyle. The neck of the femur swings anteriorly	Tensor fasciae latae, gluteus medius, gluteus minimus, iliopsoas, pectineus, adductor longus	Ischiofemoral ligament
Lateral rotation	As medial rotation but opposite direction	Obturator internus, obturator externus, piriformis, gemelli, quadratus femoris, gluteus maximus	Iliofemoral ligament

Fig. 6.18 Movements of the hip joint and the muscles responsible for these movements.

Contents of the popliteal fossa

Popliteus muscle

This has the following characteristics:

- Origin—a pit just below the lateral epicondyle of the femur and the lateral meniscus of the knee joint. It is intracapsular.
- Insertion—popliteal surface of the tibia.
- Nerve supply—tibial nerve.
- Action—unlocks the knee and draws the lateral meniscus posteriorly.

Popliteal artery

This is the continuation of the femoral artery as it passes through the adductor hiatus. It terminates at the lower border of popliteus, where it divides into the anterior and posterior tibial arteries.

It gives off superior, middle, and inferior genicular arteries and muscular branches. Anastomoses of the genicular vessels with descending branches of the femoral and profunda femoris arteries and with ascending branches of the tibial arteries form an important collateral supply if the main vessels become occluded.

Popliteal vein

This passes up the popliteal fossa medial to, and then superficial to, the artery before entering the adductor hiatus. It receives the small saphenous vein together with veins corresponding to the arterial branches.

The popliteal artery is adjacent to the distal end of the femur. In fractures of the distal femur, the popliteal artery may be damaged causing it to haemorrhage.

Common peroneal nerve

This lies beneath the biceps femoris in the popliteal fossa until it reaches the head of the fibula. Branches in the popliteal fossa include:

- The lateral sural nerve, which supplies the skin of the calf.
- A communicating branch with the medial sural nerve.
- Branches to the knee joint.

Tibial nerve

This bisects the popliteal fossa vertically. Branches in the fossa include:

- Articular branches to the knee.
- Muscular branches to soleus, gastrocnemius, plantaris, and popliteus.

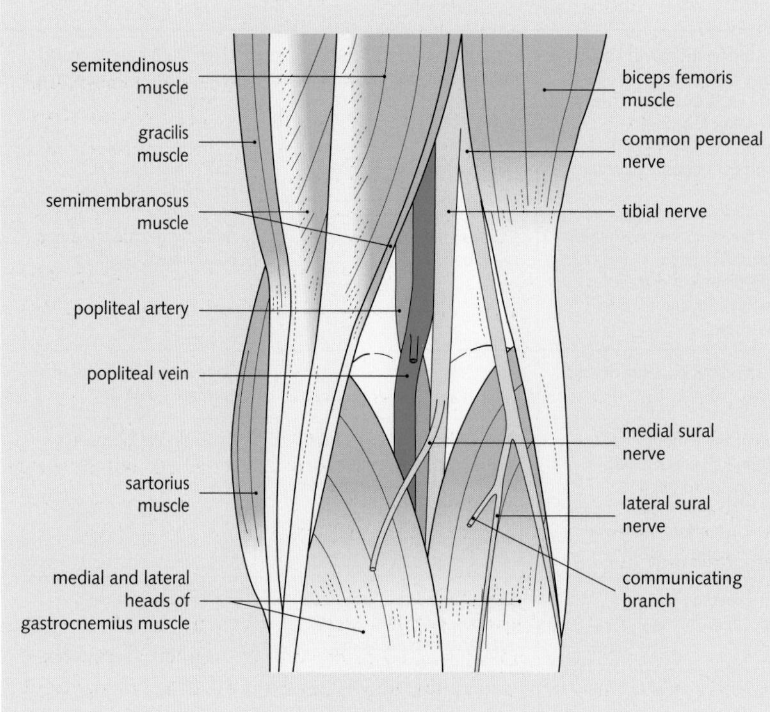

Fig. 6.19 Popliteal fossa and its contents.

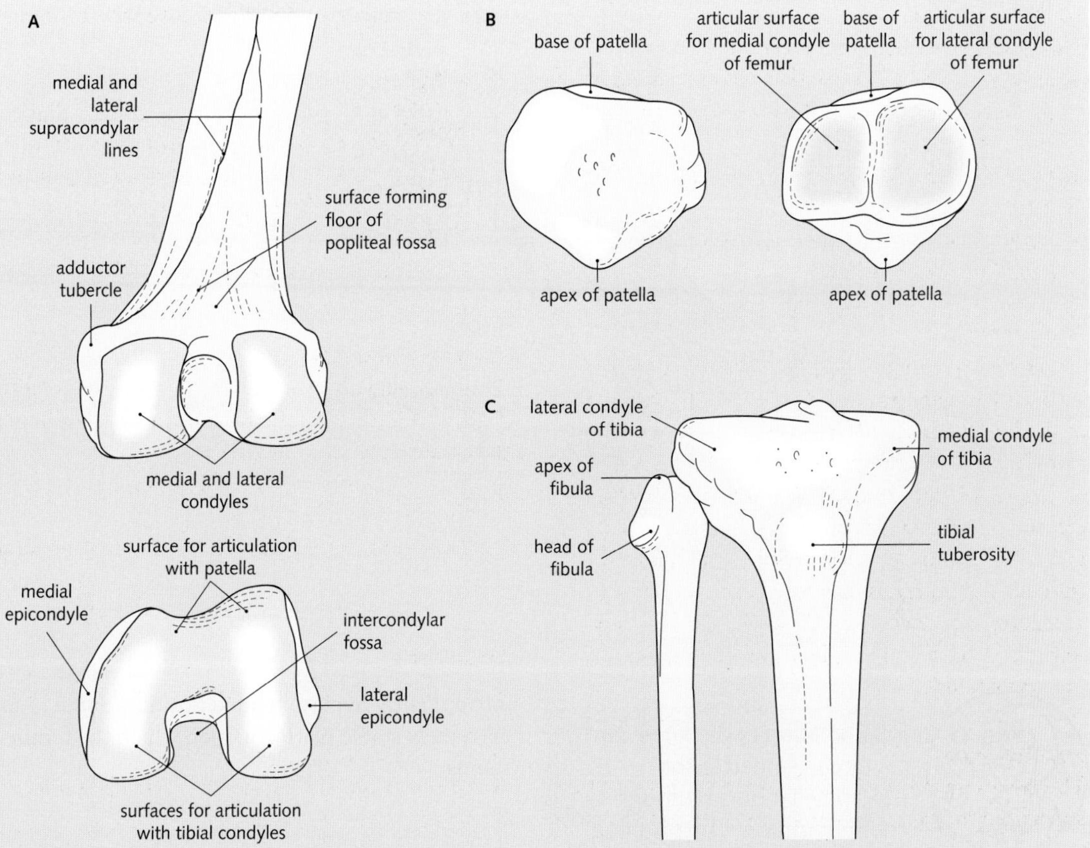

Fig. 6.20 Skeleton of the knee. (A) Posterior and inferior aspects of the lower end of the right femur. (B) Anterior and posterior aspects of the patella. (C) Anterior aspect of the upper end of the right tibia and fibula.

- The medial sural cutaneous nerve, which joins the communicating branch of the lateral sural cutaneous nerve to form the sural nerve. The sural nerve supplies the lateral side of the calf and heel.

Skeleton of the knee

The main features of the skeleton around the knee are illustrated in Fig. 6.20.

Knee joint

The knee joint is an articulation between the femur and the tibia, with the patella articulating with the femur anteriorly. It is a synovial joint.

The articular surfaces are covered by hyaline cartilage and consist of the margins of the femoral condyles, the patella, and the superior surface of the tibial condyles.

Capsule

The capsule surrounds the articular surfaces, but it is not complete: it is defective anterosuperiorly, to allow communication between the joint cavity and the suprapatellar bursa; posteroinferiorly, to allow entry of the popliteus tendon; and anteriorly it is replaced by the articular surface of the patella.

The capsule is strengthened anteriorly by the patellar retinacula—expansions of the tendons of the vastus medialis and lateralis. It is also reinforced by the quadriceps tendon, the patella, and the patellar ligament.

Synovial membrane

The synovial membrane lines the capsule (Fig. 6.21). Above the patella it becomes continuous with the lining of the suprapatellar bursa. The cruciate ligaments and popliteus tendon lie outside the synovial cavity.

Ligaments

Ligaments play a major role in stabilizing the knee joint (Fig. 6.22). The cruciate ligaments keep the articular surfaces applied to each other throughout the range of movement.

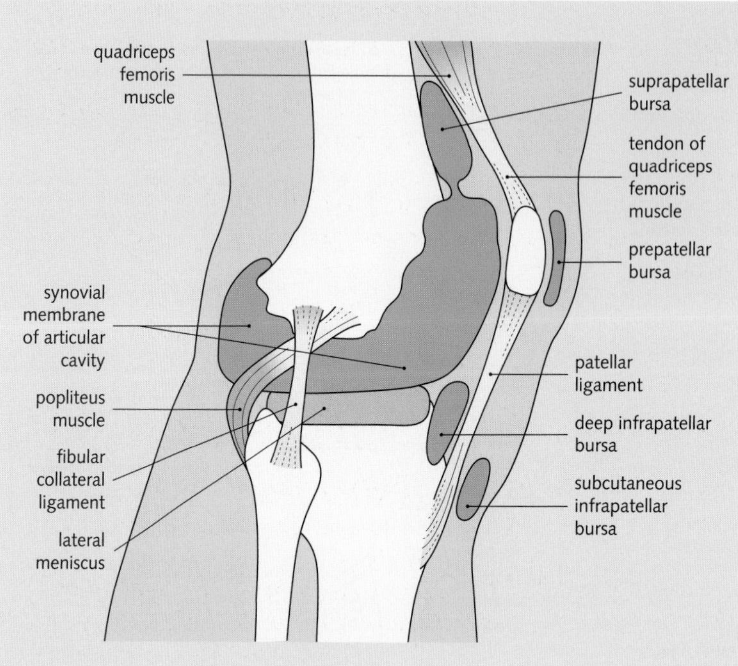

Fig. 6.21 Synovial membrane and its ligaments.

Inflammation of the knee joint cavity can spread to the suprapatellar bursa through its communication. The resulting increase in synovial fluid due to inflammation can, therefore, be palpated and aspirated from this bursa.

Menisci

The menisci are two crescentic plates of fibrocartilage (Fig. 6.23). The horns of the menisci are attached to the intercondylar area of the tibia and at the periphery to the loose coronary ligament of the capsule and to the deep part of the tibial collateral ligament. They increase the congruity of the articular surfaces.

Movements at the knee joint

The movements at the knee joint are outlined in Fig. 6.24. The 'locking of the knee joint' puts the knee into a slightly hyperextended position, and it is an extremely stable platform upon which the femur stands.

The tibial collateral ligament is attached to the medial meniscus. If the ligament is torn concomitant tearing of the medial meniscus can occur.

Ligaments of the knee joint

Ligament	Attachment
Patellar (often called patella tendon)	The termination of the quadriceps tendon running from the patella to the tibial tuberosity; its tension is controlled by the quadriceps muscle, which stabilizes the joint through its full range of movement
Tibial or medial collateral	From the medial femoral condyle to the tibia
Fibular or lateral collateral	From the lateral femoral condyle to the fibular head
Oblique popliteal	Expansion of the semi-membranosus tendon, which reinforces the capsule posteriorly
Anterior cruciate	From the anterior part of the intercondylar area of the tibia to the medial surface of the lateral femoral condyle
Posterior cruciate	From the posterior part of the intercondylar area of the tibia to the lateral surface of the medial femoral condyle

Fig. 6.22 Ligaments of the knee joint.

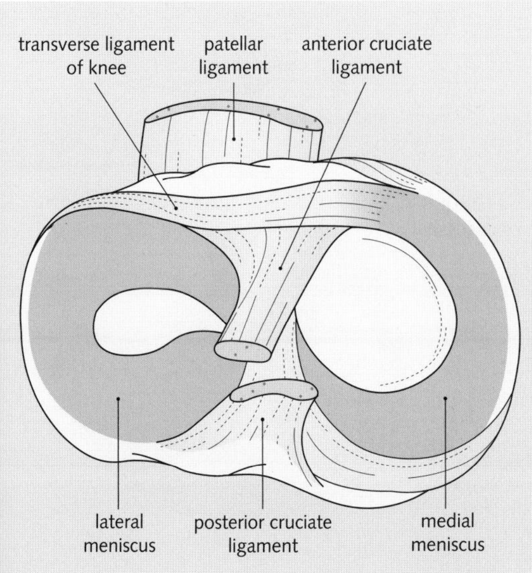

Fig. 6.23 Menisci and their ligaments.

As the knee extends, the anterior cruciate ligament becomes taut and it stops the lateral condyle of the femur from extending further. However, due to a larger surface area of the medial condyle (see Fig. 6.20) extension continues around the anterior cruciate ligament. This produces medial rotation of the femur upon the tibia, which causes the oblique, popliteal, medial collateral and lateral collateral ligaments to tighten together. At the end of this movement the knee becomes locked.

The knee joint must be unlocked before flexion can occur. This is performed by the popliteus muscle, which laterally rotates the femur. This rotation loosens the ligaments and removes the extra surface area of the femoral medial condyle.

Blood and nerve supply of the knee joint

Blood supply comes from the genicular branches of the popliteal artery.

Nerve supply comprises the articular branches of the sciatic, femoral, and obturator nerves.

Bursae around the knee joint

The bursae around the knee joint include:

- The suprapatellar bursa.
- The prepatellar bursa.
- The superficial and deep infrapatellar bursae.

The leg and foot

Skeleton of the leg

Important features of the skeleton of the leg are illustrated in Fig. 6.25.

The interosseous membrane is a tough band of tissue linking the interosseous borders of the tibia and fibula. It is pierced superiorly by the anterior tibial artery and inferiorly by branches of the peroneal artery.

Inferior tibiofibular joint

The inferior tibiofibular joint is an articulation between the lower end of the tibia and fibula. It is a fibrous joint stabilized by the anterior and posterior tibiofibular ligaments. It allows little movement, and it stabilizes the ankle joint by keeping the lateral malleolus clasped against the lateral surface of the talus.

Compartments of the leg

The leg is divided into anterior, posterior, and lateral compartments (Fig. 6.26). The muscles of the posterior compartment are divided into superficial

Movements at the knee joint	
Movement	**Muscle**
Flexion	Hamstrings, gastrocnemius, gracilis
Extension	Quadriceps femoris
Medial and lateral rotation of flexed knee	Hamstrings
Unlocking of knee	Popliteus
Locking of knee—a passive rotation of the femur upon the tibia	Is caused at the end of extension by the anterior cruciate ligament becoming taut and the femoral medial condyle moving round this ligament

Fig. 6.24 Movements at the knee joint.

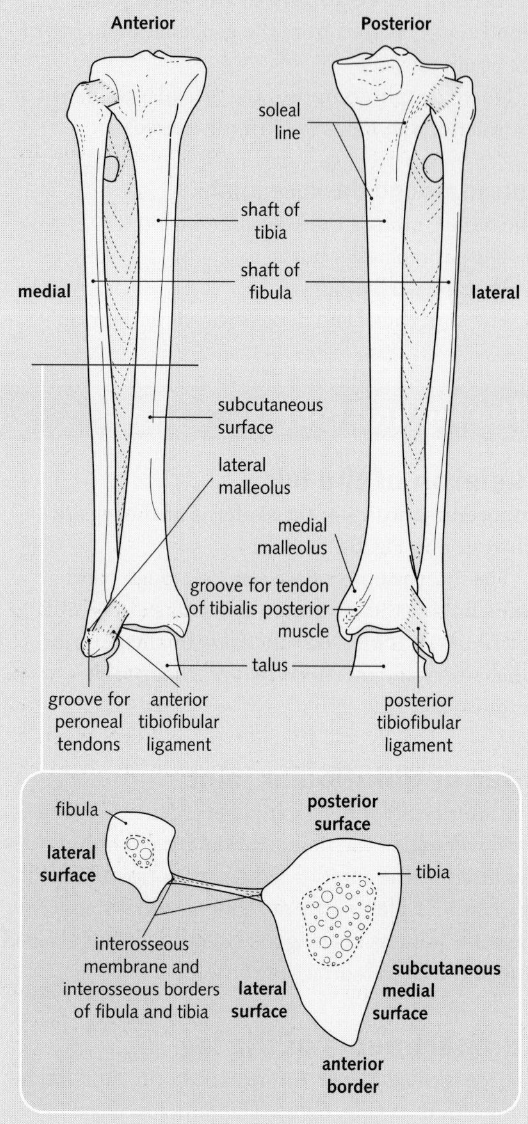

Fig. 6.25 Anterior and posterior features of the right tibia and fibula, and the relationship of the two bones and the interosseous membrane in cross-section.

and deep groups by the deep transverse crural fascia.

Anterior compartment of the leg

Muscles of the anterior compartment are shown in Figs 6.27 and 6.28.

Vessels of the anterior compartment

The anterior tibial artery is the vessel of the anterior compartment (Figs 6.29 and 6.30). It is a terminal branch of the popliteal artery.

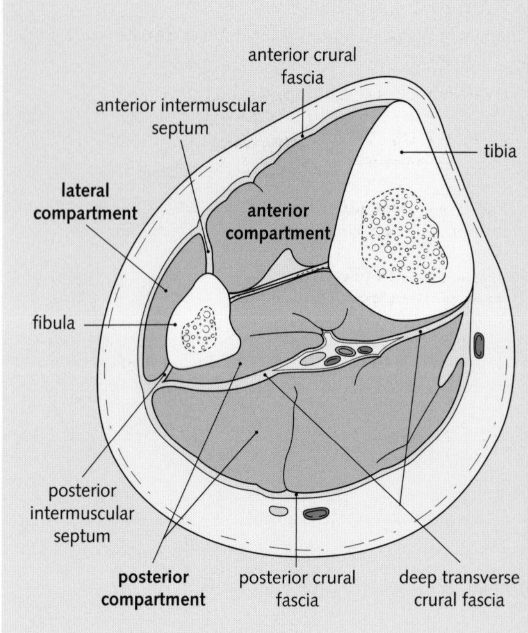

Fig. 6.26 Compartments of the leg.

Nerves of the anterior compartment

The common peroneal nerve (L4–L5, S1–S2) leaves the popliteal fossa to enter the lateral compartment of the leg by winding around the neck of the fibula (Fig. 6.31). Here, it divides into the superficial and deep peroneal nerves.

The common peroneal nerve is vulnerable to injury, e.g. by a car bumper, as it passes around the fibula. Damage results in footdrop.

Intramuscular septa of the leg are unyieldingly strong. Inflammation of the anterior compartment muscles can compress the anterior tibial artery causing pain on activity through ischaemia (anterior compartment syndrome).

Dorsum of the foot

The structures from the anterior compartment pass onto the dorsum of the foot. The extensor digitorum longus tendons pass to the lateral four digits of the

Muscles of the anterior compartment of the leg			
Name of muscle (nerve supply)	**Origin**	**Insertion**	**Action**
Extensor digitorum longus (deep peroneal nerve)	Fibula and interosseous membrane	Extensor expansion of lateral four toes	Extends toes and dorsiflexes foot at ankle joint
Extensor hallucis longus (deep peroneal nerve)	Fibula and interosseous membrane	Base of distal phalanx of great toe	Extends big toe, and dorsiflexes foot at ankle joint
Peroneus tertius (deep peroneal nerve)	Fibula and interosseous membrane	Base of 5th metatarsal bone	Dorsiflexes and everts foot
Tibialis anterior (deep peroneal nerve)	Tibia and interosseous membrane	Medial cuneiform and base of 1st metatarsal bone	Dorsiflexes and inverts foot

Fig. 6.27 Muscles of the anterior compartment of the leg.

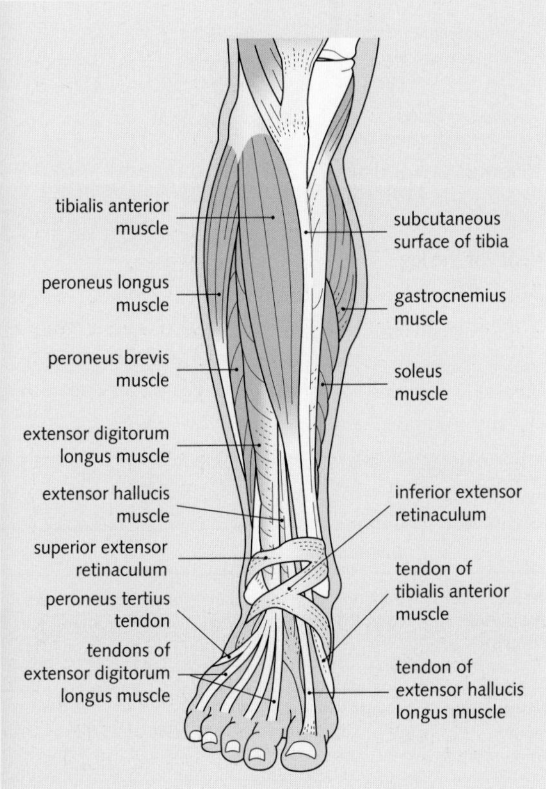

Fig. 6.28 Extensor muscles of the anterior compartment of the leg.

foot and join an extensor expansion. These expansions have the same arrangement as those found in the medial four digits of the hand. Over the proximal phalanx each expansion splits into three slips. An axial (central) slip inserts into the middle phalanx. Two collateral slips insert into the distal phalanx. Each expansion receives interossei and lumbrical muscle attachments.

The extensor digitorum brevis and extensor hallucis brevis muscles are the only muscles intrinsic to the dorsum of the foot. Arising from the calcaneum, the three tendons of extensor digitorum brevis join the lateral sides of the extensor digitorum longus tendons for the middle three toes over the metatarsophalangeal joint. From a similar origin, the extensor hallucis brevis tendon inserts into the proximal phalanx of the hallux (great toe). The muscle belly of this tendon is usually separate from the rest of the muscle. The nerve supply is by the deep peroneal nerve and the muscle extends the digits (toes).

Cutaneous innervation of the dorsum of the foot is from the superficial peroneal nerve, the deep peroneal nerve, the sural nerve, and the saphenous nerve.

Extensor retinacula

The superior and inferior extensor retinacula keep the extensor tendons firmly bound down to the dorsum of the foot (Fig. 6.32). The retinacula are derived from the anterior crural fascia.

The superior band passes from the anterior border of the tibia to the anterior border of the fibula. The inferior band is Y-shaped, and it runs from the calcaneus to the medial malleolus and plantar fascia.

Nerves and vessels of the dorsum of the foot

The deep peroneal nerve and anterior tibial artery enter the foot beneath the extensor retinacula. The anterior tibial artery continues as the dorsalis pedis artery (see Fig. 6.30).

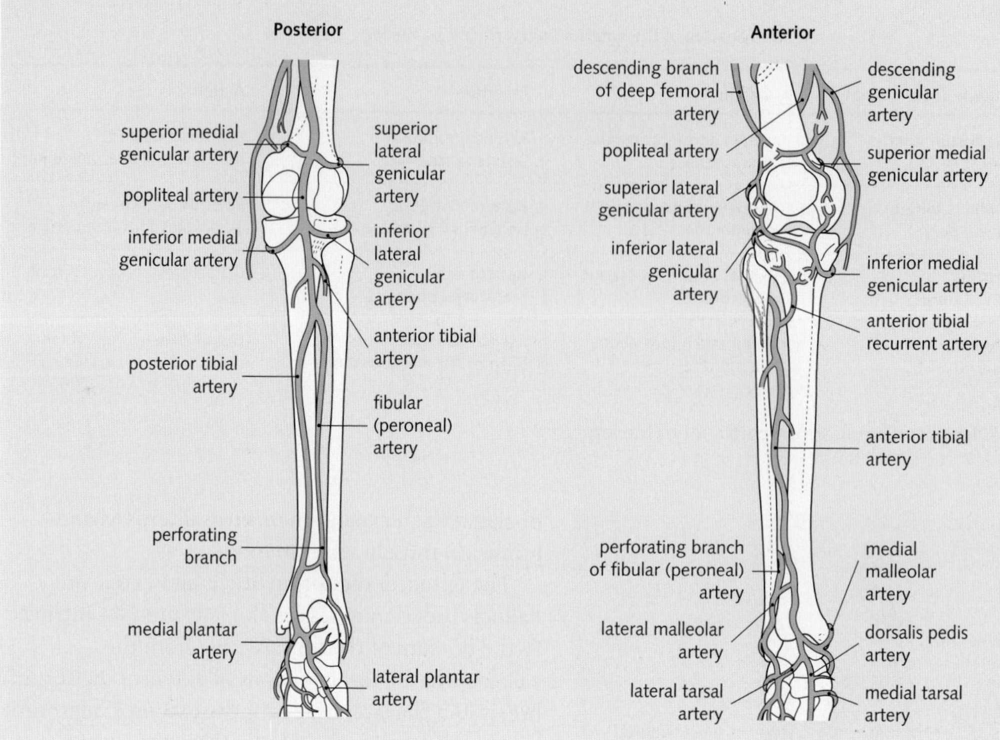

Fig. 6.29 Arterial supply to the posterior and anterior compartments of the leg.

Arterial supply to the leg		
Artery (origin)	**Course**	**Distribution**
Popliteal (continuation of femoral artery at adductor hiatus)	Passes through popliteal fossa to leg; ends at lower border of popliteus muscle by dividing into anterior and posterior tibial arteries	Superior, middle, and inferior genicular arteries to both lateral and medial aspects of knee
Anterior tibial (popliteal artery)	Passes into anterior compartment through gap in superior part of interosseous membrane and descends on this membrane	Anterior compartment of leg
Dorsalis pedis (continuation of anterior tibial artery distal to extensor retinaculum)	Runs on the dorsum of the foot and gives tarsal, arcuate and first dorsal metatarsal arteries before descending into the first interosseous space to join plantar arch	Muscles on dorsum of foot; pierces first dorsal interosseous muscle to contribute to formation of plantar arch
Posterior tibial (popliteal artery)	Passes through posterior compartment of leg and terminates distal to flexor retinaculum by dividing into medial and lateral plantar arteries	Posterior and lateral compartments of leg; nutrient artery passes to tibia, contributes to knee anastomoses
Peroneal (posterior tibial artery)	Descends in posterior compartment adjacent to posterior intermuscular septum	Posterior compartment of leg; perforating branches supply lateral compartment of leg

Fig. 6.30 Arterial supply to the leg.

Nerves of the leg	
Nerve (origin)	**Course and distribution**
Common peroneal (sciatic nerve)	Arises at apex of popliteal fossa and follows medial border of biceps femoris and its tendon; passes over posterior aspect of head of fibula and then winds around neck of fibula, deep to peroneus longus, where it divides into deep and superficial peroneal nerves; supplies skin on posterolateral part of leg via its branch —lateral sural cutaneous nerve
Deep peroneal (common peroneal nerve)	Arises between peroneus longus and neck of fibula; descends on interosseous membrane and enters dorsum of foot; supplies anterior muscles of leg, and skin of first interdigital cleft
Saphenous (femoral nerve)	Descends with femoral vessels and the great saphenous vein to supply skin on medial side of leg and foot
Superficial peroneal (common peroneal nerve)	Arises between peroneus longus and neck of fibula and descends in lateral compartment of leg; supplies peroneus longus and brevis and skin on anterior surface of leg and dorsum of foot
Sural (usually arises from both tibial and common peroneal nerves)	Descends between heads of gastrocnemius and becomes superficial at middle of leg; supplies skin on posterolateral aspects of leg and lateral side of foot
Tibial (sciatic nerve)	Descends through popliteal fossa and lies on popliteus; then runs inferiorly with posterior tibial vessels and terminates beneath flexor retinaculum by dividing into medial and lateral plantar nerves; supplies posterior muscles of leg, knee joint, skin, and muscles of the sole of the foot

Fig. 6.31 Nerves of the leg.

The dorsalis pedis artery can be palpated between the extensor hallucis longus tendon and the extensor digitorum longus tendon upon the navicular bone.

The superficial calf muscles (triceps surae) act as a venous pump on contraction, pushing the blood superiorly. The deep fascia improves the pumping action by acting as an elastic stocking.

Lateral compartment of the leg

The composition of the lateral compartment is shown in Fig. 6.33. Its muscles are outlined in Fig. 6.34.

Posterior compartment of the leg

Fig. 6.35 outlines the muscles of the posterior compartment.

Flexor retinaculum

The flexor retinaculum runs from the medial malleolus to the calcaneus and plantar fascia (Fig. 6.36). The deep flexor muscles pass beneath the retinaculum, surrounded by synovial sheaths, together with the tibial nerve and the posterior tibial artery.

For the structures passing behind the medial malleolus remember the mnemonic: Tom, Dick And Very Naughty Harry (tibialis posterior, flexor digitorum longus, artery, vein, nerve, flexor hallucis longus).

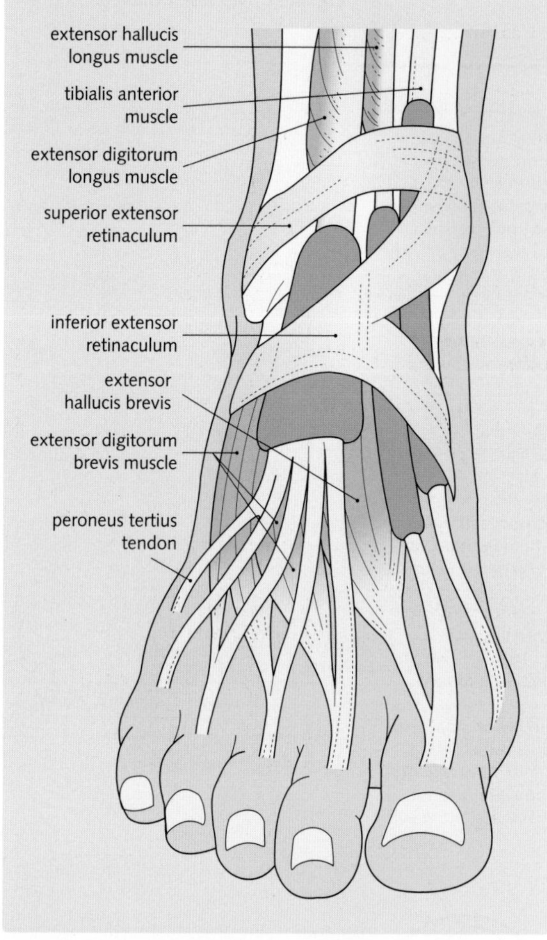

Fig. 6.32 Structures of the extensor retinacula.

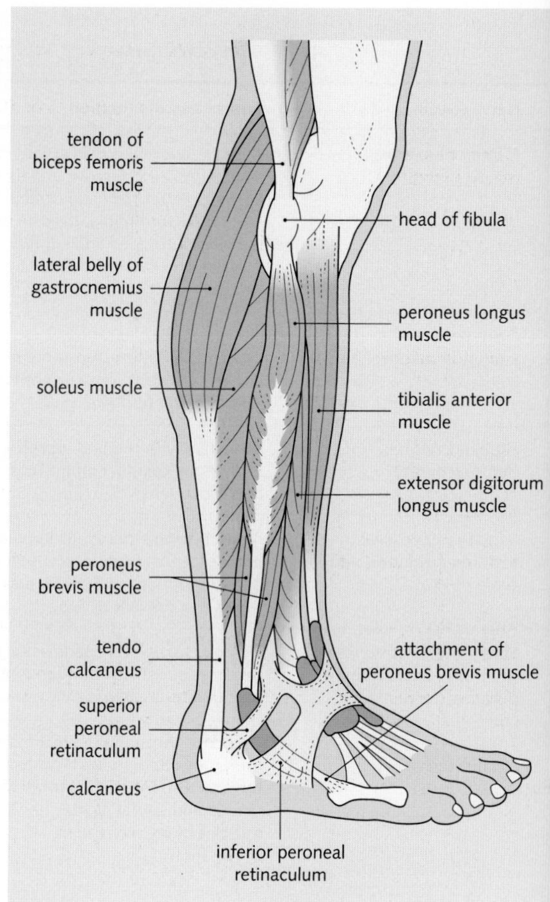

Fig. 6.33 Lateral compartment of the leg and its structures.

Muscles of the lateral compartment of the leg			
Name of muscle (nerve supply)	**Origin**	**Insertion**	**Action**
Peroneus longus (superficial peroneal nerve)	Fibula	1st metatarsal and medial cuneiform	Plantarflexes and everts the foot
Peroneus brevis (superficial peroneal nerve)	Fibula	5th metatarsal bone	Plantarflexes and everts the foot

Fig. 6.34 Muscles of the lateral compartment of the leg.

Skeleton of the foot

The skeleton of the foot consists of the tarsal bones, the metatarsal bones, and the phalanges (Fig. 6.37).

The body weight is transferred to the talus and the calcaneus, and then across the remaining tarsal and metatarsal bones. The weight is transferred to the ground at the tuber calcanei and the heads of the metatarsals. The metatarsal bones are composed of a base proximally, a body, and a distal head. The first digit (the hallux) has two phalanges; the others have three (proximal, middle, and distal).

Ankle joint

The ankle joint is the articulation between the upper surface of the talus and the lower end of the tibia and fibula, including the medial and lateral malleoli.

It is a synovial joint, and it allows only flexion and extension. Dorsiflexion (extension) involves tibialis

Muscles of the posterior compartment of the leg			
Superficial group			
Name of muscle (nerve supply)	**Origin**	**Insertion**	**Action**
Plantaris (tibial nerve)	Lateral supracondylar ridge of femur	Calcaneum	These muscles plantarflex the foot at the ankle joint
Soleus (tibial nerve)	Tibia and fibula	Via tendo calcaneus (Achilles tendon) into calcaneum	
Gastrocnemius (tibial nerve)	Medial and lateral condyles of femur	Via tendo calcaneus (Achilles tendon) into calcaneum	
Deep group			
Name of muscle (nerve supply)	**Origin**	**Insertion**	**Action**
Flexor digitorum longus (tibial nerve)	Tibia	Distal phalanges of lateral four toes	Flexes lateral four toes; plantarflexes foot
Flexor hallucis longus (tibial nerve)	Fibula	Distal phalanx of big toe	Flexes big toe; plantarflexes foot
Tibialis posterior (tibial nerve)	Tibia and fibula and interosseous membrane	Navicular bone and surrounding bones	Plantarflexes and inverts foot

Fig. 6.35 Muscles of the posterior compartment of the leg.

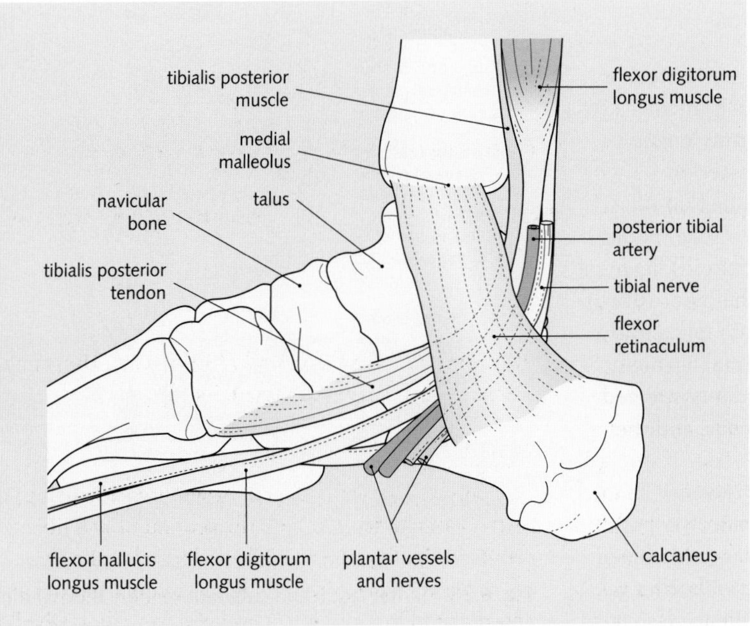

Fig. 6.36 Flexor retinaculum.

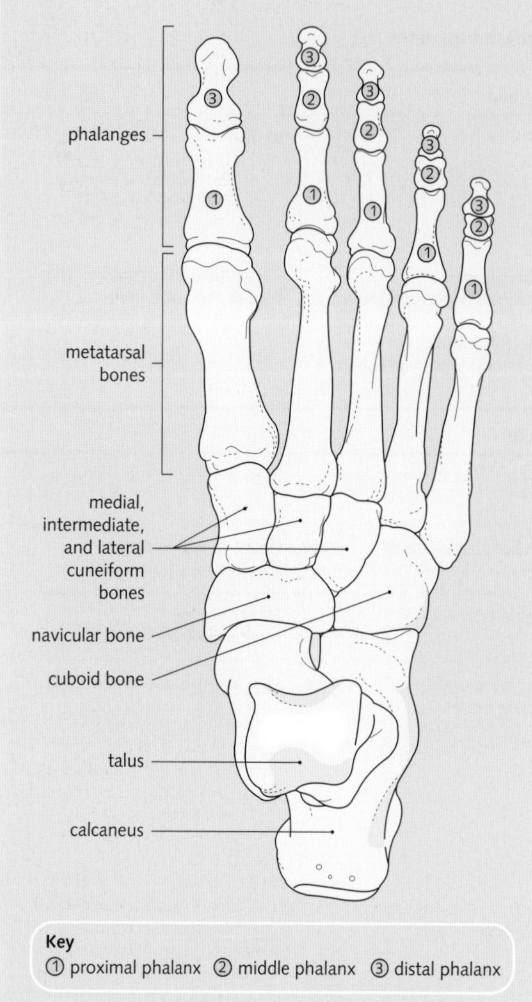

Fig. 6.37 Skeleton of the foot.

anterior, extensor digitorum longus, and extensor hallucis longus. Plantarflexion (flexion) involves gastrocnemius, soleus, flexor digitorum longus, flexor hallucis longus, and tibialis posterior.

The joint is surrounded by a capsule that is lax anteroposteriorly and reinforced by strong medial and lateral ligaments. The medial (deltoid) ligament runs from the medial malleolus to the tuberosity of the navicular bone, the sustentaculum tali, and the medial tubercle of the talus. The lateral ligament arises from the lateral malleolus, and it is inserted into the neck of the talus, the calcaneus, and the lateral tubercle of the talus.

The articular surface of the talus is wedge shaped, having a wider anterior surface and a narrower posterior surface. As a result the ankle joint is most stable in dorsiflexion because the wider anterior border is 'driven' between the two malleoli, which clasp it.

The ankle is the most frequently injured joint in the body. The lateral ligament is slightly weaker than the medial ligament and is, therefore, usually damaged. Severe damage to the ligament results in an unstable joint.

Blood supply is from the anterior and posterior tibial arteries.

Remember: the talar articular surface is wedge shaped, being wider anteriorly. Therefore, the ankle joint is most stable in dorsiflexion.

The lateral ligament of the ankle is weaker than the deltoid (medial) ligament. Therefore, the lateral ligament tears more commonly in ankle sprains.

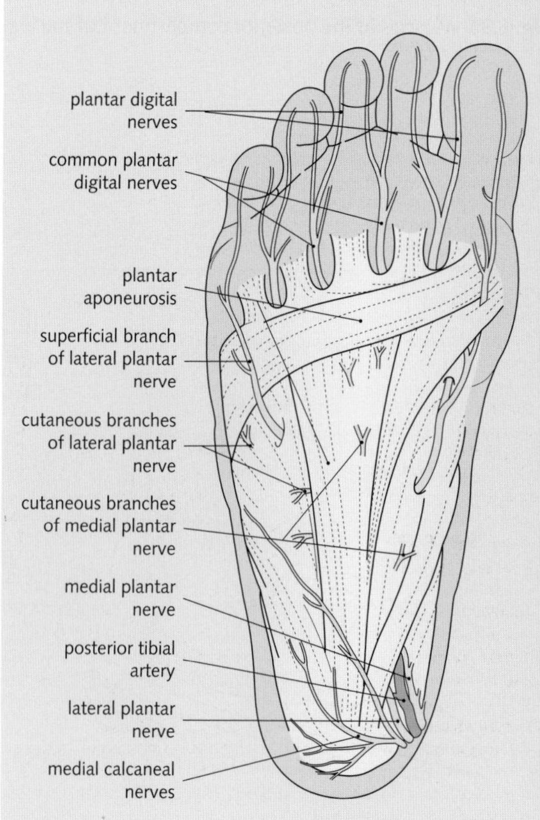

Fig. 6.38 Plantar fascia and cutaneous innervation of the sole of the foot.

Nerve supply is from the tibial and deep peroneal nerves.

Sole of the foot

The sole bears the weight of the body. The skin is thick and hairless. Fibrous septa divide the subcutaneous fat into small loculi, and they anchor the skin to the deep fascia or plantar aponeurosis (Fig. 6.38).

The plantar aponeurosis extends from the calcaneal tuberosity and divides into five bifurcating slips that insert into the flexor fibrous sheaths at the base of the toes. The bifurcation allows passage of the flexor tendons. It is very strong, and it protects the underlying muscles, vessels, and nerves. It is perforated by the cutaneous nerves supplying the sole of the foot.

Intrinsic plantar muscles of the foot

Muscle (nerve supply)	Origin	Insertion	Action
First layer			
Abductor hallucis (medial plantar nerve)	Calcaneum, flexor retinaculum, plantar aponeurosis	Proximal phalanx of hallux	Abduct hallux (great toe)
Flexor digitorum brevis (medial plantar nerve)	Calcaneum, plantar aponeurosis	Each tendon bifurcates and inserts into the middle phalanx of the lateral four digits	Flexes lateral four digits
Abductor digiti minimi (lateral plantar nerve)	Calcaneus, plantar aponeurosis	Proximal phalanx of digitus minimus	Abducts digitus minimus (little toe)
Second layer			
Flexor accessorius or Quadratus plantae (lateral plantar nerve)	Calcaneum	Tendon of flexor digitorum longus	Pulls on the tendon of flexor digitorum longus and takes up the slack of this tendon when the ankle is plantarflexed. This allows the digits to be flexed in this position
Lumbricals (first medial lumbrical—medial plantar nerve; lateral three lumbricals—lateral plantar nerve)	Tendons of flexor digitorum longus	Extensor expansions of lateral four digits	Maintains extension of the digits at DIP and PIP joints while flexor digitorum longus tendons are flexing the lateral four digits at the MTP
Third layer			
Flexor hallucis brevis (medial plantar nerve)	Cuboid and three cuneiforms	Proximal phalanx of hallux	Flexes hallux
Adductor hallucis: (lateral plantar nerve) Oblique head Transverse head	Second to fourth metatarsal bases and plantar ligament Deep transverse ligament	Both heads insert into the proximal phalanx of the hallux	Adducts hallux towards second toe
Flexor digiti minimi brevis (medial plantar nerve)	Fifth metatarsal	Proximal phalanx of digitus minimus	Flexes digitus minimus (little toe)
Fourth layer			
Palmar interossei (lateral plantar nerve)	Third, fourth, fifth metatarsals	Proximal phalanx of digits and their extensor expansions	Adduct digits towards second digit.
Dorsal interossei (lateral plantar nerve)	First and second; second and third; third and fourth; fourth and fifth metatarsals	Proximal phalanx of digits and their extensor expansions	Abduct digits away from second digit. With palmar interossei and lumbricals they extend the DIP, PIP joints and flex the MTP joint.

Fig. 6.39 Intrinsic plantar muscles of the foot. (IP, interphalangeal joint; PIP, proximal interphalangeal joint; DIP, distal interphalangeal joint; MTP, metatarsophalangeal joint.)

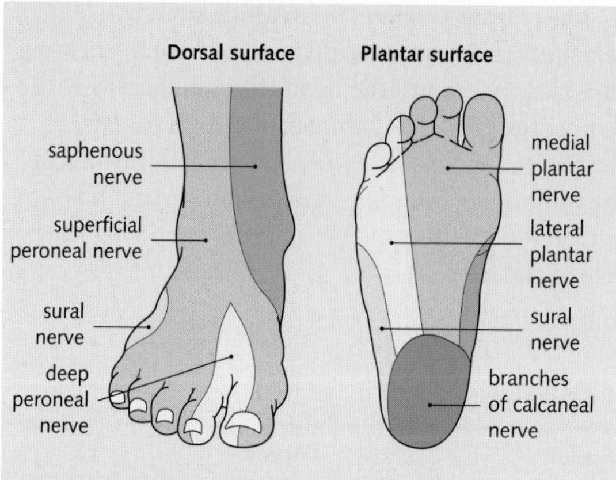

Fig. 6.40 Distribution of nerves of the foot.

The muscles of the sole of the foot are in four layers (see Fig. 6.39). The axis of abduction and adduction passes through the second digit. Therefore, digits either move towards (adduction) or away (abduction) from the second digit.

Nerves of the foot

The nerves of the foot are described in Figs 6.40 and 6.41.

Blood supply to the foot

The posterior tibial artery terminates by dividing into the medial and lateral plantar arteries under the flexor retinaculum (Fig. 6.42).

The medial plantar artery passes forward with the medial plantar nerve. It gives off muscular branches and terminates as a plantar digital branch to the medial side of the big toe and as branches that join the metatarsal branch of the palmar arch.

The lateral plantar artery crosses the sole of the foot, and it gives off muscular and cutaneous branches. At the level of the base of the fifth metatarsal, the artery passes medially and anastomoses with the dorsalis pedis artery to form the plantar arch. From this arch plantar metatarsal arteries arise and these form the plantar digital arteries for the toes.

Fibrous flexor and flexor synovial sheaths

Fibrous flexor sheaths in the foot are similar to those of the hand. They run from the head of the metatarsal bone to the base of the distal phalanx and surround the synovial sheaths. The sheath has anular (ring) fibres over the bones and cruciate

Outline of the nerves of the foot

Nerve (origin)	Distribution
Saphenous (femoral nerve)	Supplies skin on medial side of foot as far anteriorly as head of 1st metatarsal
Superficial peroneal (common peroneal nerve)	Supplies skin on dorsum of foot and all digits, except adjoining sides of first and second digits
Deep peroneal (common peroneal nerve)	Supplies extensor digitorum brevis, extensor hallucis brevis and skin on contiguous sides of first and second digits
Medial plantar (larger terminal branch of the tibial nerve)	Supplies skin of medial side of sole of foot and plantar surfaces of first three and one half digits; also supplies abductor hallucis, flexor digitorum brevis, flexor hallucis brevis, and first lumbrical
Lateral plantar (smaller terminal branch of the tibial nerve)	Supplies quadratus plantae, abductor digiti minimi and flexor digiti minimi brevis; deep branch supplies plantar and dorsal interossei, lateral three lumbricals, and adductor hallucis; supplies skin on sole lateral to a line splitting fourth digit
Sural (tibial and common peroneal nerves)	Lateral aspect of foot
Calcaneal nerves (tibial and sural nerves)	Skin of heel

Fig. 6.41 Nerves of the foot.

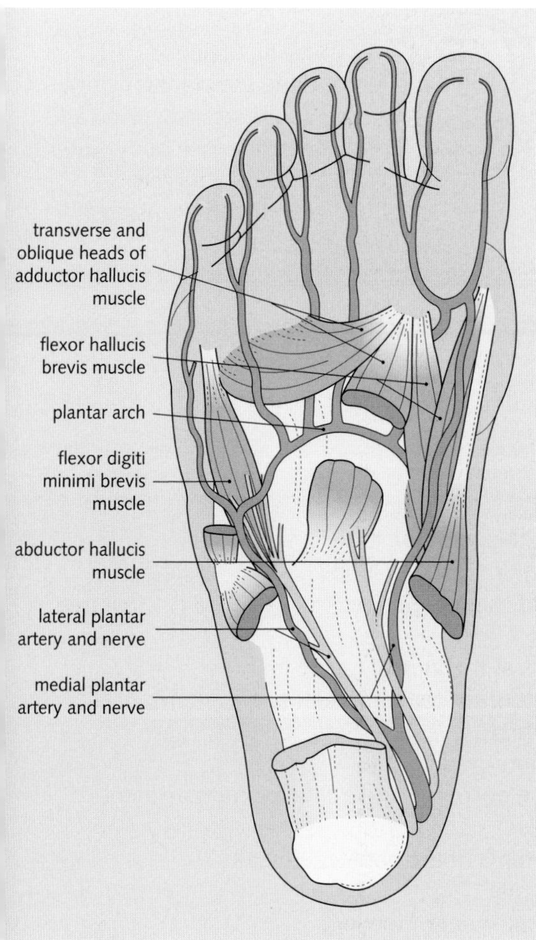

Fig. 6.42 Blood and nerve supply to the sole of the foot.

criss-cross) fibres over the joints (to allow movement).

The synovial sheaths for the flexor digitorum longus tendons surround them from just above the fibrous flexor retinaculum to the navicular bone. The tendon of flexor hallucis longus is covered by a synovial sheath from above the retinaculum to the base of the first metatarsal. As the above mentioned tendons enter their fibrous flexor sheath, they are again covered by a synovial sheath. The synovial sheath for tibialis posterior extends from the flexor retinaculum to the tendons insertion into the navicular bone.

Intertarsal joints of the foot

The intertarsal joints of the foot are formed by the articulating tarsal bones. All are synovial joints except the cuboidonavicular joint, which is a fibrous joint.

Subtalar (talocalcaneal) joint

The subtalar joint is formed by the talus articulating with the calcaneum. It is strengthened by the talocalcaneal ligaments. Inversion and eversion movements occur at this joint.

Midtarsal joints

The midtarsal joints include the talocalcaneonavicular and the calcaneocuboidal joints. The talocalcaneonavicular joint is formed by the talus, calcaneum, and navicular bones. It is strengthened by the spring (calcaneonavicular) ligament. The calcaneocuboidal joint is formed by the calcaneum and cuboid bones. It is strengthened by the bifurcate ligament, and long and short plantar ligaments. Movements at these joints are inversion and eversion.

Other tarsal joints

The cuneonavicular, cuboideonavicular, intercuneiform, and cuneocuboidal joints are strengthened by dorsal, plantar, and interosseous ligaments. There is very little movement in these joints, and it is only a slight gliding movement.

Arches of the foot

Each foot has a lateral and medial longitudinal arch. If the feet are placed together they form a transverse arch. These arches support the weight of the body, and they are maintained by bone shape, muscles and ligaments.

The medial is higher than the lateral arch, and it consists of the calcaneus, talus, navicular, and cuneiform bones and the medial three metatarsals. The talus acts as a keystone in the centre of the arch. The plantar calcaneonavicular (spring), long and short plantar ligaments, and strong dorsal ligaments tie the bones together. The plantar aponeurosis, abductor and flexor muscles in the first and third layers, and the flexor digitorum longus tendon support the arch. Finally, the tibialis muscles and deltoid ligament suspend the arch through their attachments.

The lower lateral arch consists of the calcaneus, the cuboid, and the lateral two metatarsals. This is maintained by the long and short (calcaneocuboidal) plantar ligaments, the plantar aponeurosis, muscles of the first layer of the foot, and the peroneus longus tendon.

Each foot contains half of the transverse arch. Each half consists of the metatarsal bases, cuboid, and the three cuneiforms. It is maintained by the wedge shaped cuneiform bones and metatarsal bases, the strong long and short plantar ligaments, as well as the deep transverse ligaments. The peroneus longus and brevis tendons suspend and tie the arch ends together.

In prolonged periods of standing, the plantar ligaments and aponeurosis weaken and stretch under the body weight. As a result the medial longitudinal arch 'falls' causing a flatfoot (pes planus).

Functions of the feet

The feet serve to:

- Support the body weight.
- Maintain balance.
- Act as propulsive levers, e.g. in walking and running.

- Describe the superficial venous drainage of the lower limb.
- Outline the cutaneous innervation of the lower limb.
- Outline the skeleton of the hip and thigh region.
- What is the fascia lata?
- Describe the muscles and contents of the gluteal region.
- What are the boundaries and contents of the femoral triangle?
- Discuss the arterial supply of the thigh.
- Discuss the anatomy of the hip joint.
- Discuss the muscles of the anterior, posterior, and adductor compartments of the thigh.
- What are the boundaries and contents of the popliteal fossa?
- Outline the anatomy of the knee joint.
- Outline the muscles of each compartment of the leg.
- Describe the arterial supply of the leg.
- Describe the nerve supply of the leg.
- Describe the skeleton of the foot.
- Describe the anatomy of the dorsum of the foot.
- Describe the anatomy of the ankle joint.
- Outline the muscles of the foot.
- Discuss the blood and nerve supply of the sole of the foot.
- Describe the arches of the foot.

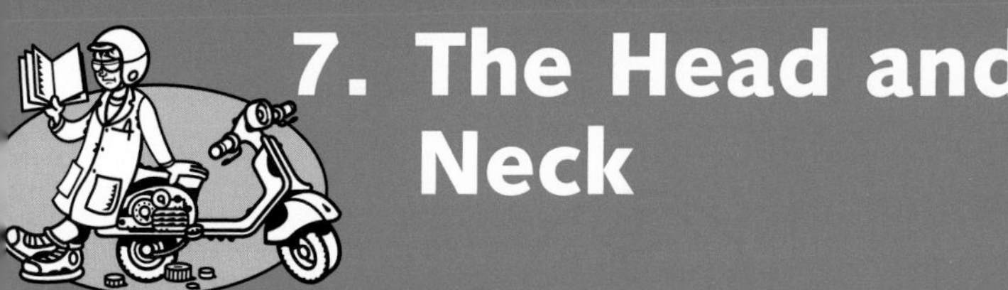

7. The Head and Neck

:egions and components of the ead and neck

kull

he skull is composed of a number of different bones ›ined at sutures. The bones of the skull may be ›ivided into:

- The cranium
- The facial skeleton.

he cranium is subdivided into:

- An upper part—the vault.
- A lower part—the base of the skull.

orma verticalis

view of the skull from above (norma verticalis) is lustrated in Fig. 7.1.

orma occipitalis

view of the skull from behind (norma ccipitalis) is shown in Fig. 7.2. Note the astoid process of the temporal bone and the xternal occipital protuberance—a midline evation from the occipital bone—from which the superior nuchal line extends laterally.

Norma frontalis

The skull is viewed from the front (norma frontalis) in Fig. 7.3. The frontal bones form the forehead and the superior margin of the orbits. They articulate with the nasal bones and the frontal process of the maxilla (upper jaw). The mandible (lower jaw) lies below the maxilla. Both mandible and maxilla bear teeth.

Norma lateralis

Fig. 7.4 shows the lateral view (norma lateralis) of the skull. The parietal bone articulates with the greater wing of the sphenoid bone and the temporal bone. These three bones meet at a point, the pterion, which is the thinnest part of the lateral aspect of the skull and vulnerable to damage. The middle meningeal artery lies just deep to this point, and this may rupture following trauma to the head.

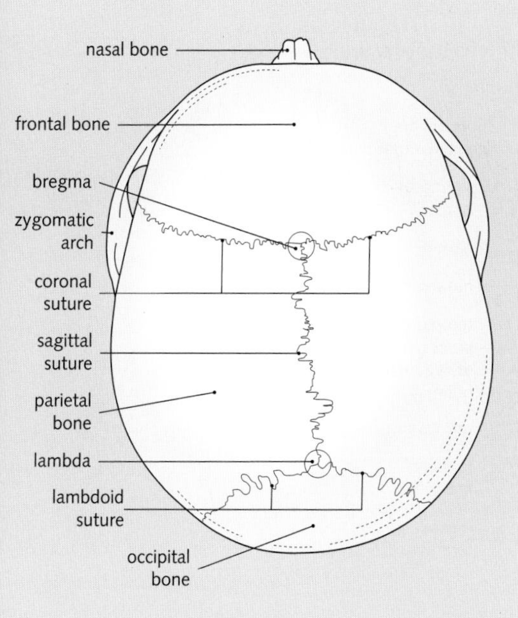

g. 7.1 Skull viewed from above.

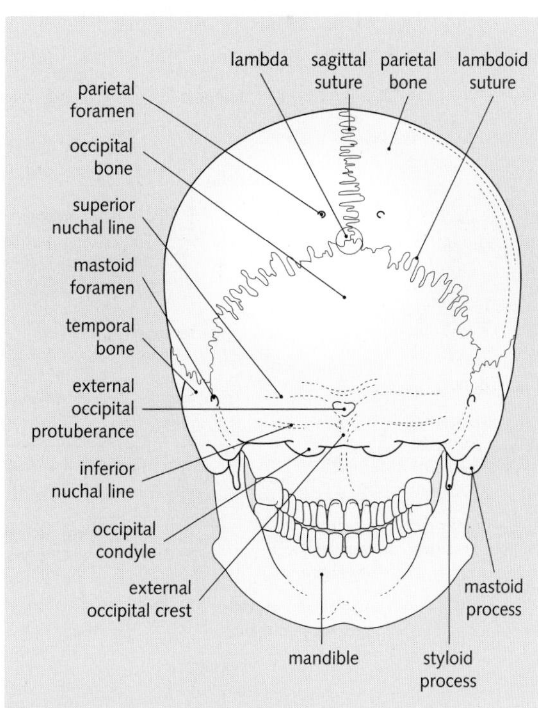

Fig. 7.2 Skull viewed from behind.

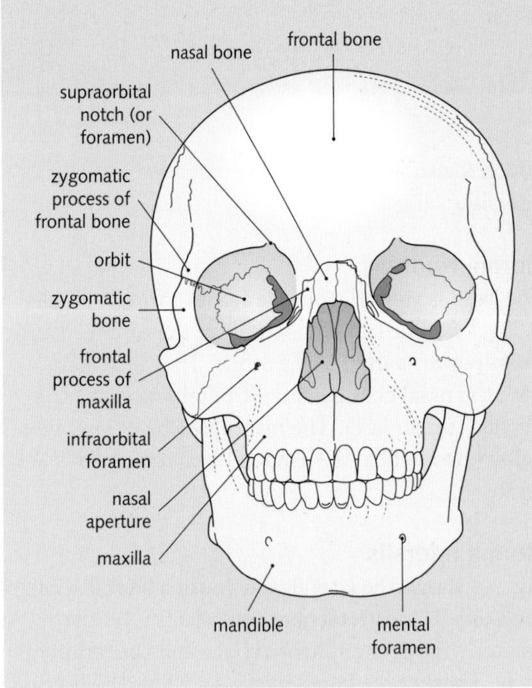

Fig. 7.3 Skull viewed from the front.

The zygomatic arch is formed by the zygomatic process of the temporal bone and the temporal process of the zygomatic bone.

Pterion damage can cut the middle meningeal artery and cause bleeding externally to the dura mater (extradural haemorrhage).

Norma basalis

In Fig. 7.5 the skull is viewed from below (norma basalis). The palate is formed by the palatine proces of the maxilla and the horizontal processes of the palatine bones. The alveolar process of the maxilla surrounds the palate.

Openings in the skull

Fig. 7.6 lists the important openings in the base of the skull and their contents.

Cervical vertebrae

There are seven cervical vertebrae forming the skeleton of the neck. All except C1 (atlas), C2 (axis), and C7 are typical vertebrae (Figs 7.7 and 7.8). C7 possesses a longer spinous process, which is usually the most superior spinous process palpable.

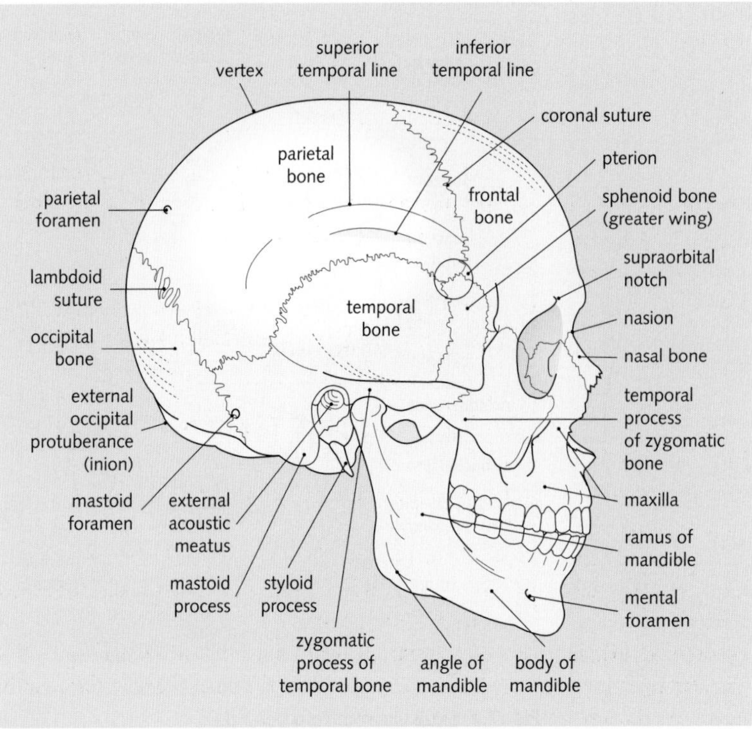

Fig. 7.4 Lateral view of the skull.

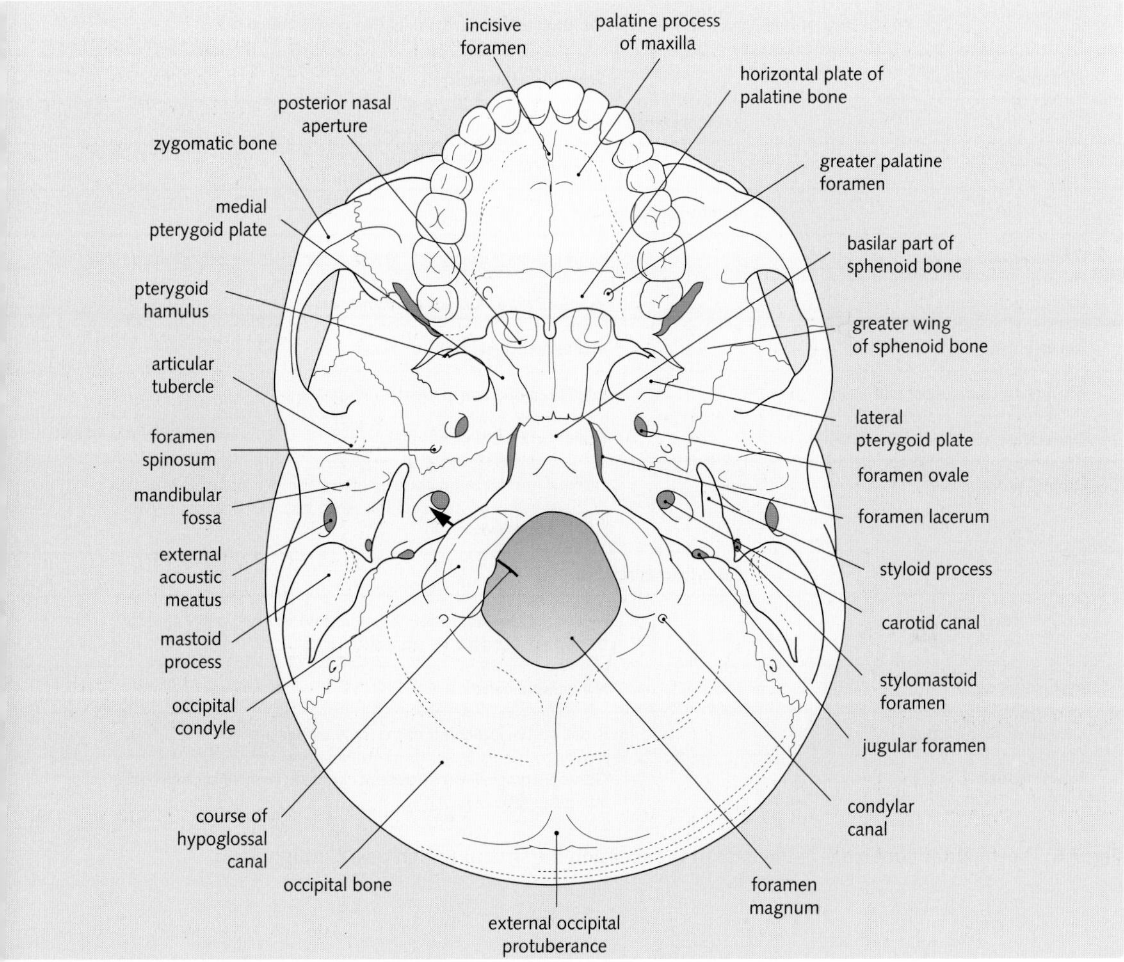

Fig. 7.5 Skull seen from below.

Atlanto-occipital joint

The atlanto-occipital joint is the articulation between the lateral masses of C1 vertebra and the occipital condyles. It is a synovial joint surrounded by a loose capsule. Flexion and extension are allowed at this joint, but no rotation.

Atlanto-axial joints

There are two lateral synovial joints between the lateral masses of the axis and atlas (Fig. 7.9). There is also a median joint between the dens (odontoid process) of the axis and the anterior arch of the atlas. These joints allow rotational movements of the head in which the skull and the atlas rotate as a unit on the axis. Alar ligaments prevent excess rotation.

The transverse ligament of the atlas holds the dens against the anterior arch of the atlas. Rupture of this ligament, e.g. during head trauma, allows the dens to impinge on the cervical spinal cord, causing paralysis of the body below the neck. If the dens compresses the medulla, the patient may die.

The face and scalp

Scalp

The scalp consists of five layers (Fig. 7.10), and it has a very rich blood supply (Fig. 7.11). Injuries to this region often result in profuse bleeding.

The scalp veins follow the arterial supply:

- The supraorbital and supratrochlear veins unite to form the facial vein.
- The superficial temporal vein joins with the maxillary vein to form the retromandibular vein in the parotid salivary gland.

The important openings in the base of the skull and the structures that pass through them	
Opening in skull	**Structures transmitted**
Anterior cranial fossa	
Cribriform plate	Olfactory nerve
Middle cranial fossa	
Foramen ovale	V_3 (mandibular division of trigeminal nerve), lesser petrosal nerve
Foramen rotundum	V_2 (maxillary division of trigeminal nerve)
Foramen spinosum	Middle meningeal artery and vein
Foramen lacerum (upper part only)	Internal carotid artery, greater petrosal nerve
Optic canal	Optic nerve, ophthalmic artery
Superior orbital fissure	Lacrimal, frontal and nasociliary branches of V_1 (ophthalmic branch of trigeminal nerve); oculomotor, abducent, and trochlear nerves; superior ophthalmic vein
Posterior cranial fossa	
Foramen magnum	Medulla oblongata, spinal part of accessory nerve, upper cervical nerves; right and left vertebral arteries
Hypoglossal canal	Hypoglossal nerve
Internal acoustic meatus	Facial, vestibulocochlear nerves; labyrinthine artery
Jugular foramen	Glossopharyngeal, vagus, accessory nerves; sigmoid sinus becomes internal jugular vein

Fig. 7.6 The important openings in the base of the skull and the structures that pass through them.

- The posterior auricular vein unites with the posterior division of the retromandibular vein to form the external jugular vein.
- Occipital veins drain into the suboccipital venous complex.

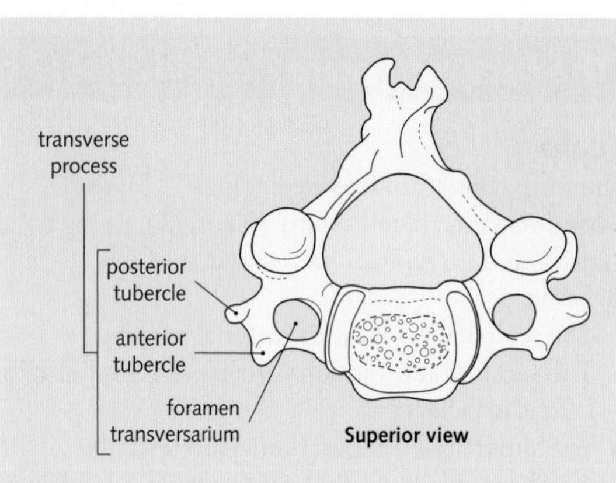

Fig. 7.7 Typical cervical vertebra.

Distinctive characteristics of a typical cervical vertebra	
Part	**Characteristics**
Body	Small; longer from side to side than anteroposteriorly; superior surface is concave, inferior surface is convex
Vertebral foramen	Large and triangular
Transverse processes	Foramina transversaria (small or absent in C7)
Articular processes	Superior facets directed superoposteriorly, inferior facets directed inferoanteriorly
Spinous processes	Short and bifid in C3 to C5, long in C6, and longer in C7

Fig. 7.8 Distinctive characteristics of a typical cervical vertebra.

The scalp veins connect with the diploic veins in the skull bones and the intracranial venous sinuses by valveless emissary veins. Infections of the scalp are, therefore, potentially very serious as they may spread intracranially.

Sensory nerve supply to the scalp is shown in Fig. 7.11. The muscles of the scalp and external ear are supplied by the facial nerve.

Face

The skin of the face is connected to the facial bones by loose connective tissue. There is no deep fascia. The muscles (of facial expression) of the face lie in this connective tissue. Like the scalp, the skin of the face is very sensitive and very vascular.

Sensory innervation of the face is from the trigeminal nerve (V). It has three divisions: the ophthalmic (V_1), the maxillary (V_2), and the mandibular (V_3) nerves. These supply the upper, middle, and lower thirds of the face, respectively (Fig. 7.12).

The order of the layers of the scalp are:
S Skin
C Connective tissue
A Aponeurosis
L Loose connective tissue
P Pericranium (periostium)

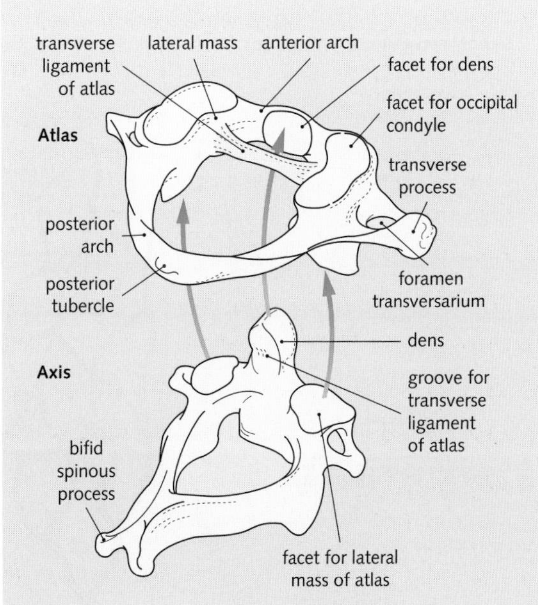

Fig. 7.9 Atlas and axis (showing their articulations).

Muscles of the face

Most of the muscles of facial expression are attached to the overlying skin (Fig. 7.13). They allow a wide variety of facial postures. All of these muscles are supplied by the facial nerve (VII).

Fig. 7.10 Layers of the scalp.

Sensory nerve supply to the scalp
Arterial supply to the scalp
supratrochlear nerve
supraorbital nerve
zygomaticotemporal nerve
auriculotemporal nerve
lesser occipital nerve
greater occipital nerve
supratrochlear artery
supraorbital artery
zygomaticotemporal artery
superficial temporal artery
posterior auricular artery
occipital artery

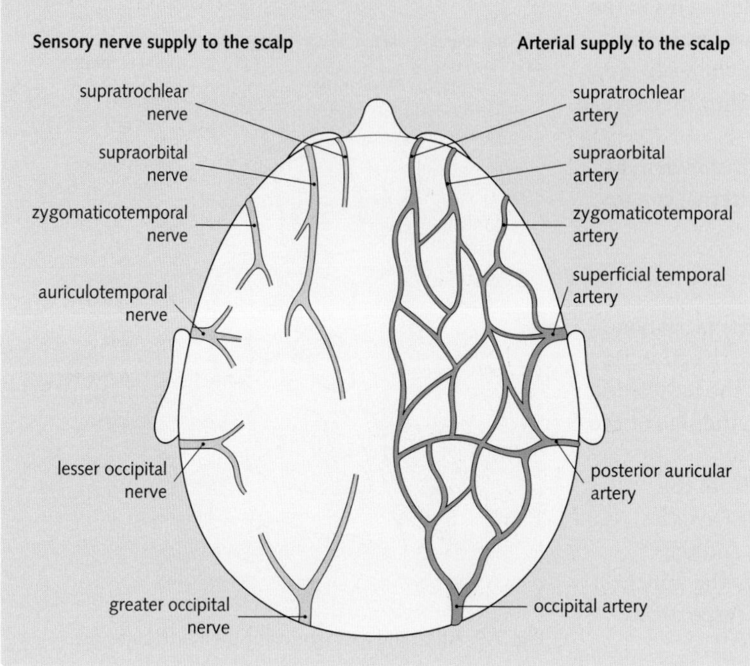

Fig. 7.11 Nerve and arterial supply to the scalp.

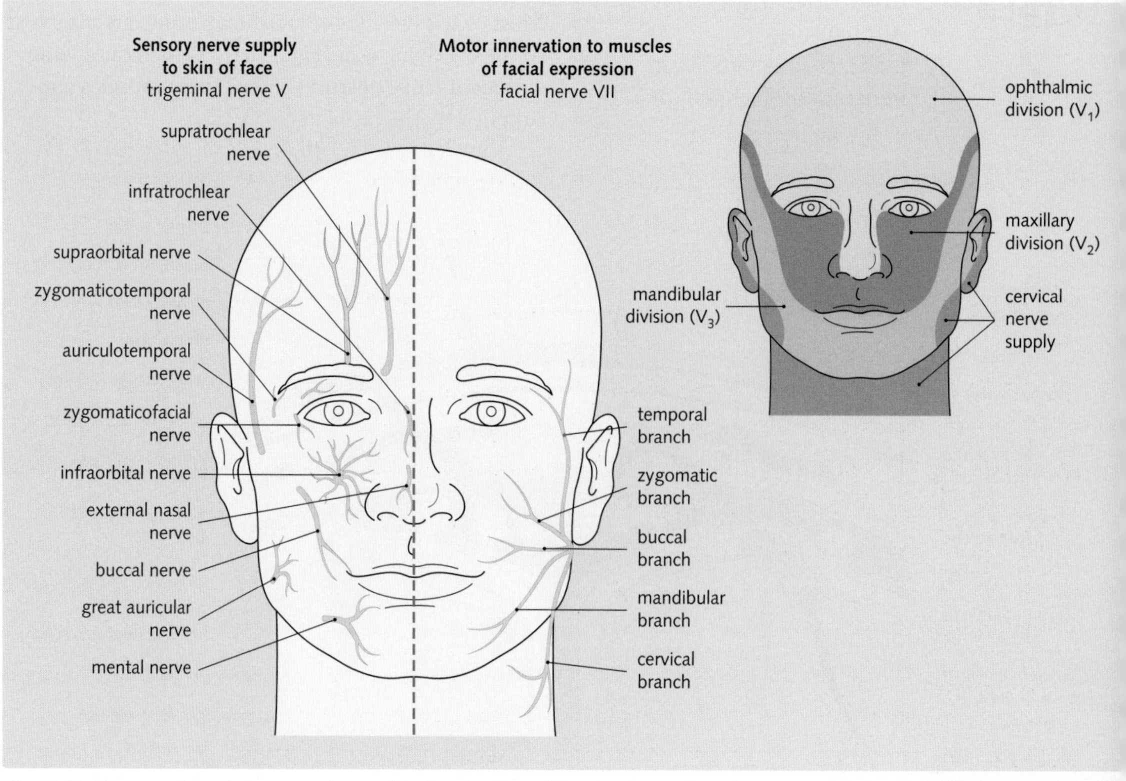

Sensory nerve supply to skin of face
trigeminal nerve V
Motor innervation to muscles of facial expression
facial nerve VII
supratrochlear nerve
infratrochlear nerve
supraorbital nerve
zygomaticotemporal nerve
auriculotemporal nerve
zygomaticofacial nerve
infraorbital nerve
external nasal nerve
buccal nerve
great auricular nerve
mental nerve
temporal branch
zygomatic branch
buccal branch
mandibular branch
cervical branch
ophthalmic division (V_1)
maxillary division (V_2)
mandibular division (V_3)
cervical nerve supply

Fig. 7.12 Nerves of the face. Inset shows the distribution of the divisions of the trigeminal nerve. Note the great auricular nerve is not part of the trigeminal nerve.

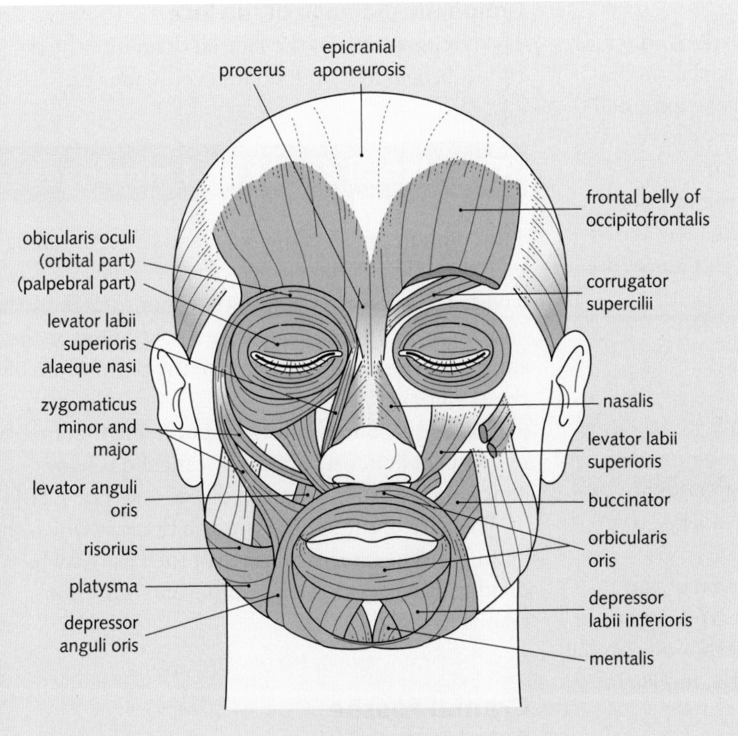

Fig. 7.13 Muscles of the face.

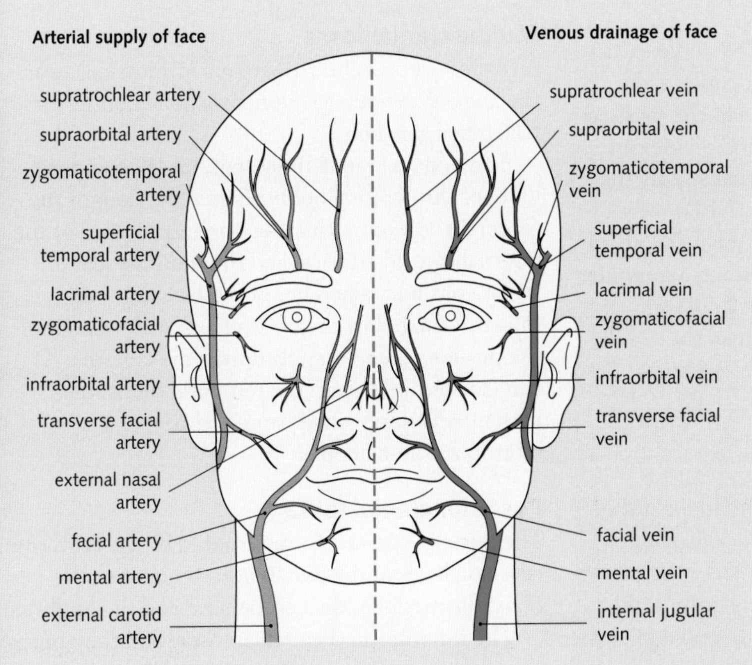

Fig. 7.14 Blood supply to the face.

Vessels of the face

The face has a very rich blood supply, derived mainly from the facial artery and the superficial temporal artery, both of which are branches of the external carotid artery (Fig. 7.14).

The facial artery ascends deep to the submandibular gland, winds around the inferior border of the mandible, and enters the face. It gives off the following branches as it ascends in the face:

- The inferior labial artery.
- The superior labial artery.
- The lateral nasal artery.
- The angular artery.

The supraorbital and supratrochlear arteries are terminal branches of the ophthalmic artery—a branch of the internal carotid artery.

The facial vein is formed by the union of the supraorbital and supratrochlear veins. It descends in the face and receives tributaries corresponding to the branches of the artery. It drains into the internal jugular vein.

The facial vein connects with the cavernous sinus in the skull via the superior ophthalmic vein, and it provides a path for spread of infection from the face to the cavernous sinus. Even minor infections in the infraorbital region may cause this grave complication.

Motor nerve supply to the face

This is from the facial nerve (VII). The nerve exits the skull through the stylomastoid foramen to lie between the ramus of the mandible and the mastoid process. It enters the parotid gland and divides into its five groups of terminal branches that supply the muscles of facial expression (see Fig. 7.12).

Before entering the parotid gland the facial nerve gives off the posterior auricular nerve and a muscular branch, which supply the occipital belly of occipitofrontalis, the stylohyoid muscle, the posterior belly of the digastric muscle, and the posterior auricular muscle.

For branches of the facial nerve to the face use the following mnemonic: Ten Zulus *Bought* My Cat (temporal, zygomatic, buccal, mandibular, cervical).

Lymphatic drainage of the face

The lymph vessels of the face all drain into the deep cervical chain of lymph nodes eventually (Fig. 7.15).

The cranial cavity and meninges

The cranium protects the brain and its surrounding meninges. The outer surface of the cranial bones is covered by the pericranium, the inner surface by the endocranium. The two layers are continuous at the sutures of the skull, and they are the periosteum of the skull bones.

The cranial bones consist of outer and inner tables of compact bones separated by cancellous bone containing red marrow—the diploë (see Fig. 7.10).

The base of the skull forms the floor on which the brain lies. The internal surface of the base may be divided into the anterior, middle, and posterior cranial fossae (Fig. 7.16).

Cranial fossae

Anterior cranial fossa

The olfactory nerves perforate the cribriform plate of the ethmoid bone. The anterior ethmoidal nerve passes through a small slit in this bone. The crista galli projects upwards from the cribriform plate.

The anterior cranial fossa contains the frontal lobes of the brain and the olfactory bulbs.

Middle cranial fossa

A shallow depression (trigeminal impression) near the apex of the petrous temporal bone houses the trigeminal ganglion.

The middle cranial fossa contains the temporal lobes of the cerebral hemispheres, the floor of the forebrain, the optic chiasma, the termination of the internal carotid arteries, and the pituitary gland.

The pituitary gland lies in the pituitary fossa or sella turcica, below the optic chiasma. A tumour of the pituitary gland may compress the chiasma. As this carries fibres from the temporal fields, the patient will complain of 'tunnel vision' or a bitemporal hemianopia.

Posterior cranial fossa

The posterior cranial fossa is roofed by the tentorium cerebelli layer of the dura mater. It contains the pons, the medulla, the cerebellum, and the midbrain.

The foramina in the cranial fossae and their main contents are outlined in Fig. 7.6.

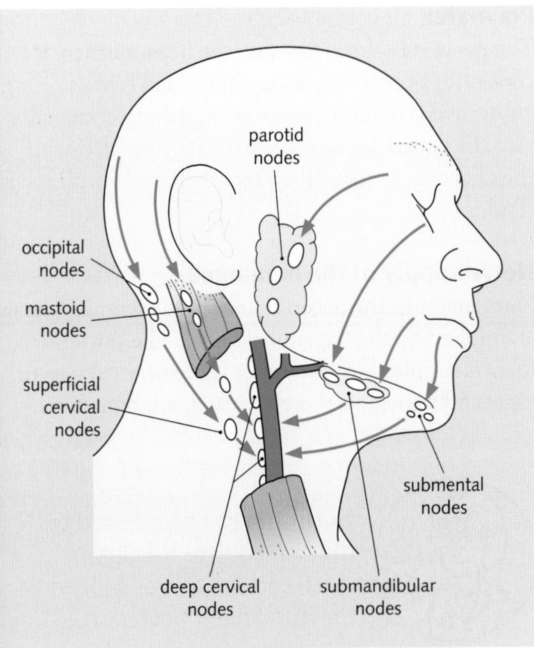

Fig. 7.15 Lymphatic drainage of the face. (Adapted from *Anatomy as a Basis for Clinical Medicine*, by E C B Hall-Craggs. Courtesy of Williams & Wilkins.)

Meninges

There are three meningeal layers surrounding the brain and spinal cord (see Fig. 7.10):

- The dura mater.
- The arachnoid mater.
- The pia mater.

Dura mater

The dura mater may be divided into two layers:

- The endosteal layer is the endosteum (periosteum) lining the inner surface of the cranial bones.
- The meningeal layer (the dura mater proper) is made up of dense strong fibrous tissue. It is continuous with the dura mater of the spinal cord through the foramen magnum. The dura sends sleeves around the cranial nerves, which fuse with the epineurium of the nerves outside the skull.

The meningeal layer of dura mater gives rise to four septa that assist in restricting rotatory displacement of the brain:

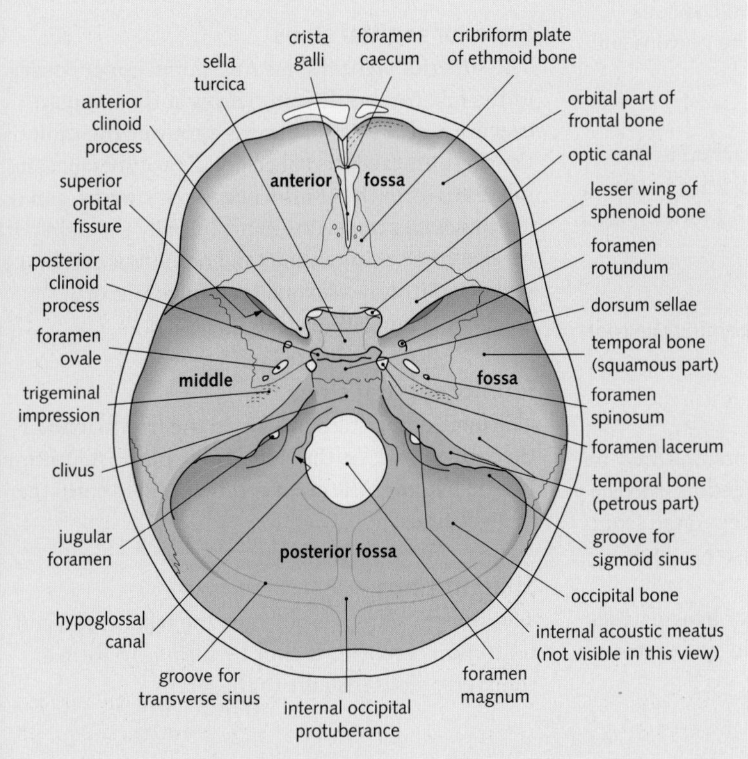

Fig. 7.16 Internal surface of the base of the skull, showing the cranial fossae.

- The falx cerebri.
- The tentorium cerebelli.
- The falx cerebelli.
- The diaphragma sellae.

Falx cerebri

This is a sickle-shaped fold of dura lying in the midline between the two cerebral hemispheres (Fig. 7.17). It is attached anteriorly to the internal frontal crest and the crista galli. Posteriorly, it blends with the tentorium cerebelli.

The superior sagittal sinus runs in its superior margin, which is its attachment to the vault of the skull. The inferior sagittal sinus runs in its inferior margin. The straight sinus runs along its attachment to the tentorium cerebelli.

Tentorium cerebelli

This is a crescent-shaped fold of dura mater that roofs the posterior cranial fossa (see Fig. 7.17). It covers the cerebellum, and it supports the occipital lobes of the cerebral hemispheres. The midbrain passes through the tentorial notch anteriorly.

The tentorium is attached to the anterior clinoid process anteriorly. Posteriorly the falx cerebri and falx cerebelli are attached to its upper and lower surfaces.

The superior petrosal and transverse venous sinuses run along its attachment to the petrous and occipital bones, respectively.

Falx cerebelli

This is a small fold of dura mater attached to the internal occipital crest. It projects forward between the two cerebellar hemispheres. Its posterior margin contains the occipital sinus.

Diaphragma sellae

This is a small circular fold of dura forming the roof of the pituitary fossa.

Arachnoid mater

The arachnoid mater surrounds the brain. It may be separated from the dura mater by bleeding into the subdural (potential) space, and it is separated from the pia mater by the subarachnoid space, which contains cerebrospinal fluid.

Where the arachnoid bridges major irregularities of the brain surface, the subarachnoid space expands to form subarachnoid cisterns.

Pia mater

The pia mater closely invests the brain surface. It continues as a sheath around the small vessels entering the brain. In some areas the pia invaginates into the ventricles to take part in the formation of the choroid plexus, which secretes cerebrospinal fluid.

Nerve supply of the meninges

Dura mater in the anterior and middle cranial fossae is supplied by the trigeminal nerve. The posterior fossa is supplied by the upper three cervical nerves, vagal and hypoglossal meningeal nerve branches.

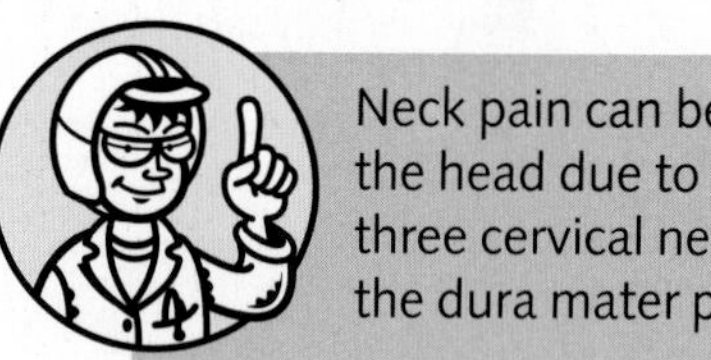

Neck pain can be referred to the head due to the upper three cervical nerves supplying the dura mater posteriorly.

Cranial venous sinuses

The cranial venous sinuses are lined by endothelium, and are valveless (see Fig. 7.17). Tributaries from the various parts of the brain and from the diploë, the orbit, and the inner ear drain into these sinuses.

Superior sagittal sinus

The superior sagittal sinus runs in the upper border of the falx cerebri. It commences at the foramen caecum and passes backwards, grooving the vault of the skull. At the internal occipital protuberance, it dilates to form the confluence of the sinuses, and it continues as a transverse sinus (usually the right). It receives the occipital sinus and numerous cerebral veins, and it shows several accumulations of arachnoid granulations (see Fig. 7.10).

Inferior sagittal sinus

The inferior sagittal sinus lies in the free margin of the falx cerebri. At the free margin of the tentorium cerebelli it joins the great cerebral vein to form the straight sinus.

Straight sinus

The straight sinus lies between the falx cerebri and tentorium cerebelli. It ends by turning to form a transverse sinus (usually to the left).

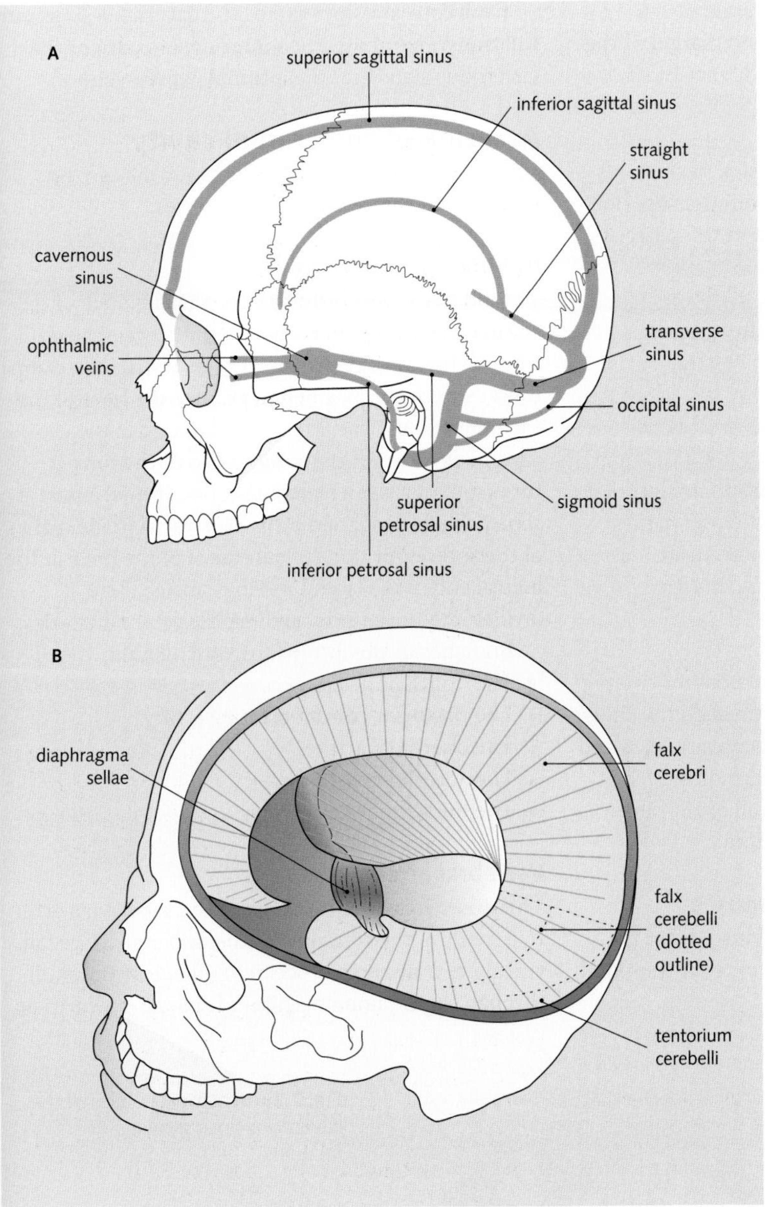

Fig. 7.17 (A) The positions of the cranial venous sinuses. (B) The falx cerebri, falx cerebelli, diaphragma sellae, and tentorium cerebelli.

Transverse sinuses

The transverse sinuses commence at the internal occipital protuberance and run in the attachment of the tentorium cerebelli. They end by turning inferiorly as the sigmoid sinuses. They receive the superior petrosal sinuses and the cerebral, cerebellar, and diploic veins.

Sigmoid sinuses

Each sigmoid sinus turns downward and medially to groove the mastoid process, lying behind the mastoid antrum. It then turns downward through the posterior part of the jugular foramen to become continuous with the superior bulb of the internal jugular vein.

Occipital sinus

The occipital sinus lies in the attached margin of the falx cerebelli. It drains into the confluence of sinuses.

Cavernous sinuses

The cavernous sinuses lie on either side of the body of the sphenoid, and they extend from the superior orbital fissure anteriorly to the apex of the petrous temporal bone posteriorly.

They receive:

- The superior and inferior ophthalmic veins.
- The cerebral veins.
- The sphenoparietal sinus.
- The central vein of the retina.

They drain posteriorly into the superior and inferior petrosal sinuses and inferiorly into the pterygoid venous plexus. The two sinuses communicate via anterior and posterior intercavernous sinuses.

Relations of the cavernous sinuses

The internal carotid artery and its sympathetic nerve plexus and the abducens nerve run through the sinus (Fig. 7.18). The oculomotor and trochlear nerves and the ophthalmic and maxillary divisions of the trigeminal nerves lie in the lateral wall of the sinus, between the endothelium and the dura.

Superior and inferior petrosal sinuses

The superior and inferior petrosal sinuses lie at the superior and inferior borders of the petrous temporal bone, respectively.

Each superior sinus drains the cavernous sinus into the transverse sinus. Each inferior sinus drains the cavernous sinus into the internal jugular vein.

Arteries of the cranial cavity

The brain is supplied by the two internal carotid arteries and the two vertebral arteries.

Internal carotid artery

The internal carotid artery is a terminal branch of the common carotid artery (Fig. 7.19). It enters the skull through the carotid canal and enters the middle cranial fossa through the foramen lacerum in the floor of the cavernous sinus. The artery runs forward in the cavernous sinus. At the anterior end of the sinus it turns superiorly and pierces the roof. It then enters the subarachnoid space and turns backwards to the region of the anterior perforated substance of the brain at the medial end of the lateral cerebral sulcus. Here, it divides into the anterior and middle cerebral arteries.

Branches of the internal carotid include:

- The ophthalmic artery.
- The posterior communicating artery.
- The choroidal artery.
- The anterior cerebral artery.
- The middle cerebral artery.

Vertebral artery

This arises from the first part of the subclavian artery. It ascends in the foramina of the upper six cervical vertebrae transverse processes and enters the skull through the foramen magnum. It passes upwards on

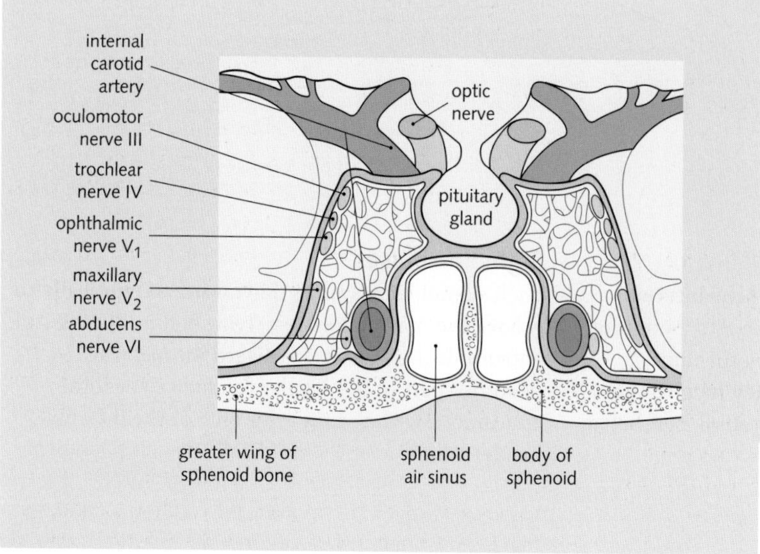

Fig. 7.18 A coronal section of the cavernous sinus showing its relations.

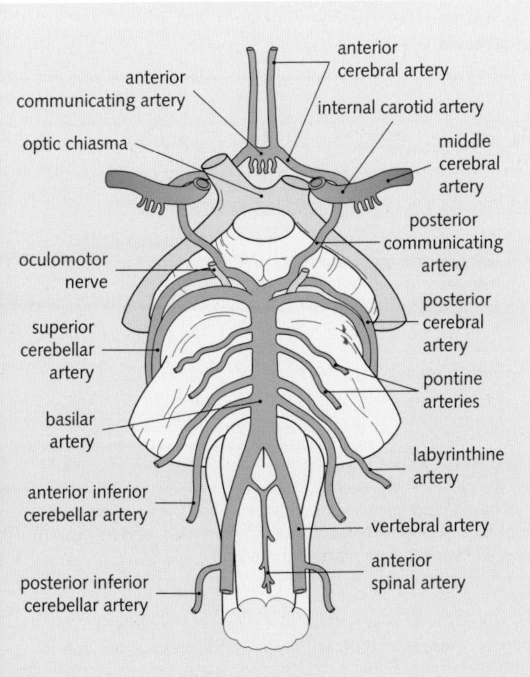

Fig. 7.19 Internal carotid and vertebral arteries on base of the brain.

the medulla oblongata (see Fig. 7.19), and it joins the vessel from the opposite side to form the basilar artery.

Cranial branches of the vertebral artery include:

- The meningeal arteries.
- The anterior and posterior spinal arteries.
- The posterior inferior cerebellar artery.
- The medullary arteries.

Basilar artery

The basilar artery ascends on the anterior border of the pons (see Fig. 7.19). At the upper border of the pons, it divides into the posterior cerebral arteries. It also gives off branches to the pons, cerebellum, and internal ear.

> Plaque formation (atheroma) in a vertebral artery can result in impaired blood flow to the brain (vertebro-basilar insuffiency). Impairing blood flow can be caused by hyperextension of the neck, which blocks the artery where these plaques are and causes fainting (syncope).

Circulus arteriosus

The circulus arteriosus is an anastomosis between branches of the internal carotid arteries and the vertebral arteries (see Fig. 7.19). It lies in the interpeduncular fossa beneath the forebrain. It allows blood entering either the carotid or vertebral artery to flow to any part of both cerebral hemispheres.

Strokes or cerebrovascular accidents are common in Western countries. They commonly arise because of blockage of one of the major arteries, usually the middle cerebral artery. The anastomosis between the cerebral vessels is not sufficient to supply the affected tissue, resulting in death of that part of the brain.

Cranial nerves

The cranial nerves are summarized in Fig. 7.20.

The orbit

The eyeball and its associated muscles, nerves, and vessels are protected by the bony orbital cavity. The mobile eyelids protect the eyes anteriorly.

Eyelids

The eyelids are two mobile folds of skin lying in front of the eye and separated by the palpebral fissure (Fig. 7.21). The superficial surface of the lids is covered by the skin; the deep surface is covered by mucosa—the conjunctiva.

Sebaceous, ciliary, and tarsal glands pour their secretions onto the eyelid.

The space between the medial part of the eyelids and the eyeball is called the lacus lacrimalis. It contains an elevation—the caruncle lacrimalis. Tears produced by the lacrimal gland are continually spread over the conjunctiva and cornea by blinking, preventing damage to the eye. The tears then pass towards the lacus lacrimalis and into the puncta lacrimalia to enter the canaliculi lacrimales, which drain into the lacrimal sac. This sac is the upper blind end of the nasolacrimal duct, which drains the tears into the inferior meatus of the nose.

The nerve supply to the lacrimal gland is parasympathetic secretomotor, which originates in the lacrimal nucleus. Parasympathetic fibres travel in

Summary of cranial nerves	
Nerve	**Distribution and functions**
Olfactory (I)	Smell from nasal mucosa of roof of each nasal cavity
Optic (II)	Vision from retina
Oculomotor (III)	Motor to all extrinsic eye muscles except superior oblique and lateral rectus; parasympathetic innervation to sphincter pupillae and ciliary muscle (constricts pupil and accommodates lens of eye) carries sympathetic nerve fibres to smooth muscle part of levator palpebrae superioris
Trochlear (IV)	Motor to superior oblique
Trigeminal (V) —ophthalmic division (V_1)	Sensation from upper third of face, including cornea, scalp, eyelids, and paranasal sinuses
Trigeminal (V) —maxillary division (V_2)	Sensation from the middle third of face, including upper lip, maxillary teeth, mucosa of nose, maxillary sinuses, and palate; supplies dura mater anteriorly
Trigeminal (V) —mandibular division (V_3)	Motor to muscles of mastication, mylohyoid, anterior belly of digastric, tensor veli palatini and tensor tympani; sensation from lower third of face, including temporomandibular joint, and mucosa of mouth and anterior two thirds of tongue, supplies dura mater anteriorly
Abducent (VI)	Motor to lateral rectus
Facial (VII)	Motor to muscles of facial expression and scalp, stapedius, stylohyoid, and posterior belly of digastric; taste from anterior two thirds of tongue, floor of mouth, and palate; sensation from skin of external acoustic meatus; parasympathetic innervation to submandibular and sublingual salivary glands, lacrimal gland, and glands of nose and palate
Vestibulocochlear (VIII)	Vestibular sensation from semicircular ducts, utricle, and saccule; hearing from spiral organ
Glossopharyngeal (IX)	Motor to stylopharyngeus, parasympathetic innervation to parotid gland; visceral sensation from parotid gland, carotid body and sinus, pharynx, and middle ear; taste and general sensation from posterior third of tongue
Vagus (X)	Motor to constrictor muscles of pharynx, intrinsic muscles of larynx, and muscles of palate (except tensor veli palatini) and superior two thirds of oesophagus; parasympathetic innervation to smooth muscle of trachea, bronchi, digestive tract, and cardiac muscle of heart; visceral sensation from pharynx, larynx, trachea, bronchi, heart, oesophagus, stomach, and intestine; taste from epiglottis and palate; sensation from auricle, external acoustic meatus, and dura mater of posterior cranial fossa
Accessory (XI) cranial root spinal root	 Motor to striated muscles of soft palate, pharynx, and larynx via fibres that join X in jugular foramen Motor to sternocleidomastoid and trapezius
Hypoglossal (XII)	Sensory to dura mater, posteriorly; motor to intrinsic and extrinsic muscles of tongue (except palatoglossus)

Fig. 7.20 Cranial nerves.

the facial nerve, greater petrosal nerve, and synapse in the pterygopalatine ganglion. The postganglionic fibres now join the zygomaticotemporal nerve (a V_2 branch), then the lacrimal nerve (a V_1 branch) before supplying the lacrimal gland.

The fibrous framework of the eyelids is formed by the orbital septum (Fig. 7.22). This is thickened at the lid margins to form the tarsal plates, which medially and laterally form the medial palpebral ligament and the lateral palpebral raphe, respectively. The levator palpebrae superioris muscle is attached to the superior tarsal plate.

The conjunctiva is the mucous membrane lining the eyelid. It is reflected at the superior and inferior fornices onto the anterior surface of the eyeball, forming the conjunctival sac when the eyes are closed (Fig. 7.23).

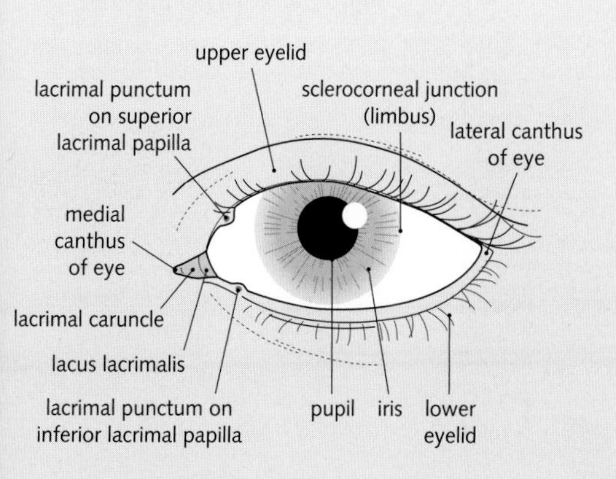

ig. 7.21 Eyelids, palpebral fissure, and eyeball.

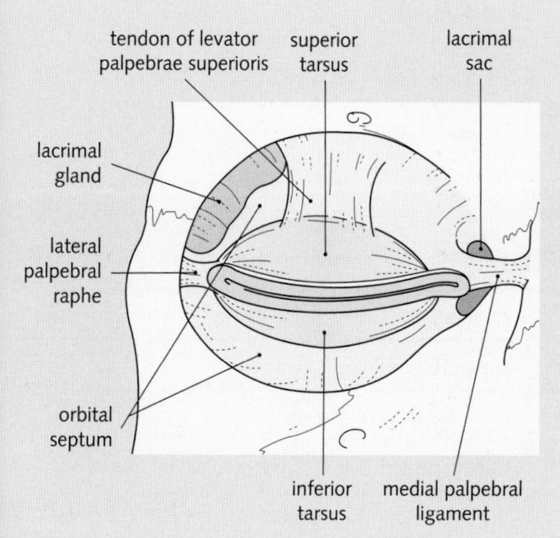

ig. 7.22 Orbital septum, tarsi, and palpebral gaments.

rbital cavity

ig. 7.24 illustrates the bony components of the rbital cavity. The orbital cavity boundaries are as llows:

- Superiorly (roof)—frontal bone (orbital part), lesser wing of sphenoid bone
- Medial wall—maxillary bone, lacrimal bone, ethmoid bone, body of sphenoid bone
- Inferiorly (floor)—maxillary bone, zygomatic bone, palatine bone
- Lateral wall—zygomatic bone (frontal process), greater wing of sphenoid bone

ig. 7.25 shows the orbital openings and their ontents.

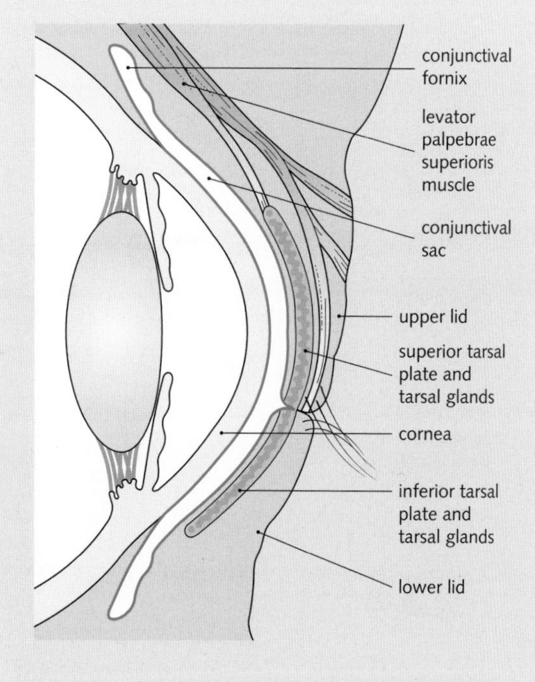

Fig. 7.23 Conjunctival sac, upper and lower lids, and cornea.

Muscles of the orbit

The muscles of the orbit are outlined in Fig. 7.26 and illustrated in Fig. 7.27.

To recall the superior oblique muscle's action remember: 'it's the poor mans muscles, always looking down and out.'

Vessels of the orbit

Fig. 7.28 shows the arterial supply to the orbit.

The superior ophthalmic vein communicates anteriorly with the facial vein and, posteriorly, it drains to the cavernous sinus. The inferior ophthalmic vein communicates via the inferior orbital fissure with the pterygoid venous plexus.

Nerves of the orbit

Optic nerve (II)

The optic nerve is surrounded by the three meningeal layers as it enters the orbit. It runs forward and laterally within the cone of rectus

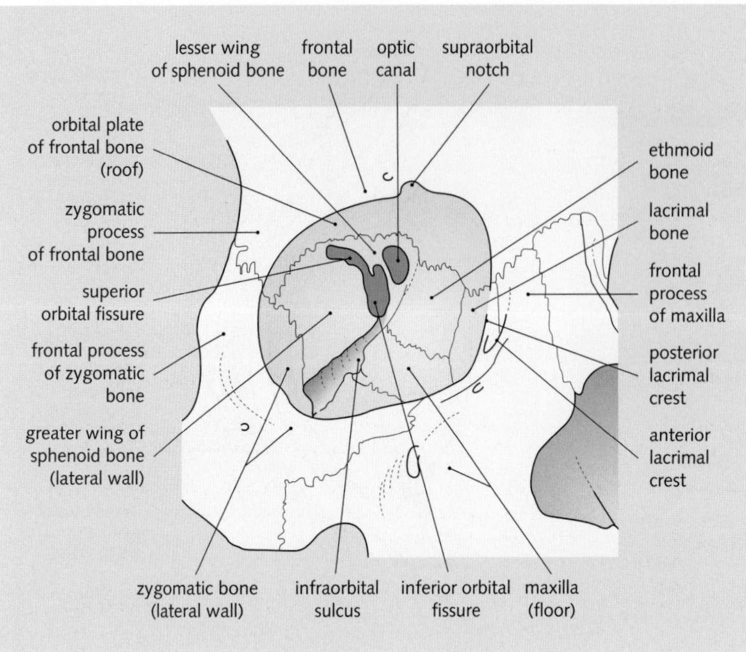

Fig. 7.24 Bones of the orbit.

Orbital openings and their contents		
Openings	**Bones**	**Contents**
Supraorbital notch (foramen)	Orbital plate of the frontal bone	Supraorbital nerve and vessels
Infraorbital groove and canal	Orbital plate of the maxilla	Infraorbital nerve and vessels
Inferior orbital fissure	Maxilla and greater wing of the sphenoid bone	Communicates with the pterygopalatine fossa and transmits the maxillary nerve and its zygomatic branch, the inferior ophthalmic vein, and sympathetic nerves
Superior orbital fissure	Greater and lesser wing of the sphenoid bone	Lacrimal, frontal, trochlear, oculomotor, abducens, and nasociliary nerves, and superior ophthalmic vein
Optic canal	Lesser wing of the sphenoid bone	Optic nerve and ophthalmic artery
Zygomaticotemporal and zygomaticofacial foramina	Zygomatic bone	Zygomaticotemporal and zygomaticofacial nerves
Anterior and posterior ethmoidal foramina	Ethmoid bone	Anterior and posterior ethmoidal nerves and vessels

Fig. 7.25 Orbital openings and their contents.

muscles and pierces the sclera. The meningeal layer fuses with the sclera here. The nerve carries afferent fibres from the retina.

Lacrimal nerve

The lacrimal nerve is a branch of the ophthalmic division of the trigeminal nerve (V_1). It passes along the upper part of the lateral rectus muscle to supply the skin of the upper lid laterally. It is joined by a branch of the zygomaticotemporal nerve carrying parasympathetic fibres to the lacrimal gland.

Frontal nerve

The frontal nerve is also a branch of V_1. It passes forward on the superior surface of levator palpebrae

Muscles of the eyeballs and eyelids			
Extrinsic muscles of eyeball (striated skeletal muscle)			
Name of muscle (nerve supply)	**Origin**	**Insertion**	**Action**
Superior rectus (III nerve)	Common tendinous ring on posterior wall of orbital cavity	Superior surface of eyeball just posterior to corneoscleral junction	Raises cornea upward and medially
Inferior rectus (III nerve)	Common tendinous ring on posterior wall of orbital cavity	Inferior surface of eyeball just posterior to corneoscleral junction	Depresses cornea downard and medially
Medial rectus (III nerve)	Common tendinous ring on posterior wall of orbital cavity	Medial surface of eyeball just posterior to corneoscleral junction	Rotates eyeball so that cornea looks medially
Lateral rectus (VI nerve)	Common tendinous ring on posterior wall of orbital cavity	Lateral surface of eyeball just posterior to corneoscleral junction	Rotates eyeball so that cornea looks laterally
Superior oblique (IV nerve)	Body of sphenoid bone	Passes through trochlea and is attached to superior surface of eyeball beneath superior rectus, behind the equator	Rotates eyeball so that cornea looks downward and laterally
Inferior oblique (III nerve)	Floor of orbital cavity	Lateral surface of eyeball deep to lateral rectus	Rotates eyeball so that cornea looks upward and laterally
Intrinsic muscles of eyeball (smooth muscle)			
Name of muscle (nerve supply)	**Origin**	**Insertion**	**Action**
Sphincter pupillae of iris (parasympathetic via III nerve)	Ring of smooth muscle passing circumferentially around pupil	–	Constricts pupil
Dilator pupillae of iris (sympathetic)	Ciliary body	Sphincter pupillae	Dilates pupil
Ciliary muscle (parasympathetic via III nerve)	Corneoscleral junction	Ciliary body	Controls shape of lens; in accommodation, makes lens more globular
Muscles of eyelids			
Name of muscle (nerve supply)	**Origin**	**Insertion**	**Action**
Orbicularis oculi (VII nerve)	Medial palpebral ligament, lacrimal bone	Skin around orbit, tarsal plates	Closes eyelids
Levator palpebrae superioris (striated muscle: III nerve; smooth muscle: sympathetic)	Lesser wing of sphenoid bone	Superior tarsal plate	Raises upper lid

g. 7.26 Muscles of the eyeballs and eyelids.

perioris. Just before it reaches the orbital margin, it vides into the supraorbital and supratrochlear rves, which supply the skin of the forehead and alp, and the frontal sinus.

Nasociliary nerve

Another branch of V_1, the nasociliary nerve enters the orbit and crosses above the optic nerve with the ophthalmic artery to reach the medial wall of the

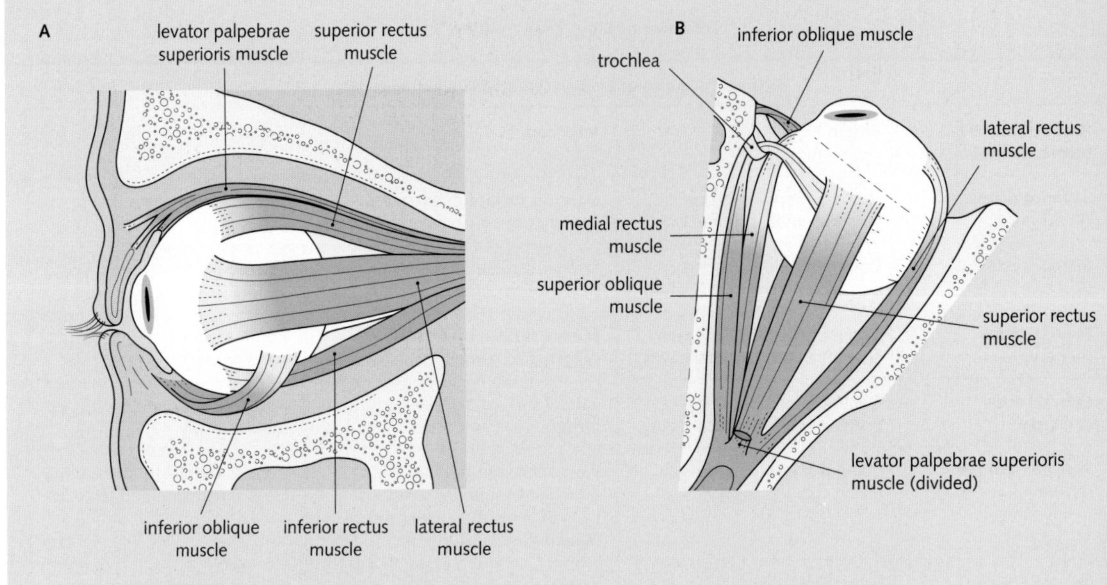

Fig. 7.27 Muscles of the orbit seen laterally (A) and from above (B).

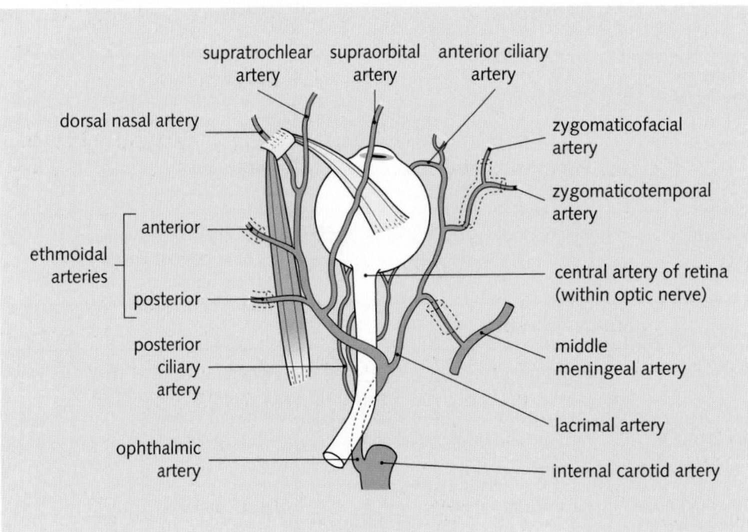

Fig. 7.28 Arterial supply to the orb
Branches of the ophthalmic artery.

orbit. It runs forward on the upper margin of medial rectus, and it ends by dividing into the anterior ethmoidal and infratrochlear nerves (Fig. 7.29).

Trochlear nerve (IV)

The trochlear nerve leaves the lateral wall of the cavernous sinus to enter the orbit. It runs forward and medially across the origin of levator palpebrae superioris to the superior oblique muscle, which it supplies.

Oculomotor nerve (III)

The oculomotor nerve is divided into superior and inferior divisions:

- The superior division supplies the superior rectu and levator palpebrae superioris.
- The inferior division supplies the inferior rectus, medial rectus, and inferior oblique.

The nerve to inferior oblique sends a branch to the ciliary ganglion. This carries parasympathetic fibres to the sphincter pupillae and ciliary muscle.

Branches of the nasociliary nerve	
Branch	**Action**
Communicating branch	Communicates with the ciliary ganglion—general sensory fibres from the eyeball pass to the ciliary ganglion via the short ciliary nerves and then to the nasociliary nerve via the communicating branch
Long ciliary nerve	2–3 branches containing sympathetic fibres for the dilator pupillae—runs with the short ciliary nerves and pierces the sclera to reach the iris
Posterior ethmoidal nerve	Exits through the posterior ethmoidal foramen to supply the ethmoidal and sphenoidal air sinuses
Infratrochlear nerve	Passes below the trochlea to supply the skin over the upper eyelid
Anterior ethmoidal nerve	Exits via the anterior ethmoidal foramen and enters the anterior cranial fossa on the cribriform plate of the ethmoid; then enters the nasal cavity via an opening opposite the crista galli to supply the mucosa of the nose; then supplies the skin of the nose as the external nasal nerve

Fig. 7.29 Branches of the nasociliary nerve.

Abducent nerve (VI)

The abducent nerve enters the orbit and supplies the lateral rectus muscle.

Ciliary ganglion

The ciliary ganglion is a parasympathetic ganglion situated posteriorly in the orbit, lateral to the optic nerve.

Preganglionic fibres from the Edinger–Westphal nucleus pass to the ganglion via the oculomotor nerve.

Postganglionic parasympathetic fibres pass to the back of the eyeball via the short ciliary nerves.

Sympathetic fibres (from the internal carotid plexus) pass through the ganglion to enter the short ciliary nerves. General sensory fibres leave the ganglion via the nasociliary nerve. The long ciliary nerve also carries sympathetic and sensory fibres to the eyeball.

The parasympathetic nerves supply the constrictor pupillae and ciliary muscle; the sympathetic fibres supply the dilator pupillae.

Motor innervation of the orbit is from the oculomotor nerve except for SO 4 and LR 6 (superior oblique—IV nerve; lateral rectus—VI nerve).

The parotid region

Parotid gland

The parotid gland is the largest of the major salivary glands. It lies between the ramus of the mandible and the sternocleidomastoid muscle (Fig. 7.30).

The gland is surrounded by a capsule derived from the investing layer of deep cervical fascia. The stylomandibular ligament is part of the fascia, running from the mandibular angle to the styloid process. It separates the parotid and submandibular glands.

The parotid duct emerges from the anterior border of the gland. It crosses the masseter muscle superficially and, at the anterior border of this muscle, it turns medially to pierce the buccal fat pad and buccinator to open into the oral cavity opposite the upper second molar tooth.

Structures within the parotid gland

The structures of the parotid gland are shown in Fig. 7.31.

Facial nerve (VII)

The facial nerve emerges from the stylomastoid foramen to enter the gland after giving off the posterior auricular nerve and muscular branches. It divides into its five terminal branches in the parotid gland.

A parotid tumour compresses the facial nerve weakening the facial muscles ipsilaterally (Bell's palsy). The corner of the mouth and eye may drop.

Retromandibular vein

The retromandibular vein is formed in the gland by the union of the superficial temporal and maxillary veins. It divides into anterior and posterior divisions, which leave the lower border of the gland. The anterior division joins the facial vein; the posterior division joins with the posterior auricular vein to form the external jugular vein.

External carotid artery

The external carotid artery enters the parotid gland by passing up from the carotid triangle. At the neck of the mandible it divides into its two terminal branches—the maxillary and superficial temporal arteries.

Blood supply, lymphatic drainage, and innervation of the parotid gland

Blood supply is from the external carotid artery and its terminal branches.

Parasympathetic secretomotor fibres from the inferior salivary nucleus of the glossopharyngeal nerve (IX) pass to the otic ganglion via the tympanic branch of the IX nerve and the lesser petrosal nerve. Postganglionic fibres pass to the parotid via the auriculotemporal nerve. The great auricular nerve

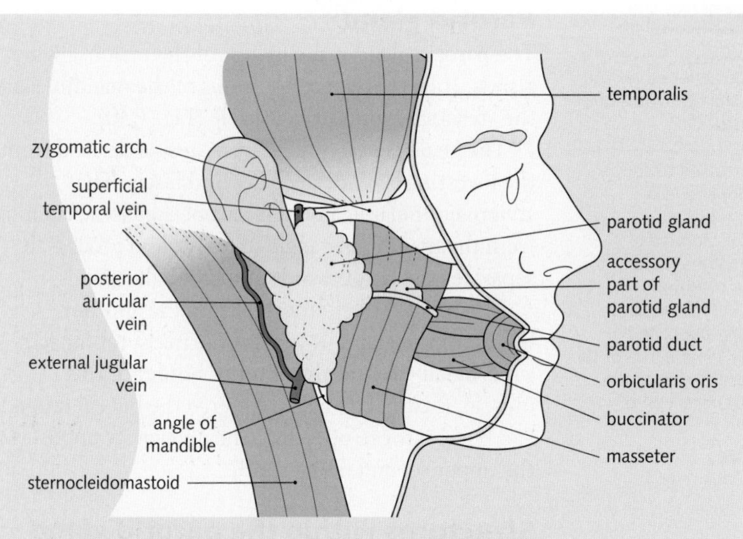

Fig. 7.30 Parotid gland and its relations. (Adapted from *Clinical Anatomy For Medical Students*, 4th edn, by R S Snell. Little Brown & Co.)

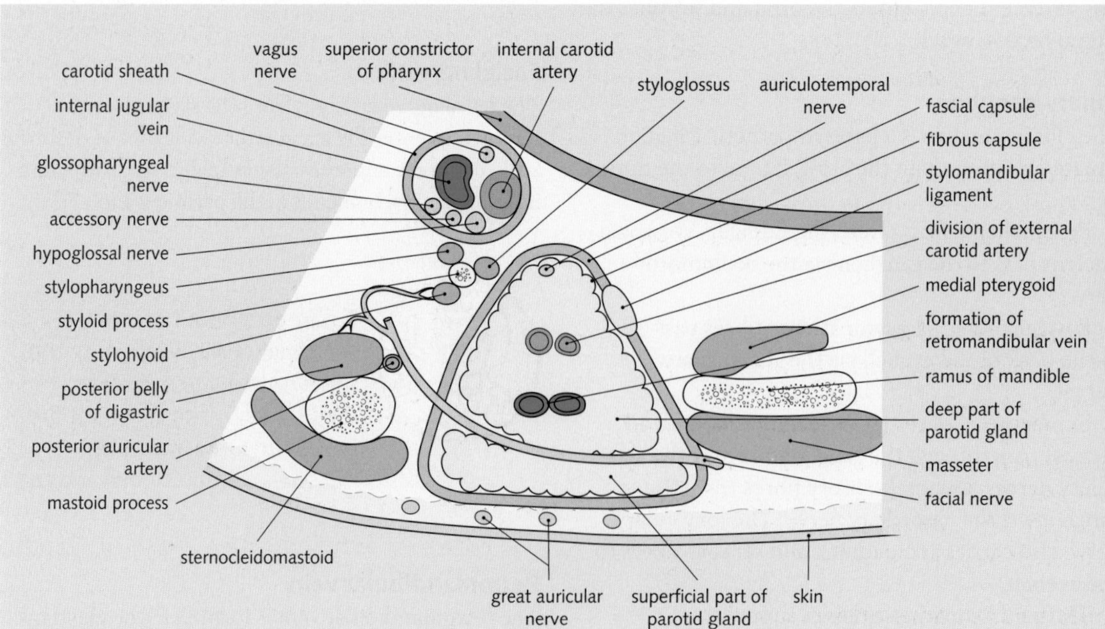

Fig. 7.31 Horizontal section of the neck, showing structures within the parotid gland. (Adapted from *Clinical Anatomy For Medical Students*, 4th edn, by R S Snell. Little Brown & Co.)

upplies sensory fibres to the gland capsule. The uriculotemporal nerve supplies sensory fibres to the land itself.

Parotid nodes drain the gland to the deep cervical odes.

The temporal and infratemporal fossae

Temporal fossa

The temporal fossa lies on the lateral aspect of the kull. It is bounded by the superior temporal line of he temporal bone superiorly, by the frontal process f the zygomatic bone anteriorly, and by the ygomatic arch inferiorly.

Contents of the temporal fossa are:

- The temporalis muscle and temporal fascia.
- The deep temporal nerves and vessels.
- The auriculotemporal nerve.
- The superficial temporal artery.

Temporal fascia

The temporal fascia covers the temporalis muscle bove the zygomatic arch. It is attached inferiorly to he zygomatic arch and superiorly to the superior emporal line.

Deep temporal nerves

Two or three nerves on each side arise from the nandibular nerve and emerge from the upper border f lateral pterygoid to enter and supply the emporalis muscle.

Deep temporal arteries

The deep temporal arteries are branches of the naxillary artery. They accompany the deep temporal erves.

Auriculotemporal nerve

The auriculotemporal nerve is a branch of the osterior division of the mandibular nerve. It merges from behind the temporomandibular joint, nd it crosses the root of the zygomatic arch behind he superficial temporal artery. It supplies the skin of he auricle, the external auditory meatus, and the calp over the temporal region.

Superficial temporal artery

The superficial temporal artery emerges from behind he temporomandibular joint, crosses the zygomatic rch, and ascends to the scalp.

Infratemporal fossa

The infratemporal fossa lies beneath the base of the skull between the pharynx and the ramus of the mandible (Fig. 7.32). It communicates with the temporal region deep to the zygomatic arch.

The infratemporal fossa contains (Fig. 7.33):

- The medial and lateral pterygoid muscles.
- Branches of the mandibular nerve.
- The otic ganglion.
- The chorda tympani.
- The maxillary artery.
- The pterygoid venous plexus.

Muscles of mastication

There are four large and powerful muscles of mastication (see Fig. 7.34). They can be clinically tested by asking the patient to:

- Clench their teeth (masseter and temporalis muscles).
- Move their chin (mental protuberance) from side to side (lateral and medial pterygoid muscles).

Mandible

Important features of the mandible are shown in Fig. 7.35. The two halves of the mandible unite at the midline symphysis menti.

Temporomandibular joint

The temporomandibular joint is the articulation between the condylar head of the mandible and the mandibular fossa of the temporal bone (Figs 7.36 and 7.37).

It is a synovial joint. The joint space is divided into upper and lower compartments by an articular disc attached to the lateral pterygoid muscle anteriorly, and to the capsule of the joint.

Boundaries of the infratemporal fossa

Boundary	Components
Anterior	Posterior surface of the maxilla
Posterior	Styloid process
Superior	Infratemporal surface of the greater wing of the sphenoid bone
Medial	Lateral pterygoid plate
Lateral	Ramus of the mandible

Fig. 7.32 Boundaries of the infratemporal fossa.

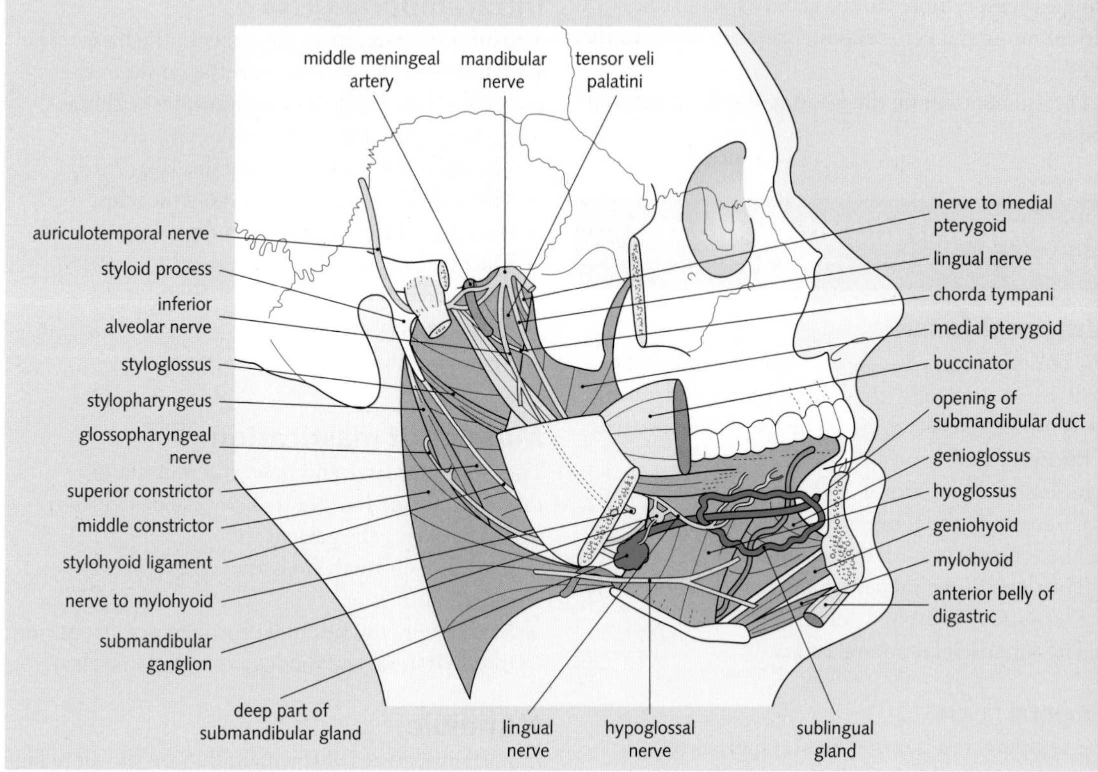

Fig. 7.33 Infratemporal fossa and its relations.

Muscles of mastication			
Muscle (nerve supply)	**Origin**	**Insertion**	**Action**
Temporalis (V_3 nerve)	Temporal fossa floor	Coronoid process	Elevates mandible; posterior fibres retract a protruded mandible
Masseter (V_3 nerve)	Zygomatic arch	Ramus of mandible	Elevates and protrudes mandible
Lateral pterygoid (V_3 nerve)			
Superior head	Infratemporal surface of sphenoid bone	Neck of the mandible	Acting together they protrude the mandible and pull the articular disc anteriorly; acting alone on one side produces deviation of mandible to contralateral side
Inferior head	Lateral surface of lateral pterygoid plate	Articular disc	
Medial pterygoid (V_3 nerve)			
Superficial head	Tuberosity of maxilla	Ramus and angle of the mandible	Acting together they elevate the mandible; acting alone on one side produces deviation of mandible to contralateral side
Deep head	Medial surface of lateral pterygoid plate	Ramus and angle of the mandible	

Fig. 7.34 Muscles of mastication.

The capsule surrounds the joint, and it is attached to the margins of the mandibular fossa and the neck of the mandible. It is strengthened by the lateral temporomandibular ligament. The sphenomandibular and stylomandibular ligaments are also functionally associated with the joint.

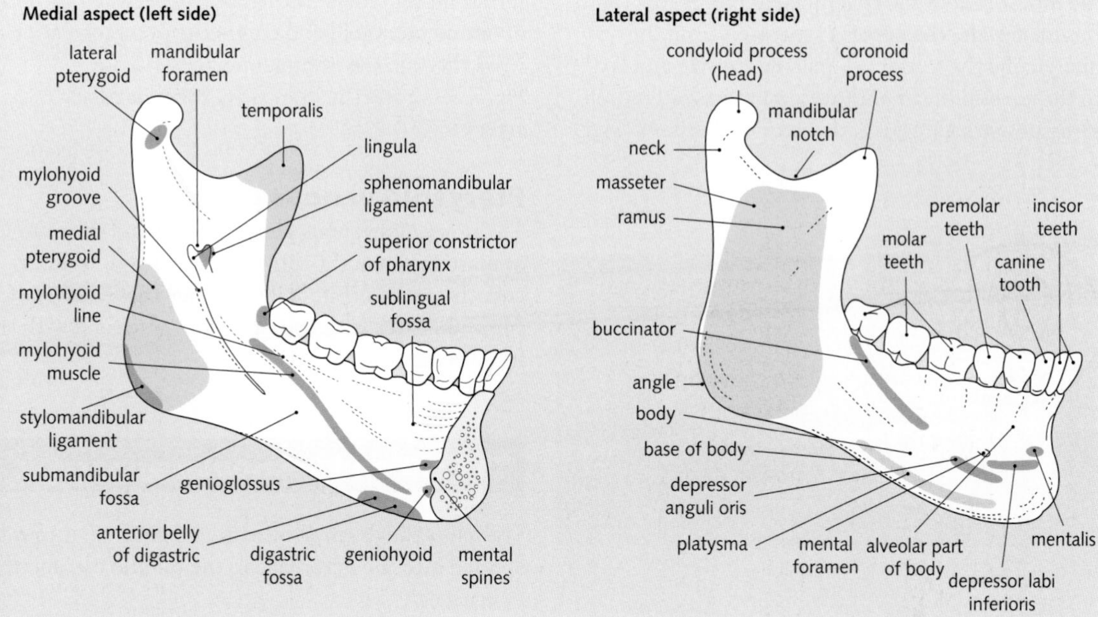

Fig. 7.35 Features of the mandible.

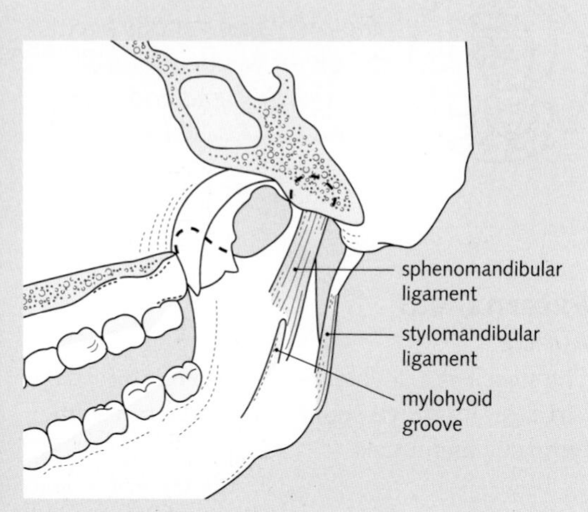

Fig. 7.36 Temporomandibular ligaments.

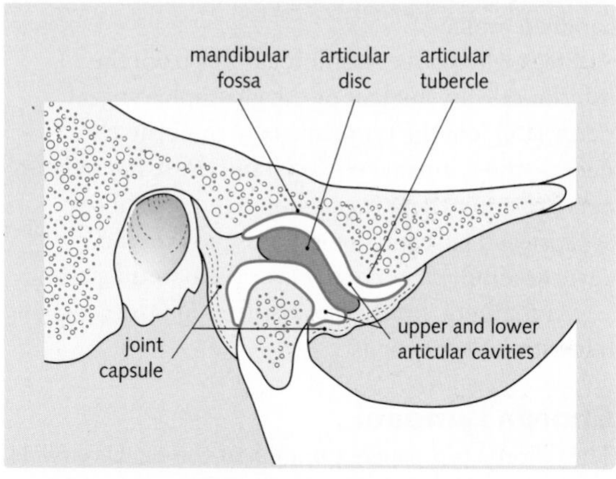

Fig. 7.37 Temporomandibular joint.

Hinge-like movements (elevation and depression) take place in the lower joint space (between the condyle and the articular disc). Gliding movements occur in the upper joint space (between the articular disc and the mandibular fossa).

When yawning, if the lateral pterygoid muscle contracts too much the mandibular head may pass anteriorly over the articular tubercle and dislocate the temporomandibular joint. An affected individual is unable to close their mouth.

Mandibular nerve

V_3 exits the skull through the foramen ovale to enter the infratemporal fossa, where it immediately joins

the motor root of the trigeminal nerve. Below the foramen ovale, the nerve is separated from the pharynx by the tensor veli palatini muscle and is deep to the superior head of the lateral pterygoid muscle. It divides into anterior and posterior divisions (Fig. 7.38).

Remember, the anterior division of the mandibular nerve is concerned with supplying the muscles of mastication except for its buccal branch which is sensory to the cheek skin, mucosa, and gingivae.

Otic ganglion

This is a parasympathetic ganglion lying below the foramen ovale.

Preganglionic secretomotor fibres from the inferior salivary nucleus of the glossopharyngeal nerve (IX) join the tympanic branch of the IX nerve—the tympanic plexus—and the lesser petrosal nerve to enter the otic ganglion. Here the fibres synapse and postganglionic fibres pass via the auriculotemporal nerve to enter the parotid gland.

Sympathetic fibres also pass through the ganglion, but without synapsing.

Chorda tympani

The Chorda tympani is a branch of the facial nerve in the temporal bone. It enters the infratemporal fossa via the petrotympanic fissure and joins the lingual nerve.

It transmits preganglionic parasympathetic secretomotor fibres to the submandibular ganglion and taste fibres from the anterior two thirds of the tongue. Cell bodies of the taste fibres are in the geniculate ganglion of the facial nerve and end by synapsing with cells of the nucleus solitarius in the pons.

Maxillary artery

The maxillary artery is the large terminal branch of the external carotid artery in the parotid gland. It runs forward medial to the neck of the mandible to the lower border of lateral pterygoid, entering the infratemporal fossa. It then passes between the heads of lateral pterygoid and enters the pterygopalatine fossa through the pterygomaxillary fissure.
Fig. 7.39 shows the branches of the maxillary artery.

Pterygoid venous plexus

The pterygoid venous plexus lies around the muscles of mastication in the infratemporal fossa. It drains veins from the orbit, oral cavity, and nasal cavity. It communicates with the cavernous sinus and with the facial vein.

The ear and vestibular apparatus

The ear is the organ of hearing and balance. It may be divided into the external ear, the middle ear, and the internal ear.

The pterygoid venous plexus is devoid of valves, as are all veins of the head and neck.

External ear

Auricle

The auricle is a double layer of skin reinforced by cartilage. It collects sound and conducts it to the tympanic membrane.

External auditory (acoustic) meatus

The external auditory meatus extends from the auricle to the tympanic membrane (Fig. 7.40). The lateral third is cartilaginous and the medial two thirds are bony. It is lined by a layer of thin skin. Ceruminous and sebaceous glands produce cerumen (wax).

Tympanic membrane

The tympanic membrane is a thin membrane lying between the external and middle ears (see Fig. 7.40). It is covered by skin externally and by mucous membrane internally. The membrane shows a

Branches of the mandibular nerve	
Branch	**Area supplied**
Main trunk	
Meningeal branch	Re-enters cranial cavity via foramen spinosum
Nerve to medial pterygoid	Medial pterygoid and a branch that passes through otic ganglion to supply tensor tympani and tensor veli palatini
Anterior division (motor except the buccal nerve)	
Deep temporal nerves	Temporalis muscle (see temporal region)
Masseteric nerve	Passes through mandibular notch to supply masseter muscle
Nerve to lateral pterygoid	Enters deep surface of lateral pterygoid and supplies it
Buccal nerve	Passes anteriorly between heads of lateral pterygoid to appear at anterior border of masseter; is sensory to skin of cheek and underlying buccal mucosa and gingiva
Posterior division (mainly sensory)	
Lingual nerve	Appears at lower border of lateral pterygoid and runs over superior surface of medial pterygoid to lie just beneath mucosa lining inner aspect of mandible adjacent to 3rd molar tooth (its subsequent course is described later); deep to lateral pterygoid, the nerve receives the chorda tympani
Inferior alveolar nerve	Runs parallel with lingual nerve over medial pterygoid; enters mandibular foramen and supplies teeth of lower jaw; at mental foramen, a branch of the nerve, the mental nerve, exits mandible to supply lower lip and chin region; mylohyoid nerve arises from inferior alveolar nerve just above mandibular foramen to supply mylohyoid and anterior belly of digastric
Auriculotemporal nerve	See temporal region

Fig. 7.38 Branches of the mandibular nerve and the areas they supply.

concavity towards the meatus, with a central depression—the umbo.

The membrane moves in response to air vibration. Movements are transmitted by the ossicles through the middle ear to the internal ear.

The auriculotemporal nerve supplies the external surface of the tympanic membrane. The glossopharyngeal nerve supplies the internal surface.

Middle ear

The middle ear lies in the petrous temporal bone. It consists of the tympanic cavity and the epitympanic recess, which lies superior to the tympanic cavity.

It is connected to the nasopharynx via the auditory tube and to the mastoid air cells via the mastoid antrum. The mucosa lining the tympanic cavity is continuous with that of the auditory tube, mastoid cells, and the mastoid antrum.

The middle ear contains:

- The ossicles (malleus, incus, and stapes).
- Stapedius and tensor tympani muscles.
- The chorda tympani.
- The tympanic plexus of nerves.

Fig. 7.41 describes the walls of the middle ear.

Mastoid antrum

The aditus to the antrum connects the mastoid antrum to the epitympanic recess of the tympanic cavity. The tegmen tympani separates the antrum from the middle cranial fossa. The floor of the antrum communicates with the mastoid air cells via

Branches of the maxillary artery		
Branch	**Site of origin**	**Area supplied**
Deep auricular artery	Behind neck of mandible	External auditory meatus and outer surface of eardrum
Anterior tympanic artery	Behind neck of mandible	Inner surface of eardrum via petrotympanic fissure
Middle meningeal artery	Infratemporal fossa	Enters cranial cavity via foramen spinosum to supply meninges
Inferior alveolar artery	Infratemporal fossa	Follows inferior alveolar nerve into mandibular canal and supplies lower jaw and teeth, and surrounding mucosa
Deep temporal arteries Masseteric artery Pterygoid branches	Infratemporal fossa	Muscles of mastication
Posterior superior alveolar artery	Pterygopalatine fossa	Enters posterior aspect of maxilla to supply molar and premolar teeth of maxilla
Infraorbital artery	Pterygopalatine fossa	Accompanies infraorbital nerve through infraorbital foramen onto face; reaches foramen by passing forward in infraorbital canal in orbital floor
Anterior superior alveolar artery	Infraorbital canal	Incisor and canine teeth
Palatine Sphenopalatine Pharyngeal branches	Pterygopalatine fossa	Described with the nasal cavity

Fig. 7.39 Branches of the maxillary artery.

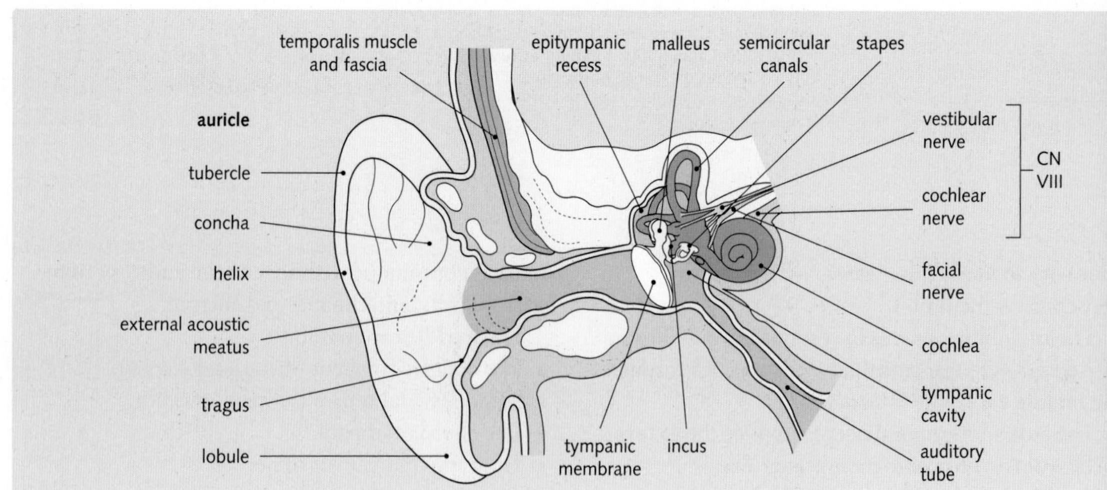

Fig. 7.40 External auditory meatus.

several openings. The antrum and air cells are lined by mucosa. Anteroinferiorly the antrum is related to the canal for the facial nerve.

Auditory tube

The auditory tube connects the tympanic cavity to the nasopharynx. The posterior third is bony and the

emainder is cartilaginous. The mucosa is ontinuous with that of the tympanic cavity and asopharynx.

It equalizes pressure in the middle ear with tmospheric pressure, allowing free movement f the tympanic membrane. Pressure changes, e.g. uring flying, can be equalized by swallowing or hewing—these movements open the auditory ubes.

Nerve supply is from the tympanic plexus formed y the glossopharyngeal nerve.

The auditory tube provides a passage for infection to spread from the nasopharynx to the tympanic cavity (middle ear).

Ossicles

The ossicles are the incus, malleus, and stapes. The nalleus is attached to the tympanic membrane. The ncus connects the malleus to the stapes, which is ttached to the oval window (Fig. 7.42).

Walls of the middle ear

Wall	Components
Roof (tegmental wall)	Tegmen tympani (thin plate of bone): separates cavity from dura in floor of middle cranial fossa
Floor (jugular wall)	A layer of bone separates tympanic cavity from superior bulb of internal jugular vein
Lateral wall (membranous)	Tympanic membrane with epitympanic recess superiorly
Medial wall (labyrinthine)	Separates tympanic cavity from inner ear
Anterior wall (carotid)	Separates tympanic cavity from carotid canal; superiorly lies opening of auditory tube and canal for tensor tympani
Posterior wall	Connected by aditus to mastoid antrum and air cells

Fig. 7.41 Walls of the middle ear.

The ossicles transmit vibration from the tympanic membrane to the oval window.

There are two muscles associated with the ossicles: tensor tympani dampens vibration of the tympanic membrane, and stapedius dampens vibration of the stapes.

They are innervated by the medial pterygoid nerve (V_3) and VII cranial nerves, respectively.

Internal ear

This lies in the petrous temporal bone (Fig. 7.43). It consists of a bony labyrinth and a membranous labyrinth. The two are separated by a space containing fluid called perilymph, which resembles cerebrospinal fluid (CSF).

Bony labyrinth

Cochlea

This contains the cochlear duct, and it is concerned with hearing. It makes 2.5 turns about a bony core—the modiolus. The large basal turn of the cochlea produces the promontory on the medial wall of the tympanic cavity.

Vestibule

The vestibule contains the utricle and saccule, components of the balance system. It is continuous with the cochlea anteriorly, with the semicircular canals posteriorly, and with the posterior cranial fossa by the aqueduct of the vestibule. The aqueduct extends to the posterior surface of the petrous

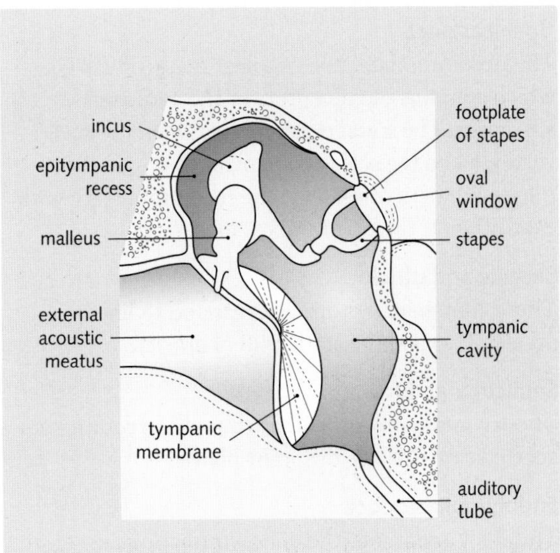

Fig. 7.42 Coronal section of the tympanic cavity showing the ossicles in situ.

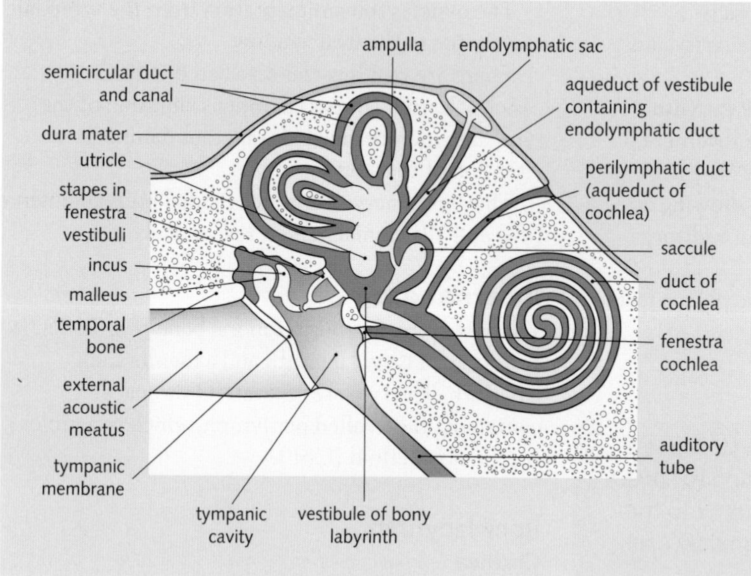

Fig. 7.43 Internal ear.

temporal bone to open into the internal auditory meatus. It contains the endolymphatic ducts and blood vessels.

Semicircular canals

These three communicate with the vestibule, and they are each perpendicular to one other. At one end of each canal is a swelling—the ampulla. The semicircular ducts lie in the canals.

Membranous labyrinth

This is a series of ducts and sacs in the bony labyrinth, which contain endolymph.

Cochlear duct

This accommodates the cochlear duct (of Corti), which contains the receptors of the auditory apparatus. The spiral organ lies between the scala vestibuli and the scala tympani, both of which are filled with perilymph and which communicate with each other at the tip of the cochlea.

Saccule and utricle

These contain receptors that respond to linear acceleration and the static pull of gravity.

Semicircular ducts

These contain receptors that respond to rotational acceleration in three different planes.

Endolymphatic duct

This duct opens into the endolymphatic sac. Endolymph has a similar composition to intracellular fluid.

Vestibulocochlear nerve (VIII)

Near the lateral end of the internal auditory meatus, the VIII nerve divides into an anterior cochlear nerve (hearing) and a posterior vestibular nerve (balance), as shown in Fig. 7.44. The vestibular nerve enlarges to form the vestibular ganglion, and its fibres supply receptors in the semicircular ducts, the saccule, and the utricle. The cochlear nerve forms the spiral ganglion and supplies the spiral organ.

Facial nerve in the temporal bone

The facial nerve (VII) and its sensory root—the nervus intermedius—enter the internal auditory meatus together with the VIII nerve. The two roots fuse and enter the facial canal to pass above the internal ear to reach the medial wall of the middle ear. The nerve then turns sharply posteriorly above the promontory and passes posteriorly to the posterior wall, where it turns downwards to leave the temporal bone through the stylomastoid foramen. The sensory geniculate ganglion lies at the sharp bend that the nerve makes on entering the middle ear.

Branches in the temporal bone

Branches in the temporal bone comprise:

- The greater petrosal nerve. This branches off at the geniculate ganglion and enters the middle cranial fossa. It is joined by the deep petrosal nerve to form the nerve of the pterygoid canal.

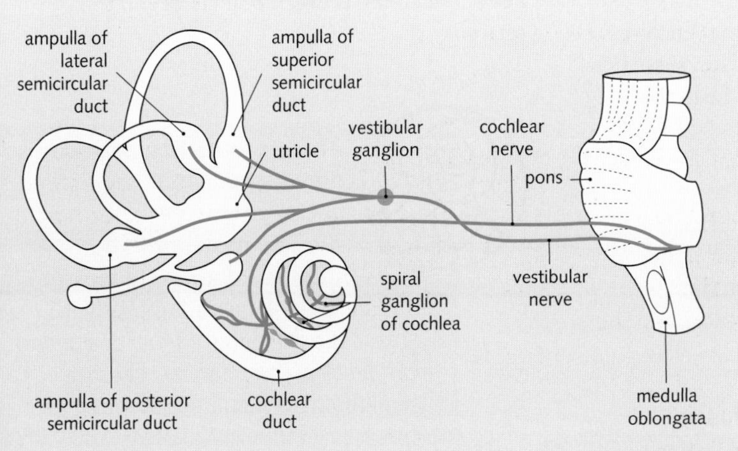

Fig. 7.44 Vestibulocochlear nerve.

- The nerve to stapedius.
- The chorda tympani. This is given off just above the stylomastoid foramen. It passes to the lateral wall of the middle ear, crosses the deep surface of the tympanic membrane, and enters a canal leading to the petrotympanic fissure. It joins the lingual nerve in the infratemporal fossa.

The soft tissues of the neck

The neck is the region between the head and the thorax.

Fascial layers of the neck

The fascial layers of the neck are illustrated in Fig. 7.45.

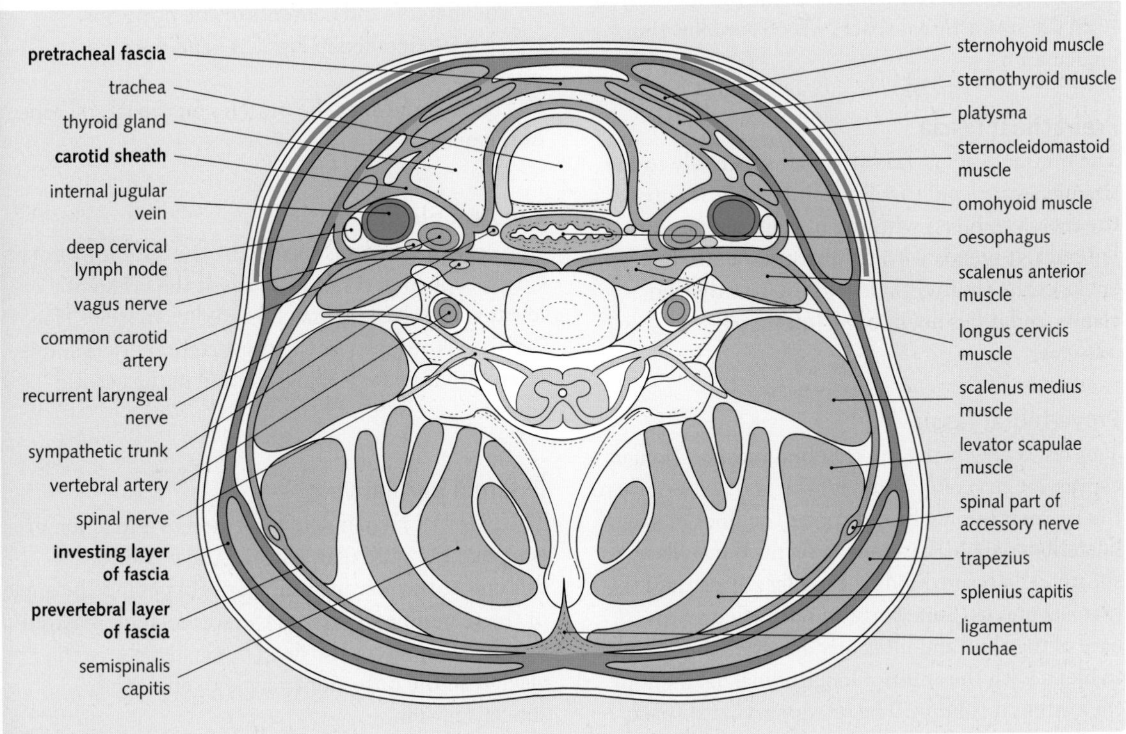

Fig. 7.45 Fascial layers of the neck.

Superficial fascia

The superficial fascia is a thin layer that encloses the platysma muscle. The cutaneous nerves, superficial vessels, and superficial lymph nodes lie in the fascia.

Deep fascia

The deep fascia lies beneath the superficial fascia. It is condensed in certain regions to form the investing layer of deep cervical fascia, the pretracheal fascia, the prevertebral fascia, and the carotid sheath.

Investing layer of deep cervical fascia

This completely encircles the neck, splitting to enclose the sternocleidomastoid and trapezius muscles. Posteriorly it is attached to the ligamentum nuchae. Superiorly it is attached to the hyoid bone. Above this it splits to enclose the submandibular gland, and it then attaches to the lower border of the mandible. The fascia also splits to enclose the parotid gland and is attached to the zygomatic arch and the base of the skull. The stylomandibular ligament is a thickening of the fascia between the angle of the mandible and the styloid process.

Inferiorly the fascia is attached to the acromion, the clavicle, and the manubrium. It attaches to the anterior and posterior borders of the manubrium to form the suprasternal space, which contains the jugular arch.

Pretracheal fascia

The pretracheal fascia is attached superiorly to the thyroid and cricoid cartilages. Inferiorly, it enters the thorax to blend with the fibrous pericardium. Laterally, it blends with the carotid sheath. It encloses the thyroid gland and the parathyroid glands, and it lies deep to the infrahyoid muscles.

Prevertebral fascia

This fascia covers the prevertebral muscles (longus capitis, longus colli), and it is attached posteriorly to the ligamentum nuchae. It forms the floor of the posterior triangle. Laterally, it forms the axillary sheath, which surrounds the axillary artery and the brachial plexus. Superiorly, it is attached to the base of the skull and inferiorly, it enters the thorax to blend with the anterior longitudinal ligament of the vertebral column. The retropharyngeal space lies between the prevertebral fascia and the pharynx.

Abscess formation behind the prevertebral fascia can extend laterally in the neck, forming a swelling posterior to the sternocleidomastoid muscle. If it pierces the fascia anteriorly it enters the retropharyngeal space and can narrow the pharynx, causing swallowing difficulties (dysphagia).

Carotid sheath

The carotid sheath is a condensation of the fascia surrounding the common and internal carotid arteries, the internal jugular vein, the deep cervical chain of nodes, and the vagus nerve. It extends from the base of the skull to the root of the neck.

Posterior triangle of the neck

The inferior belly of omohyoid divides the posterior triangle into a large occipital triangle and a small supraclavicular triangle (Fig. 7.46).

The margins and contents of the posterior triangle are detailed in Figs 7.47 and 7.48, respectively.

Fig. 7.49 outlines the muscles on the lateral aspect of the neck.

Cervical plexus

The cervical plexus is formed by the anterior rami of C1–C4 spinal nerves, and it lies at the origin of levator scapulae and scalenus medius muscles (Fig. 7.50). It is covered by the prevertebral fascia and is related to the internal jugular vein in the carotid sheath.

External jugular vein

The external jugular vein is formed by the union of the posterior auricular vein and the posterior division of the retromandibular vein behind the angle of the mandible. It crosses the sternocleidomastoid muscle and pierces the deep fascia just above the clavicle in the posterior triangle to enter the subclavian vein.

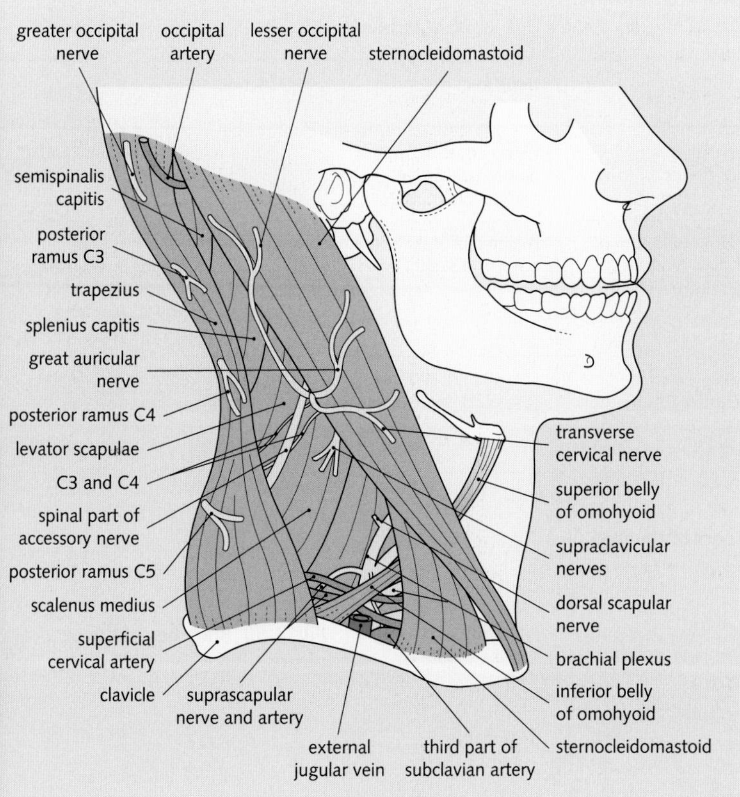

Fig. 7.46 Posterior triangle of the neck.

Margins of the posterior triangle	
Margin	**Components**
Anterior	Posterior border of sternomastoid
Posterior	Anterior border of trapezius
Inferior	Middle third of clavicle
Roof	Skin, superficial fascia, platysma, investing layer of deep fascia
Floor	Prevertebral fascia covering muscles of floor

ig. 7.47 Margins of the posterior triangle.

ᴬnterior triangle of the neck

ᵀhe anterior triangle is formed by the anterior border ◦f sternocleidomastoid muscle, the midline of the ◦eck, and the inferior border of the mandible. It is ◦ubdivided by the anterior and posterior belly of ◦igastric and the superior belly of omohyoid into the

Contents of the posterior triangle	
Structure	**Origin**
Third part of subclavian artery	Enters anterior inferior angle of triangle
Superficial cervical artery	Branch of thyrocervical trunk of subclavian artery
Suprascapular artery	Branch of thyrocervical trunk
Brachial plexus	Roots of plexus enter posterior triangle by emerging between scalenus anterior and medius; trunks and divisions also lie in posterior triangle before entering the axilla
Accessory nerve	Spinal part of accessory nerve enters posterior triangle by emerging from deep to posterior border of sternocleidomastoid
Cervical plexus	The four cutaneous branches emerge from posterior border of sternocleidomastoid

Fig. 7.48 Contents of the posterior triangle.

Major muscles of the lateral aspect of the neck			
Name of muscle (nerve supply)	**Origin**	**Insertion**	**Action**
Platysma (VII nerve)	Inferior border of mandible; skin and subcutaneous tissues of lower part of the face	Fascia covering superior parts of pectoralis major and deltoid muscles	Used to express sadness and fright by pulling angles of mouth down
Sternocleidomastoid [XI nerve (spinal part), C2, C3]	Anterior surface of manubrium of sternum; medial third of clavicle	Mastoid process of temporal bone and superior nuchal line	Individually each muscle laterally flexes neck and rotates it so face is turned upwards toward opposite side; both muscles act together to flex neck
Trapezius [XI nerve (spinal part), C2, C3]	Superior nuchal line; external occipital protuberance; ligamentum nuchae; spinous processes of C7–T12 vertebrae	Lateral third of clavicle; acromion; spine of scapula	Elevates, retracts, and rotates scapula

Fig. 7.49 Major muscles of the lateral aspect of the neck.

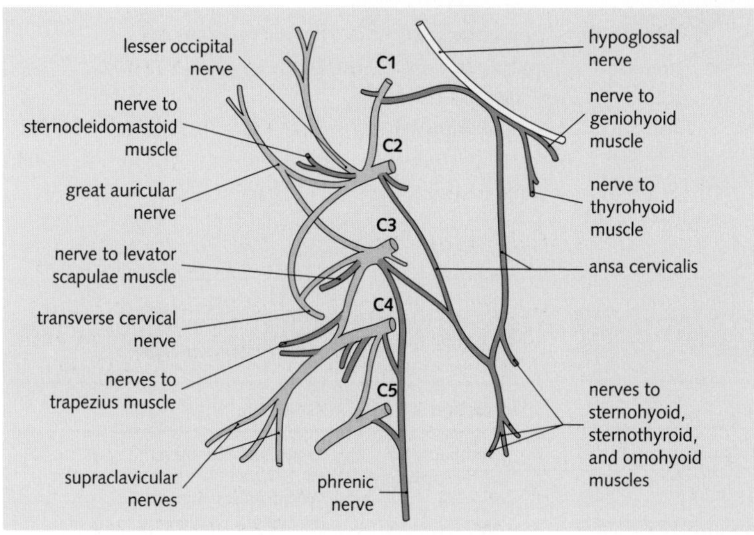

Fig. 7.50 Branches of the cervical plexus.

digastric (submandibular), carotid, and muscular triangles (Figs 7.51 and 7.52). The boundaries of these triangles are as follows:

- Carotid triangle-sternocleidomastoid, posterior belly of digastric, inferior belly of omohyoid muscles
- Digastric (or submandibular) triangle-mandible, anterior and posterior bellies of digastric muscle
- Submental-midline, hyoid bone and anterior belly of the digastric muscle
- Muscular-hyoid bone, sternocleidomastoid, and superior belly of omohyoid muscles

The muscles of the anterior triangle are shown in Fig. 7.53, and the contents are detailed in Figs 7.51 and 7.52.

Vessels of the anterior triangle

Common carotid artery

The left common carotid artery arises from the aortic arch, the right from the brachiocephalic trunk. Both ascend in the neck deep to the sternocleidomastoid muscle behind the sternoclavicular joint. At the level of the upper border of the thyroid cartilage the arteries divide into the external and internal carotid arteries (Fig. 7.54).

Contents of the anterior triangle of the neck	
Triangle	**Main contents**
Carotid	External carotid artery; larynx and pharynx, and internal and external laryngeal nerves
Muscular	Sternothyroid and sternohyoid muscles, superior belly of omohyoid; thyroid gland, trachea, and oesophagus
Digastric (submandibular)	Submandibular gland and lymph nodes; facial artery and vein; external carotid artery; internal carotid artery, internal jugular vein, glossopharyngeal (IX), vagus (X), and hypoglossal (XII) nerves
Submental	Submental lymph nodes

ig. 7.51 Contents of the anterior triangle of the eck.

At the terminal part of the common carotid artery nd the origin of the internal carotid artery there is a ocalized dilatation, the carotid sinus. The sinus ontains baroreceptors that respond to changes in rterial pressure.

The carotid body is embedded in the tunica dventitia of the artery. It contains chemoreceptors hat monitor blood carbon dioxide levels.

Both the carotid sinus and the carotid body are nnervated by the carotid sinus branch of the IX erve.

The common carotid pulse can be palpated at the upper border of the thyroid cartilage (C3, C4 vertebral levels), anterior to the sternocleidomastoid muscle.

xternal carotid artery

his commences at the upper border of the hyroid cartilage and ascends to enter the arotid. Its branches comprise the following rteries:

- Ascending pharyngeal.
- Superior thyroid.
- Lingual.
- Facial.
- Occipital.
- Posterior auricular.
- Superficial temporal.
- Maxillary.

Internal carotid artery

This commences at the upper border of the thyroid cartilage, and it ascends in the carotid sheath to the carotid canal in the base of the skull. It supplies the cerebral hemispheres and the orbital contents.

Unlike the external carotid artery, the internal carotid has no branches outside the skull.

Internal jugular vein

This commences at the end of the sigmoid sinus, and it leaves the cranial cavity through the jugular foramen. It descends through the neck in the carotid sheath, at first posterior to the carotid artery and then lateral to it. It unites with the subclavian vein to form the brachiocephalic vein behind the sternoclavicular joint.

The vein has dilatations at its upper and lower ends—the superior and inferior bulbs, respectively.

Tributaries include the inferior petrosal sinus, the facial vein, the pharyngeal vein, the lingual vein, and the superior and middle thyroid veins.

Deep cervical nodes

These form a chain along the internal jugular vein in the carotid sheath. They drain the entire head and neck region. Efferent vessels join to form the jugular lymph trunk, which in turn drains into the thoracic duct, the right lymph duct, or the subclavian trunk (see Fig. 7.15).

Nerves of the triangles of the neck

The following nerves are all found in the anterior triangle except the accessory nerve, which is in the posterior triangle.

Glossopharyngeal nerve (IX)

This runs between the two carotid arteries and passes between the superior and middle constrictors to supply sensory and taste fibres to the posterior third of the tongue and oropharynx. Its motor branch supplies the stylopharyngeus muscle.

Vagus nerve (X)

The vagus exits the skull through the jugular foramen. It has superior and inferior sensory ganglia. Below the superior ganglion the cranial part of the accessory nerve joins the

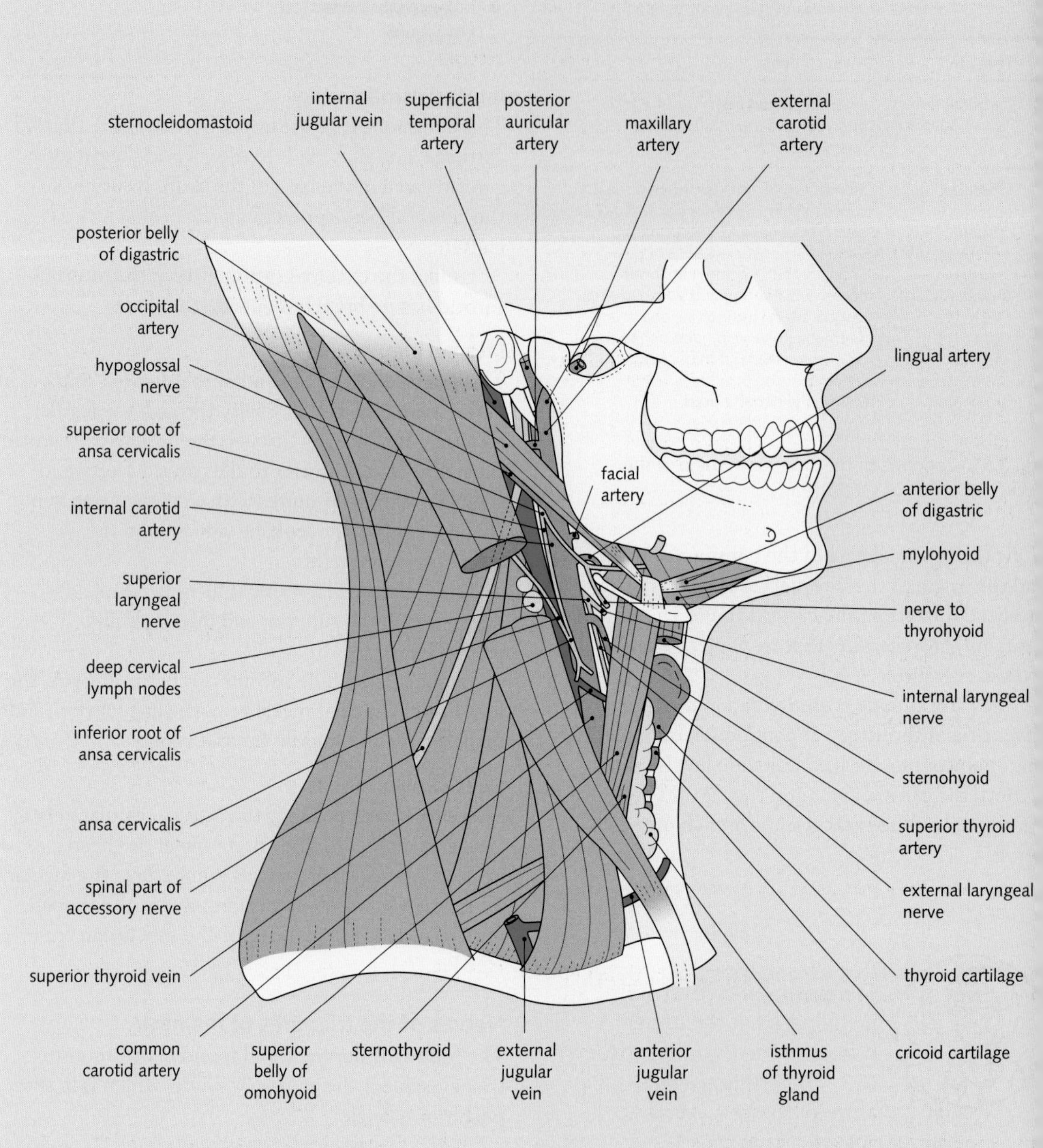

Fig. 7.52 Anterior triangle of the neck.

vagus, and it is distributed to the pharyngeal and recurrent laryngeal nerves (Fig. 7.55). The X nerve descends in the neck in the carotid sheath between the internal carotid artery and internal jugular vein. At the root of the neck the nerve passes anterior to the first part of the subclavian artery to enter the thorax.

Accessory nerve (XI)

The spinal part of the accessory nerve arises from th upper five or six cervical segments. The roots ascend in the vertebral canal and enter the skull via the foramen magnum.

The cranial root arises from the medulla oblongata. The two roots unite and exit the skull via the jugular foramen.

Suprahyoid and infrahyoid muscles			
Suprahyoid muscles			
Name of muscle (nerve supply)	**Origin**	**Insertion**	**Action**
Posterior belly of digastric (VII nerve)	Mastoid process	Intermediate tendon bound to hyoid bone	Depresses mandible and elevates hyoid bone
Anterior belly of digastric (inferior alveolar V_3 nerve)	Lower border of mandible near midline	Intermediate tendon bound to hyoid bone	Depresses mandible and elevates hyoid bone
Stylohyoid (VII nerve)	Styloid process of temporal bone	Body of hyoid bone	Elevates hyoid bone
Mylohyoid (inferior alveolar V_3 nerve)	Mylohyoid line on medial surface of mandible	Body of hyoid bone and mylohyoid raphe	Elevates floor of mouth and hyoid bone, and depresses mandible
Geniohyoid (C1 through XII nerve)	Inferior mental spine	Body of hyoid bone	Elevates hyoid bone and depresses mandible
Infrahyoid muscles			
Name of muscle (nerve supply)	**Origin**	**Insertion**	**Action**
Sternohyoid (ansa cervicalis C1–C3)	Manubrium sterni and clavicle	Body of hyoid bone	Depresses hyoid bone
Sternothyroid (ansa cervicalis C1–C3)	Manubrium sterni	Oblique line on lamina of thyroid cartilage	Depresses larynx
Thyrohyoid (C1 through XII nerve)	Oblique line on lamina of thyroid cartilage	Body of hyoid bone	Depresses hyoid bone and elevates larynx
Omohyoid—inferior belly (ansa cervicalis C1–C3)	Upper margin of scapula	Intermediate tendon bound to clavicle and first rib	Depresses hyoid bone
Omohyoid—superior belly (ansa cervicalis C1–C3)	Body of hyoid bone	Intermediate tendon bound to clavicle and first rib	Depresses hyoid bone

ig. 7.53 Suprahyoid and infrahyoid muscles. (Adapted from *Anatomy as a Basis for Clinical Medicine*, by E C B Hall-Craggs. Williams & Wilkins.)

The cranial root joins the vagus; the spinal oot supplies sternocleidomastoid and rapezius.

Remember CN XI supplies trapezius and sternocleidomastoid. If damaged, patients cannot shrug their shoulders or turn their head against resistance.

Hypoglossal nerve (XII)

This descends in the neck between the internal carotid artery and the internal jugular vein. At the lower border of the digastric muscle, the nerve loops around the occipital artery and crosses the internal and external carotid arteries to enter the submandibular region. It is motor to the muscles of the tongue.

The XII nerve is joined by fibres of C1 spinal nerve. These are given off in the superior root of the ansa cervicalis, which joins the inferior root of the ansa cervicalis (C2–C3) to form the ansa cervicalis. This is motor to omohyoid, sternohyoid, and sternothyroid. Other C1 branches from the XII nerve are:

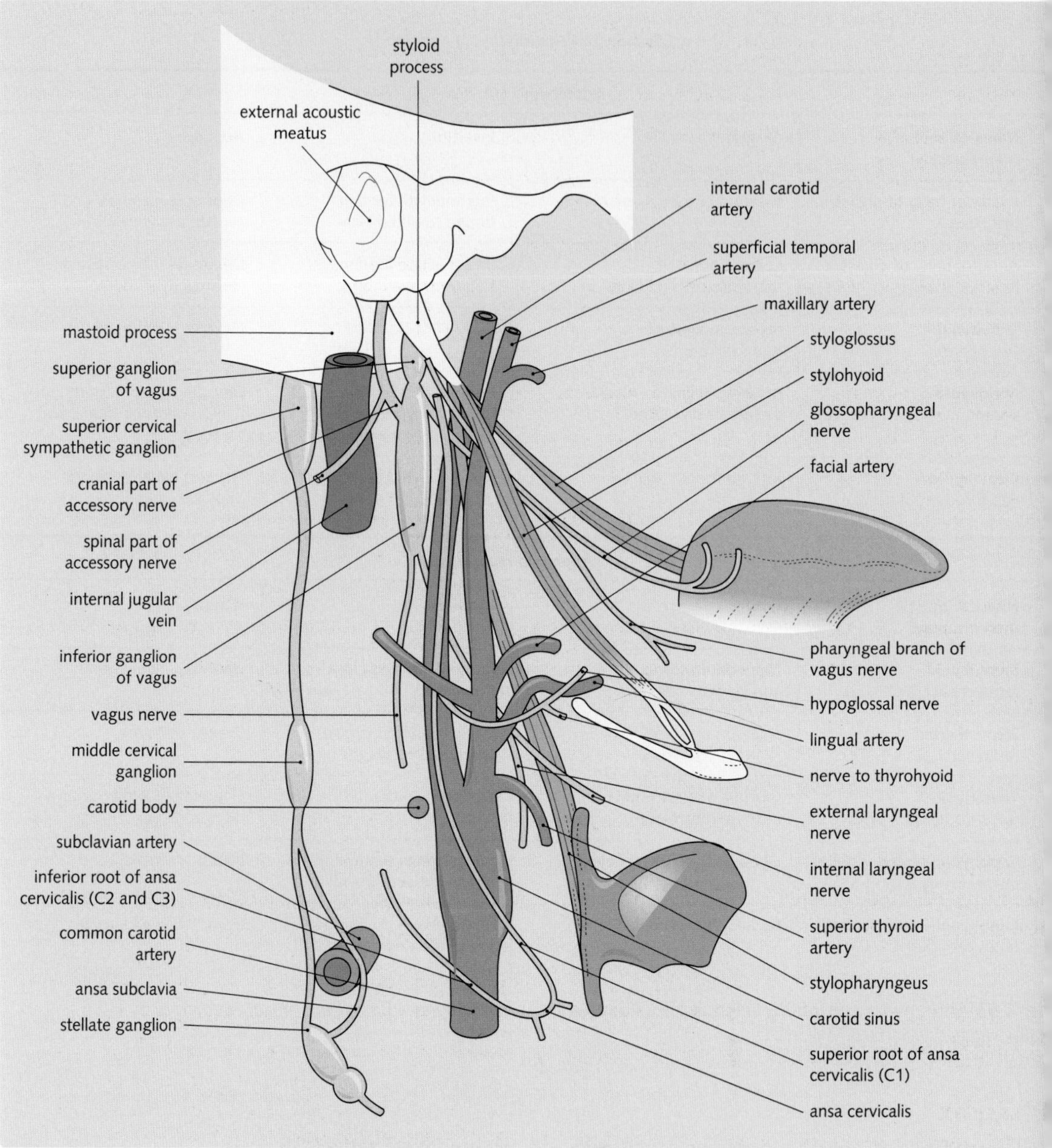

Fig. 7.54 The common carotid artery and the lower cranial nerves. (Adapted from *Clinical Anatomy For Medical Students*, 4th edn, by R S Snell. Little Brown & Co.)

- Nerve to thyrohyoid.
- Nerve to geniohyoid.

Sympathetic trunk

The trunk lies deep in the neck, between the carotid sheath and prevertebral fascia. It has superior, middle, and inferior ganglia. The inferior ganglion usually fuses with the first thoracic ganglion to form the stellate ganglion. Postganglionic fibres form plexuses around the

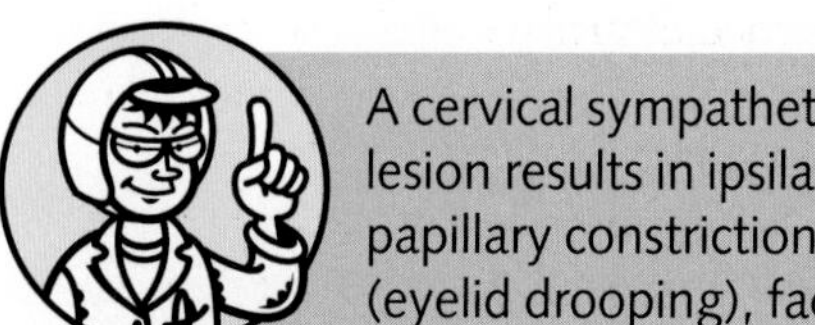

A cervical sympathetic trunk lesion results in ipsilateral papillary constriction, ptosis (eyelid drooping), facial and neck vasodilation, and lack of sweating. This is due to an interrupted sympathetic nerve supply; it is called Horner's syndrome.

Branches of the vagus nerve	
Branch	**Course and distribution**
Meningeal branch	Dura mater of posterior cranial fossa
Auricular branch	Medial surface of auricle, external auditory meatus, and adjacent tympanic membrane
Pharyngeal branch	Contains motor fibres from XI nerve (cranial part); combines with pharyngeal branches of IX nerve (sensory fibres) to form pharyngeal plexus, which supplies all pharyngeal muscles except stylopharyngeus (IX) and all soft-palate muscles except tensor veli palatini (V_3)
Superior laryngeal nerve divides into internal and external laryngeal nerves	Internal laryngeal nerve is sensory to piriform fossa and mucosa of larynx above vocal folds; external laryngeal nerve is motor to cricothyroid muscle
Cardiac branches	Assist in forming cardiac plexus in thorax
Right recurrent laryngeal nerve	Arises from X nerve as it crosses subclavian artery; hooks backwards and upwards behind artery and ascends in a groove between trachea and oesophagus; supplies all laryngeal muscles (except cricothyroid) and laryngeal mucosa below vocal folds, trachea, and oesophagus
Left recurrent laryngeal nerve	Arises from X nerve as it crosses aortic arch; hooks beneath arch behind ligamentum arteriosum and passes into neck between trachea and oesophagus; has a similar distribution to right nerve

Fig. 7.55 Branches of the vagus nerve.

najor vessels and supply the structures of the head ınd neck, e.g. blood vessels and glands. The trunks ılso give off cardiac branches.

Midline structures of the face and neck

Pharynx

The pharynx is a fibromuscular tube lying behind he nasal cavity (nasopharynx), the oral cavity oropharynx), and the larynx (laryngopharynx). It extends from the base of the skull to the inferior border of the cricoid cartilage (C6 vertebra level), where it is continuous with the oesophagus. There are three layers in the pharyngeal wall:

- The muscular layer is formed by the pharyngeal constrictors and longitudinal muscles (Figs 7.56 and 7.57).
- The pharyngobasilar fascia separates the mucosa and the muscle layer. It blends with the periosteum of the base of the skull.
- The mucous membrane (Fig. 7.58).

Nasopharynx

The nasopharynx lies behind the nasal cavity above the soft palate. During swallowing, the soft palate elevates and the pharyngeal wall is pulled forward to form a seal, preventing food entering the nasopharynx. The pharyngeal tonsil lies in the posterior wall. Tubal elevations in the lateral wall bear tonsillar tissue, and they are the sites of opening of the auditory tube. The tubal recess is a small depression in the lateral wall, behind the tubal elevation. Anteriorly the nasopharynx is continuous with the nose through choanae, superiorly and bordered by the soft palate, inferiorly.

Pharyngeal mucosa can bulge between the thyropharyngeus and cricopharyngeal muscles to form a pharyngeal pouch (Killian's dehiscence).

Muscles of the pharynx			
Name of muscle (nerve supply)	**Origin**	**Insertion**	**Action**
Superior constrictor (pharyngeal plexus)	Medial pterygoid plate, pterygoid hamulus, pterygomandibular raphe, mylohyoid line of mandible	Pharyngeal tubercle of occipital bone, midline pharyngeal raphe	Assists in separating oro- and nasopharynx and propels food bolus downward
Middle constrictor (pharyngeal plexus)	Stylohyoid ligament, lesser and greater cornua of hyoid bone	Pharyngeal raphe	Propels food bolus downward
Inferior constrictor (pharyngeal plexus)			
Thyropharyngeus	Lamina of thyroid cartilage	Pharyngeal raphe	Propels food bolus downward
Cricopharyngeus	Cricoid cartilage	Contralateral cricopharyngeus	Upper oesophageal sphincter
Palatopharyngeus (pharyngeal plexus)	Palatine aponeurosis Horizontal plate of palatine bone	Thyroid cartilage	Elevates pharyngeal wall and pulls palatopharyngeal folds medially
Salpingopharyngeus (pharyngeal plexus)	Auditory tube	Merges with palatopharyngeus	Elevates pharynx and larynx
Stylopharyngeus (IX)	Styloid process of temporal bone	Thyroid cartilage	Elevates larynx during swallowing

Fig. 7.56 Muscles of the pharynx.

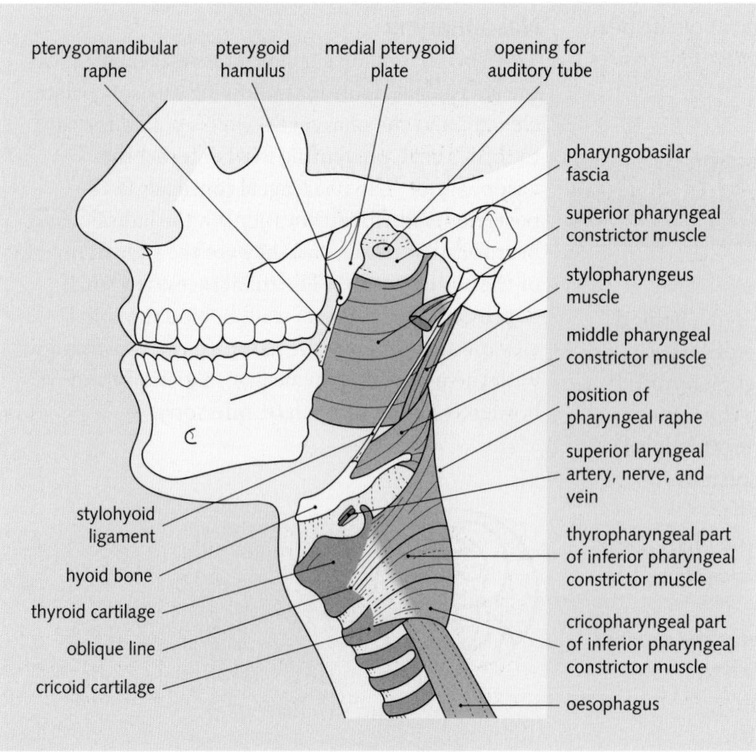

Fig. 7.57 Muscles of the pharynx.

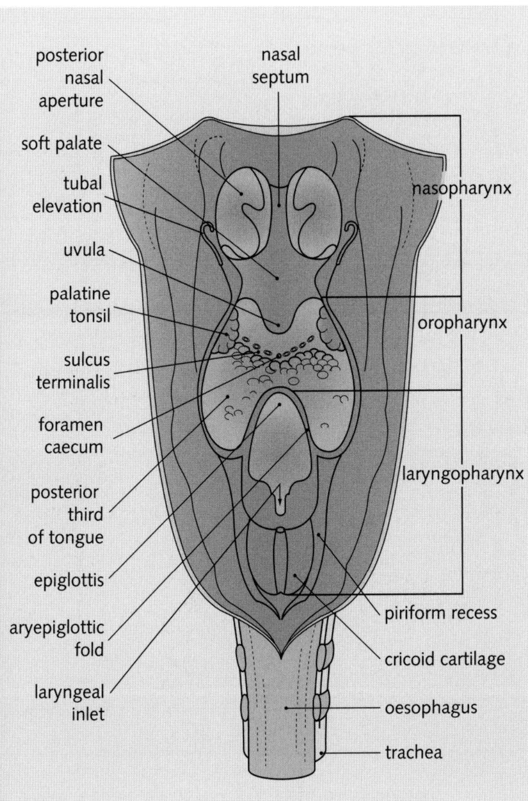

Fig. 7.58 Mucous membrane and the interior of the pharynx.

Oropharynx

The oropharynx extends from the soft palate to the upper border of the epiglottis. The palatine tonsils lie in its lateral walls. The posterior third of the tongue forms the anterior wall of the oropharynx. It has an irregular surface owing to the presence of the underlying lingual tonsils.

The mucosa is reflected from the tongue onto the epiglottis to form a median and two lateral glossoepiglottic folds. The depression on each side of the median fold is the vallecula.

Laryngopharynx

The laryngopharynx lies behind the laryngeal opening and the posterior surface of the larynx.

The piriform fossa is a groove on either side of the laryngeal inlet. It leads from the back of the tongue to the oesophagus.

Vessels of the pharynx

Blood supply is from branches of the ascending pharyngeal, ascending palatine, facial, maxillary, and lingual arteries.

Veins drain via the pharyngeal venous plexus to the internal jugular vein.

Lymphatics drain into the deep cervical nodes either directly or indirectly via the retropharyngeal or paratracheal nodes.

Nerve supply of the pharynx

The motor nerve supply to the pharynx is the cranial part of XI via X and the pharyngeal plexus.

The sensory nerve supply is as follows:

- nasopharynx—maxillary nerve (V_2)
- oropharynx—IX nerve
- laryngopharynx—internal laryngeal nerve (X).

Nose

The nose consists of:

- The external nose—this has a bony (nasal bones, frontal process of the maxilla) and cartilaginous skeleton, separated by the nasal septum.
- The nasal cavities—these communicate with the exterior via the nares or nostrils, and with the nasopharynx via the choanae.

Nasal cavity

The walls of the nasal cavity are listed in Fig. 7.59.

The openings in the lateral wall are listed in Fig. 7.60.

The nerve and blood supply of the lateral wall are illustrated in Fig. 7.61.

The nasal septum and the external nose both have cartilaginous and bony components.

Paranasal sinuses

The paranasal sinuses lie around the nasal cavity, in the bones of the face and skull. They are the sphenoidal, ethmoidal, frontal, and maxillary sinuses. They drain into the nasal cavity.

Mucous membrane of the nose

The vestibule lies just inside the anterior nares and is lined by hairy skin. The remainder of the nasal cavity is lined by ciliated columnar epithelium. There is a rich vascular plexus in the submucosa, together with numerous serous and mucous glands.

Walls of the nasal cavity	
Surface	**Components**
Floor	Palatine process of maxilla, horizontal process of palatine bone—i.e. the hard palate
Roof	Nasal, frontal, sphenoid, and ethmoid bones; above lies the anterior cranial fossa and the sphenoidal sinus
Lateral wall	Maxillary, palatine, sphenoid, lacrimal, and ethmoid bones and the inferior concha; the superior and middle conchae are projections of the ethmoid bone; the three conchae divide the lateral wall into the superior, middle, and inferior meatus and the sphenoethmoidal recess; the last lies above the superior concha
Medial wall (nasal septum)	The perpendicular plate of the ethmoid, the vomer and the septal cartilage

Fig. 7.59 Walls of the nasal cavity.

Openings in the lateral wall of the nose	
Region of lateral wall	**Features and openings**
Sphenoethmoidal recess	Sphenoidal sinus
Superior meatus	Posterior ethmoidal air cells
Middle meatus	The hiatus semilunaris lies below the middle concha; the frontal sinus, anterior ethmoidal cells, and maxillary sinus open into the hiatus; the bulla ethmoidalis is formed by the underlying middle ethmoidal cells which open onto it
Inferior meatus	Nasolacrimal duct

Fig. 7.60 Openings in the lateral wall of the nose.

Dust from the inspired air is removed by the nasal hairs and the mucus of the nasal cavity. The air is also warmed by the vascular plexus and moistened before it enters the lower airway.

The roof and superior part of the lateral wall contain olfactory epithelium, which receives the distal processes of the olfactory nerve cells. These fibres play a role in both smell and taste sensations.

The sphenopalatine artery anastomoses with the septal branch of the superior labial artery around the vestibule of the nose. This is a very common site for a nosebleed (epistaxis).

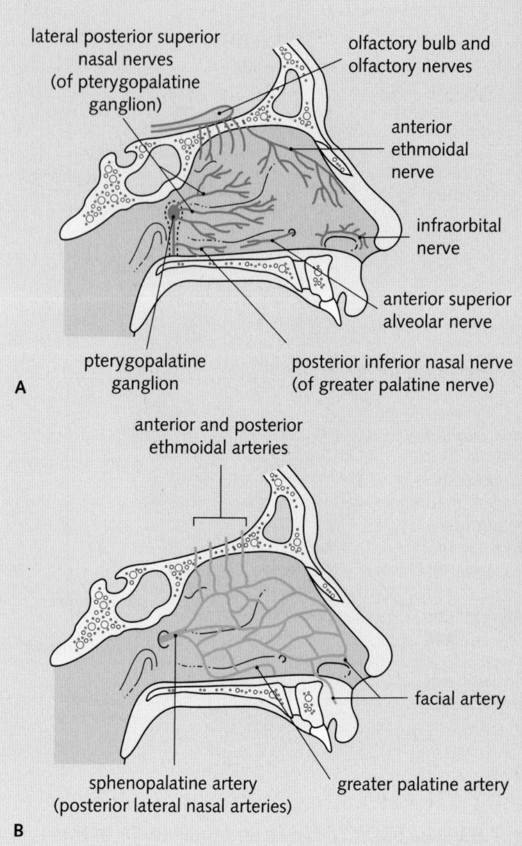

Fig. 7.61 Nerve (A) and blood (B) supplies of the lateral wall of the nose.

Pterygopalatine fossa

The pterygopalatine fossa is a small pyramidal space lying inferior to the apex of the orbit. It contains the terminal branches of the maxillary artery, the maxillary nerve, the nerve of the pterygoid canal, and the pterygopalatine ganglion. The communications of the fossa are listed in Fig. 7.62.

Pterygopalatine ganglion

The pterygopalatine ganglion is a parasympathetic ganglion lying in the pterygopalatine fossa, just lateral to the sphenopalatine foramen.

Preganglionic fibres from the superior salivary nucleus of the VII nerve enter the greater petrosal nerve. This joins the deep petrosal (sympathetic) nerve to form the nerve of the pterygoid canal, which joins the ganglion. Here, the parasympathetic fibres synapse and sympathetic fibres pass uninterrupted through the ganglion. Fibres of common sensation enter the ganglion via ganglionic branches of the maxillary nerve.

Communications of the pterygopalatine fossa	
Surface	**Communicates with**
Lateral	Infratemporal fossa
Medial	Nasal cavity via the sphenopalatine foramen
Anterior	Orbit via the inferior orbital fissure
Posterosuperior	Middle cranial fossa via the foramen rotundum and pterygoid canal

Fig. 7.62 Communications of the pterygopalatine fossa.

The branches of the ganglion are shown in Fig. 7.63.

Oral cavity

The oral cavity is divided into the vestibule and the oral cavity proper:

The vestibule lies between the lips and cheeks externally and the gums and teeth internally.

The oral cavity proper is bounded by the teeth and gums anteriorly and laterally. The palate forms the roof; the floor is formed by the anterior two thirds of the tongue and the floor of the mouth. A midline fold of mucosa—the frenulum—lies beneath the tongue (Fig. 7.64).

The submandibular ducts open onto the sublingual papillae on either side of the frenulum. The sublingual fold extends back from the papilla and overlies the sublingual glands. There are also numerous minor salivary glands that open in the oral cavity.

Nerve supply is as follows:

- Roof—greater and lesser palatine and nasopalatine nerves.
- Floor—lingual nerve.
- Cheek—buccal nerve (from V_3 nerve).

Lips

The two lips seal the oral cavity anteriorly, and they also assist in speech. The lips are covered by mucosa internally and by skin externally. The orbicularis oris muscle, the superior and inferior labial vessels and nerves, and numerous minor salivary glands lie in the substance of the lips.

Tongue

The tongue is a mobile muscular organ covered by mucous membrane. The anterior two thirds lie in the mouth, the posterior third in the oropharynx (Fig. 7.65).

The muscles of the tongue are listed in Fig. 7.66.

Branches of the pterygopalatine ganglion	
Branch	**Course and distribution**
Nasopalatine nerve	Passes through the sphenopalatine foramen to supply the nasal septum and incisive gum of the hard palate
Lateral posterior superior nasal nerve	Exits via the sphenopalatine foramen to supply the lateral wall of the nose
Greater palatine nerve	Passes through the greater palatine canal and foramen to supply the mucosa of the palate and the lateral wall of the nose
Lesser palatine nerve	Exits through the lesser palatine foramina to supply the soft palate and the mucosa over the palatine tonsil
Pharyngeal nerve	Passes via the palatovaginal canal to supply the nasopharynx
Lacrimal fibres	Parasympathetic fibres to the lacrimal gland join the zygomaticotemporal nerve of V_2 then the lacrimal nerve before supplying the gland

Fig. 7.63 Branches of the pterygopalatine ganglion.

Mucous membrane of the tongue

The sulcus terminalis divides the tongue into the anterior two thirds and the posterior third. The foramen caecum lies at the apex of the sulcus. It is the remnant of the upper end of the thyroglossal duct. Between 10 and 12 vallate papillae lie anterior to the sulcus.

The mucosa of the anterior two thirds of the tongue is relatively smooth, and it has numerous filiform and fungiform papillae on the dorsal surface. The frenulum connects it to the floor of the mouth. Lateral folds of mucosa, the plica fimbriata, are also seen on the ventral surface of the tongue.

The irregular surface of the posterior third of the tongue is caused by the underlying lingual tonsils.

Blood and nerve supply to the tongue

Vessels of the tongue comprise the lingual arteries and veins.

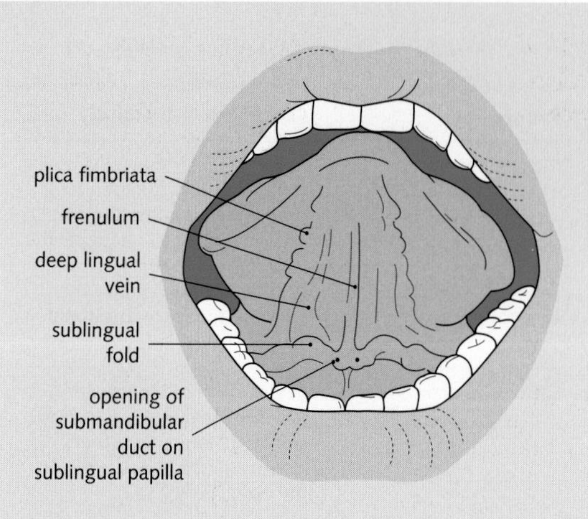

Fig. 7.64 Oral cavity.

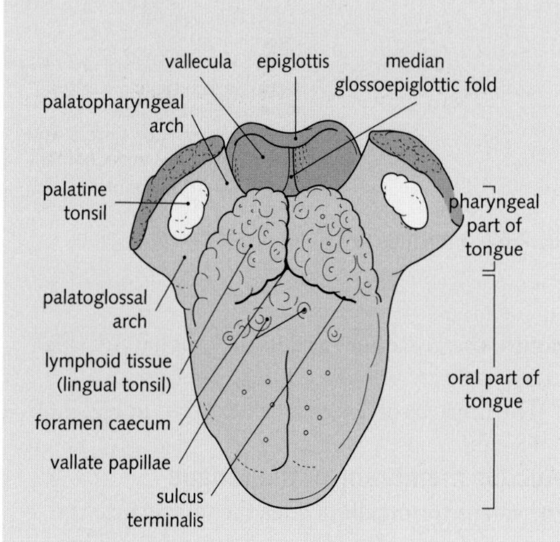

Fig. 7.65 Tongue.

Lymphatic drainage is to the deep cervical, the submandibular, and the submental nodes. Carcinoma of the tongue may spread via the lymphatics to both sides of the neck, dramatically worsening its prognosis.

The nerve supply to the tongue is shown in Fig. 7.67.

Remember, the XII nerve is motor to all the muscles of the tongue except palatoglossus (pharyngeal plexus).

Floor of the mouth and submandibular region

This region lies between the mandible and hyoid bone. It contains the following:

- Muscles—digastric, mylohyoid, hyoglossus, geniohyoid, genioglossus, and styloglossus.
- Salivary glands—submandibular and sublingual.
- Nerves—lingual, glossopharyngeal, and hypoglossal; submandibular ganglion.
- Blood vessels—facial and lingual.
- Lymph nodes—submandibular.

Submandibular gland

This consists of two parts—a large superficial part and a small deep part—that are continuous around the posterior border of mylohyoid. The deep part of the gland lies between the mylohyoid superficially and the styloglossus and hyoglossus.

Blood supply is from the facial and lingual arteries.

Nerve supply is from the submandibular ganglion, a parasympathetic ganglion with the following features:

- Preganglionic parasympathetic fibres originate in the superior salivary nucleus of the VII nerve and pass to the ganglion via the nervus intermedius, the chorda tympani, and the lingual nerve.
- Sympathetic and sensory fibres from the superior cervical ganglion and the lingual nerve pass through the ganglion.
- Postganglionic parasympathetic secretomotor fibres pass to the submandibular and sublingual glands via the lingual nerve or directly.

Lingual nerve

From the mandibular third-molar region, the lingual nerve passes across styloglossus to the lateral surface of hyoglossus and across the submandibular duct, to break up into its terminal branches supplying the mucosa of the tongue.

Hypoglossal nerve

In the submandibular region, the hypoglossal nerve runs forward below the deep part of the submandibular gland, the submandibular duct, and the lingual nerve. It divides into its terminal branches and supplies all the muscles of the tongue except palatoglossus.

Sublingual gland

The sublingual gland lies superficially under the mucosa of the floor of the mouth. The lingual nerve

Muscles of the tongue			
Intrinsic muscles			
Name of muscle (nerve supply)	**Origin**	**Insertion**	**Action**
Longitudinal (XII nerve)	Mucous membrane	Mucous membrane	Shortens tongue
Transverse (XII nerve)	Mucous membrane and median septum	Mucous membrane	Narrows tongue
Vertical (XII nerve)	Mucous membrane	Mucous membrane	Lowers tongue
Extrinsic muscles			
Name of muscle (nerve supply)	**Origin**	**Insertion**	**Action**
Palatoglossus (pharyngeal plexus)	Palatine aponeurosis	Lateral aspect of tongue	Pulls tongue upward and backward and narrows oropharyngeal isthmus
Genioglossus (XII nerve)	Superior mental spine (genial tubercle) of mandible	Merges with other tongue muscles	Draws tongue forward and pulls tip backward
Hyoglossus (XII nerve)	Body and greater cornu of hyoid bone	Merges with other tongue muscles	Depresses tongue
Styloglossus (XII nerve)	Styloid process of temporal bone	Merges with other tongue muscles	Draws tongue upward and backward

Fig. 7.66 Muscles of the tongue.

Nerve supply to the tongue		
	Posterior third	**Anterior two thirds**
General sensory	Glossopharyngeal nerve (IX)	Lingual nerve (V_3)
Taste	Glossopharyngeal nerve (IX) (also vallate papillae)	Chorda tympani (VII) (via the lingual nerve)
Motor	Hypoglossal nerve, pharyngeal plexus (palatoglossus) (XII, XI)	Hypoglossal nerve (XII)

Fig. 7.67 Nerve supply to the tongue.

and submandibular duct lie medially. It is supplied by the submandibular ganglion.

Palate and tonsils

The palate forms the roof of the mouth and the floor of the nose. It is divided into two components:

- The hard palate is composed of the palatine process of the maxilla and the horizontal process of the palatine bone. It is covered by mucous membrane.
- The soft palate is a mobile fibromuscular fold lying posteriorly. It is composed of muscles (Fig. 7.68) and the palatine aponeurosis—the expanded tendon of tensor veli palatini.

Blood supply to the palate is from the greater and lesser palatine arteries. Nerve supply is from the pterygopalatine ganglion.

The palatine tonsils are masses of lymphoid tissue lying in the tonsillar fossae between the palatoglossal and palatopharyngeal arches. They are covered by mucous membrane. The surface is pitted by many openings that lead to the tonsillar crypts. Lymphatics drain to the deep cervical nodes.

Muscles of the soft palate			
Name of muscle (nerve supply)	**Origin**	**Insertion**	**Action**
Tensor veli palatini (nerve to medial pterygoid V_3)	Spine of sphenoid, auditory tube, scaphoid fossa of pterygoid process	With muscle of other side, forms palatine aponeurosis	Tenses soft palate
Levator veli palatini (pharyngeal plexus)	Petrous part of temporal bone, auditory tube	Palatine aponeurosis	Elevates soft palate
Musculus uvulae (pharyngeal plexus)	Posterior border of hard palate	Mucous membrane of uvula	Elevates uvula
Palatopharyngeus (pharyngeal plexus)	Palatine aponeurosis horizontal plate of palatine bone	Posterior border of thyroid cartilage	Elevates pharyngeal wall and pulls palatopharyngeal folds medially and depresses soft palate
Palatoglossus (pharyngeal plexus)	Palatine aponeurosis	Lateral aspect of tongue	Pulls tongue upward and backward and narrows oropharyngeal isthmus and depresses soft palate

Fig. 7.68 Muscles of the soft palate.

Larynx

The larynx is continuous with the laryngopharynx superiorly and with the trachea inferiorly. It acts as a sphincter, separating the lower respiratory system from the alimentary system, and it is responsible for voice production.

The laryngeal cartilages are shown in Fig. 7.69. The laryngeal membranes link these cartilages together, and they join the larynx to the hyoid bone and the trachea (Fig. 7.70). The membranes thicken in places to form ligaments.

Mucous membrane of the larynx

The mucosa is tucked under the vestibular ligament to form the laryngeal ventricle between the vestibular and vocal folds. Above the vocal fold the mucosa is supplied by the internal laryngeal nerve and the superior laryngeal artery. Below the vocal fold, it is supplied by the recurrent laryngeal nerve and the inferior laryngeal artery (from the inferior thyroid artery).

Laryngeal cavity

The laryngeal inlet allows communication between the pharynx and the larynx. It is bounded by the epiglottis and the aryepiglottic and interarytenoid folds (Fig. 7.71).

The inlet leads to the vestibule, which extends to the vestibular folds. The laryngeal ventricle lies between the vestibular and vocal folds. The rima glottis is the space between the vocal folds. The infraglottic cavity lies below the vocal folds and is continuous with the trachea.

Intrinsic muscles of the larynx

The intrinsic muscles of the larynx are described in Fig. 7.72. All intrinsic muscles are paired except the transverse arytenoid muscle. They can alter the tension and length of the vocal folds and the rima glottis's size and shape (see Fig. 7.73).

All intrinsic muscles are supplied by the recurrent laryngeal nerve except for the cricothyroid muscle, which is supplied by the external laryngeal nerve.

Joints of the larynx

The synovial cricothyroid joint is formed by the inferior cornu (horn) of the thyroid cartilage articulating with the facet of the cricoid cartilage. Around a transverse axis (passing) through the joint one cartilage can tilt back and forth upon the other. This alters the vocal fold tension and length.

The synovial cricoarytenoid joint has a lax capsule. This allows rotation and gliding movements of the arytenoid cartilages upon the cricoid cartilage. Arytenoid cartilage gliding widens or narrows a V-

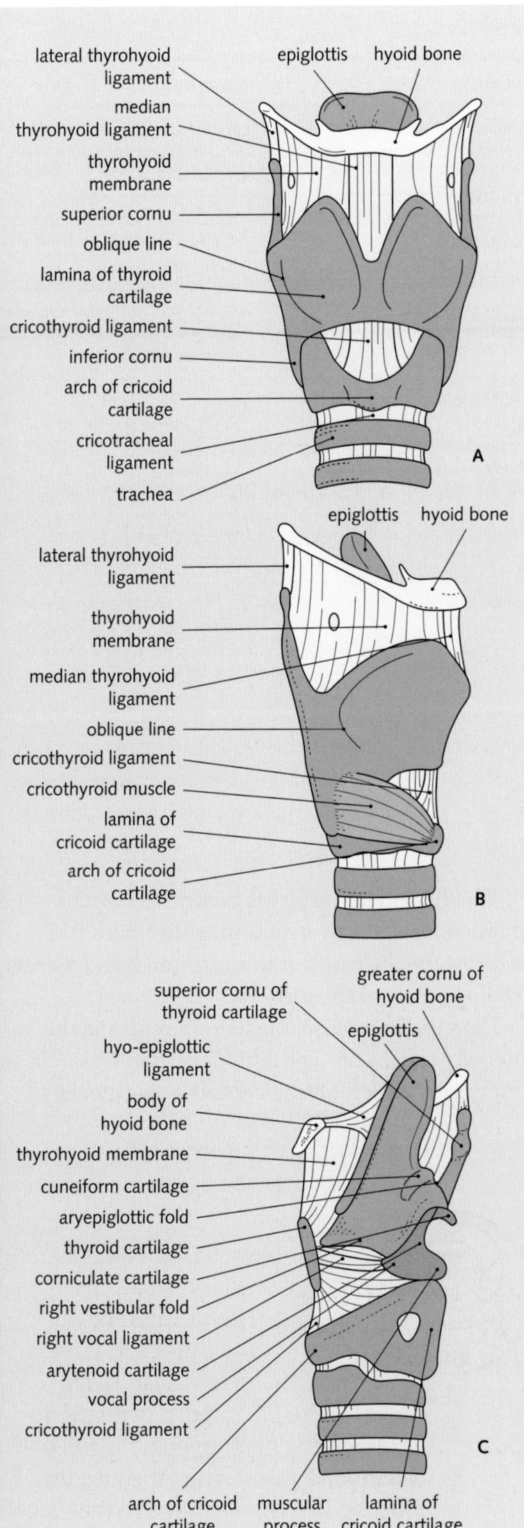

Fig. 7.69 Laryngeal cartilages from the front (A), from the right (B), and from the left without the left lamina of thyroid cartilage (C).

Laryngeal membranes	
Membrane	**Attachments**
Thyrohyoid	Runs between the thyroid cartilage and hyoid bone; has a midline thickening and two lateral thickenings, the median thyrohyoid ligament and lateral thyrohyoid ligaments, respectively
Quadrangular	Runs between the epiglottis and the arytenoid cartilage; its lower free border is the vestibular ligament
Cricothyroid	Joins the cricoid, thyroid, and arytenoid cartilages; its upper free border is the vocal ligament; there is also a midline thickening, the median cricothyroid ligament
Cricotracheal	Runs from the cricoid cartilage to the trachea

Fig. 7.70 Laryngeal membranes.

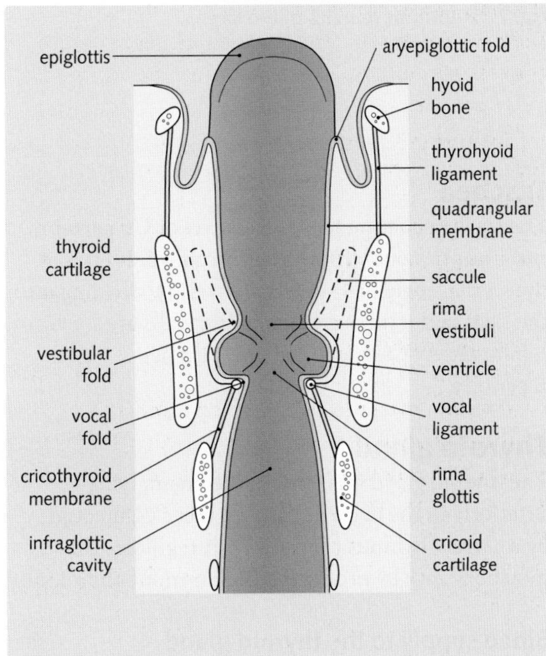

Fig. 7.71 Coronal section of the laryngeal cavity. (Adapted from *Gray's Anatomy* 38th edn, edited by L H Bannister et al. Harcourt Brace and Co.)

shaped rima glottis. Arytenoid cartilage rotation can open the rima glottis into a diamond shape or close it.

Intrinsic muscles of the larynx			
Muscle (nerve supply)	**Origin**	**Insertion**	**Action**
Cricothyroid (external laryngeal nerve)	Cricoid cartilage arch	Inferior border of thyroid cartilage and inferior cornu	Lengthens and tenses vocal cords by tilting cricoid and thus arytenoid cartilages
Posterior cricothyroid (recurrent laryngeal nerve)	Cricoid cartilage lamina	Arytenoid cartilage muscular process	Abducts vocal cords by laterally rotating arytenoid cartilages on cricoid cartilage
Lateral cricothyroid (recurrent laryngeal nerve)	Cricoid cartilage arch	Arytenoid cartilage muscular process	Adducts vocal cords by medially rotating arytenoid cartilages on cricoid cartilage
Thyroarytenoid (recurrent laryngeal nerve)	Posterior surface of thyroid cartilage	Arytenoid cartilage muscular process	Shortens vocal cord
Transverse arytenoid (recurrent laryngeal nerve)	Body of arytenoid cartilage	Body of arytenoid cartilage	Closes rima glottis by adducting arytenoid cartilages
Oblique arytenoid (recurrent laryngeal nerve)	Muscular process of arytenoid cartilage	Apex of arytenoid cartilage	Closes rima glottis by drawing arytenoid cartilages together
Vocalis (recurrent laryngeal nerve)	Vocal process of arytenoid cartilage	Vocal ligament	Maintains/increases tension in anterior part of vocal ligament; relaxes posterior part of vocal ligament

Fig. 7.72 Intrinsic muscles of the larynx.

Trachea

The trachea commences at the level of C6 vertebra and is continuous with the larynx above. It ends at the sternal angle (T4 vertebral level) by dividing into the right and left main bronchi. Its walls are reinforced by C-shaped hyaline cartilage anteriorly.

Thyroid gland

The thyroid gland is an endocrine organ, lying anteriorly in the neck. It has two lobes connected by a narrow isthmus (Fig. 7.74). It regulates the metabolic rate by producing the hormone thyroxine.

Blood supply to the thyroid gland

The superior thyroid artery (the external carotid artery's first branch) has the external laryngeal nerve running with it. The artery at the upper pole of the gland branches to supply it. The inferior thyroid artery (arising from the thyrocervical trunk of the subclavian artery) has the recurrent laryngeal nerve running with it. The inferior artery anastomoses with its superior counterpart.

The thyroid ima artery is present in only 3% of individuals. It can arise from either the brachiocephalic trunk or the aortic arch and it enters the lower part of the isthmus.

The superior and middle thyroid veins join the internal jugular vein. The inferior thyroid veins join and empty into the left brachiocephalic vein (usually).

The inferior thyroid artery travels with the recurrent laryngeal nerve, and the superior thyroid artery travels with the external laryngeal nerve. These are important relations to remember in removing part/all of the thyroid gland when the arteries must be tied off. Damage to the nerves results in a weak and hoarse voice

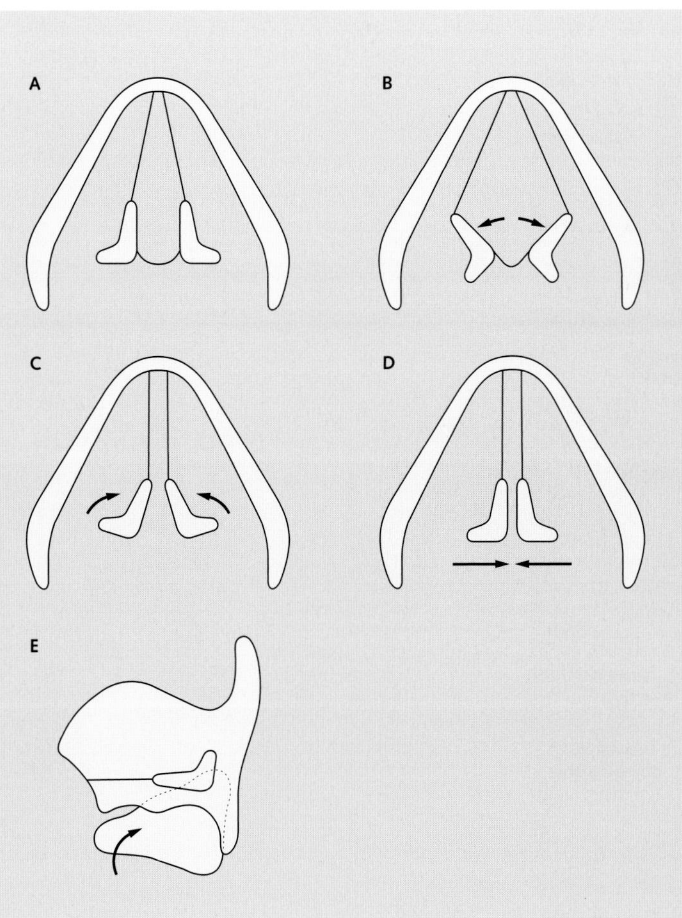

Fig. 7.73 Movements of the vocal folds, arytenoid and cricoid cartilages.

(A) Vocal cord position during quiet respiration i.e. at rest.

(B) Vocal cord abduction by posterior cricoarytenoid muscles.

(C) Vocal cord adduction by lateral cricoarytenoid muscles.

(D) Vocal cord adduction by transverse arytenoid muscle

(E) Increase in vocal cord tension by cricoid cartilage tilting through cricothyroid muscle contraction and drawing arytenoid cartilages posteriorly.

(Adapted from *Gray's Anatomy* 38th edn, edited by L H Bannister et al. Harcourt Brace and Co.)

Parathyroid glands

The parathyroid glands are four small glands related to the posterior border of the thyroid gland. They are important in the regulation of calcium metabolism. Note, the parathyroids may be damaged during thyroid surgery.

Blood supply to the parathyroid glands

The upper and lower parathyroid glands are supplied by the inferior thyroid artery. Small veins join the thyroid veins.

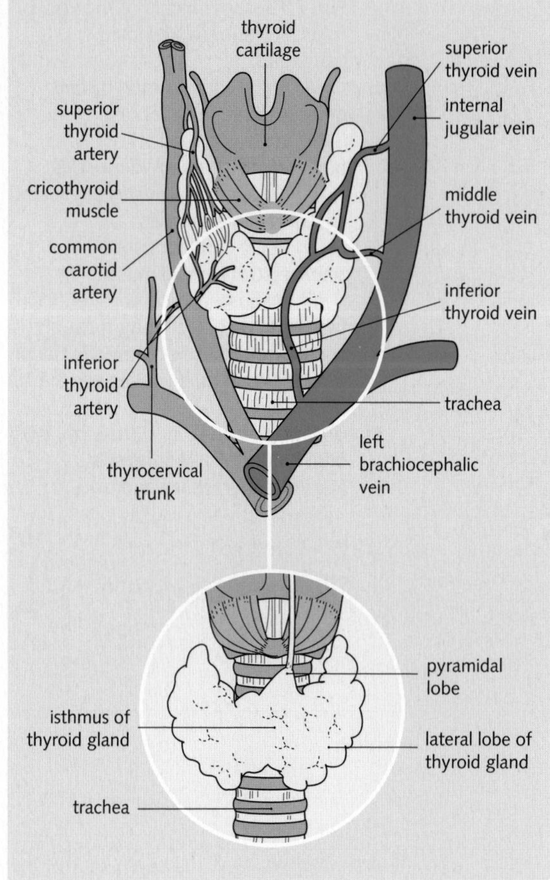

Fig. 7.74 Anterior view of the thyroid gland. The left side of the figure shows the arterial supply, and the right side shows the venous drainage. Inset shows thyroid gland anatomy.

- List the individual bones of which the skull is composed.
- Describe the atlanto-occipital and atlanto-axial joints.
- List the openings in the base of the skull and their main contents.
- Describe the anatomy of the scalp, including its blood and nerve supply.
- Define the areas of the face supplied by the divisions of the trigeminal nerve.
- Describe the anatomy of the dura mater and its reflections.
- Outline the anatomy of the cavernous sinus.
- Summarize the circulus arteriosus and the main arteries of the brain.
- List the cranial nerves and their functions.
- Describe the boundaries and contents of the orbital cavity.
- Discuss the actions of the muscles of the eye.
- Describe the anatomy of the parotid gland and the structures passing through it.
- Discuss the boundaries and contents of the infratemporal fossa.
- Describe the temporomandibular joint.
- List the branches of the mandibular nerve and what they supply.
- List the walls of the middle ear.
- Outline the components of the internal ear.
- Describe the deep cervical fascia of the neck.
- Describe the boundaries and contents of the posterior triangle.
- Outline the cervical plexus.
- What muscles are classed as suprahyoid, and what is their nerve supply?
- What is the blood and nerve supply of the pharynx?
- Outline the walls of the nasal cavity.
- Discuss the nerve supply of the tongue.
- Describe the laryngeal membranes.
- List the actions and nerve supply of the intrinsic muscles of the larynx.

8. The Back

Regions and components of the back

The back consists of the vertebral column, the spinal cord, the roots of the spinal nerves, and associated muscles.

The vertebral column extends from the skull to the coccyx. It supports the weight of the body above the pelvic girdle. There is limited movement between adjacent vertebrae, but the total movement of the entire column is considerable.

There are 33 vertebrae arranged in five regions (seven cervical, twelve thoracic, five lumbar, five sacral, and four coccygeal). The sacral and coccygeal vertebrae fuse to form the sacrum and coccyx, respectively (see Fig. 8.2).

Back pain is the consequence of a bipedal gait. Three factors combine to cause back pain. The nucleus pulposus dehydrates with age, weakened ligaments due to poor posture and lifting with a flexed vertebral column.

There are four curvatures of the vertebral column in adults. The cervical and lumbar curvatures (secondary curvatures) are concave posteriorly; the thoracic and sacrococcygeal curvatures (primary curvatures) are concave anteriorly. The primary curvatures develop during the fetal period, whereas the secondary curvatures begin to appear before birth, but they become obvious only during infancy.

Abnormal lateral curvature and vertebral rotation is called scoliosis. It can be caused by contralateral weakness in the intrinsic back muscles.

Surface anatomy and superficial structures

Visible and palpable features of the back are shown in Fig. 8.1. Note the following:

- The first easily palpable spine when passing a finger down the back of the neck is that of C7.
- The inferior angle of the scapula lies at the angle of the spine of T7 vertebra.
- A line passing through the highest point of the iliac crest passes through the spine of L4 vertebra.

Cutaneous innervation of the back

The skin and muscles of the back are supplied segmentally by the posterior rami of the 31 pairs of spinal nerves. All posterior rami of the spinal nerves, except the first cervical nerve, divide into a medial and lateral branch. The posterior ramus of the 1st (suboccipital) cervical nerve supplies the deep muscles of the back of the neck (in the suboccipital region), and it does not supply the skin.

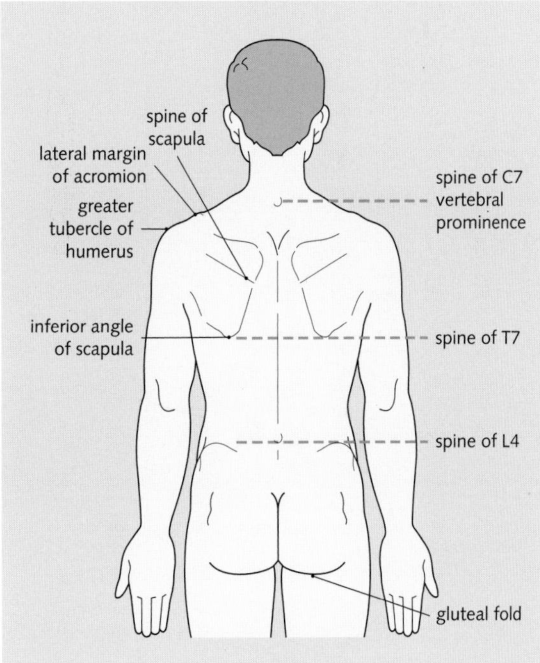

Fig. 8.1 Surface features of the back.

The vertebral column

Skeleton of the vertebral column

The vertebral column consists of 33 vertebrae lying in five regions (Fig. 8.2). Individual vertebrae articulate with each other via intervertebral discs and articular facet (zygapophyseal) joints.

The vertebral column supports the weight of the upper body. The weight is transferred to the lower limb via the pelvic girdle. The column also transmits and protects the spinal cord.

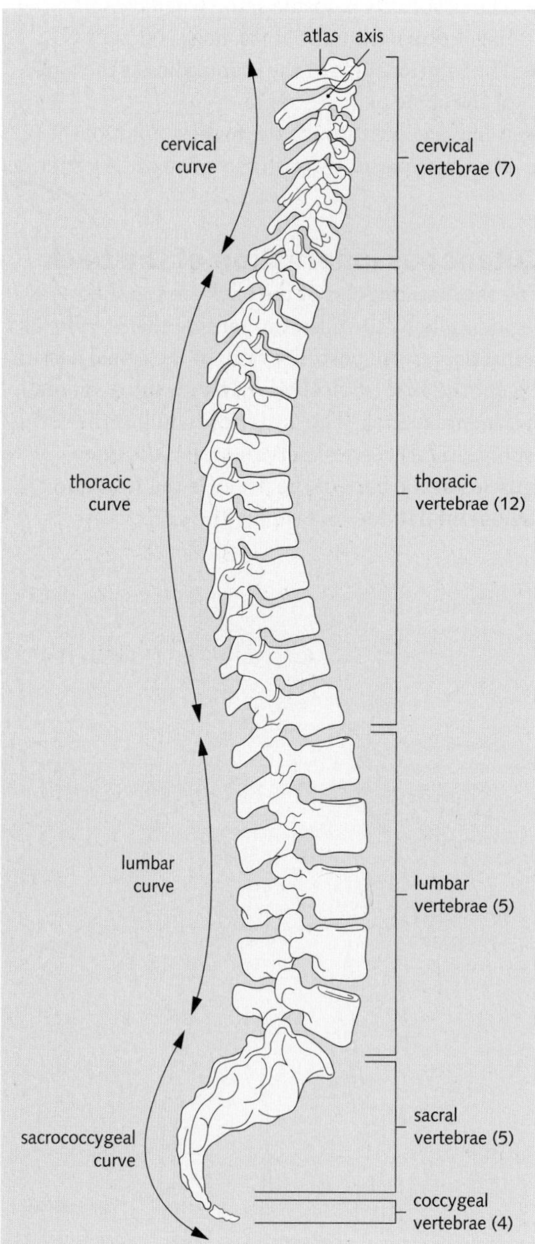

Fig. 8.2 Lateral view of the vertebral column.

Features of individual vertebrae

The cervical and thoracic vertebrae, together with the sacrum and coccyx, are discussed in Chapters 3, 5, and 7.

Fig. 8.3 illustrates the features of typical lumbar vertebrae.

Joints of the vertebral column

The craniovertebral joints are discussed in Chapter 7.

Intervertebral discs

These are the joints between the bodies of adjacent vertebrae (Fig. 8.4). They are secondary cartilaginous joints. The joints provide strength and weight-bearing capacity, and they absorb compressive forces.

The disc is composed of:

- The anulus fibrosus—an outer ring made up of concentric layers of fibrous tissue.
- The nucleus pulposus—a gelatinous core.

As individuals age, the nuclei pulposi lose their water content, becoming less turgid and thinner due to the compression forces upon the vertebral column, and

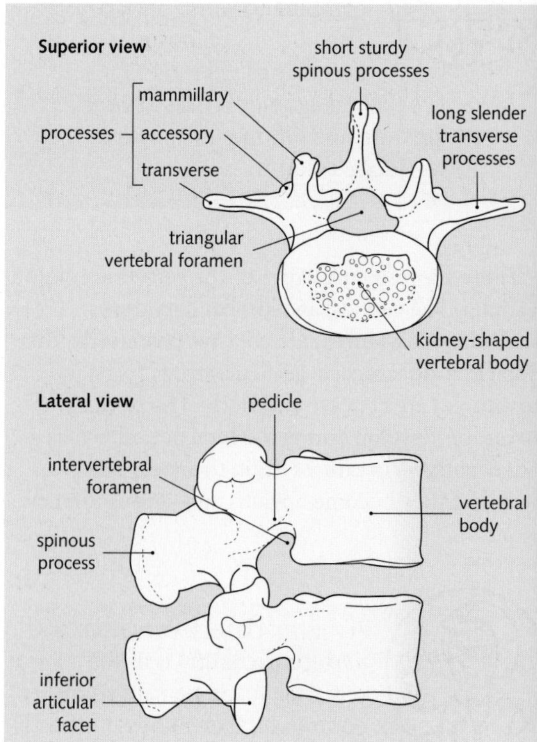

Fig. 8.3 Features of typical lumbar vertebrae.

is change can herniate through a worn anulus rosus. This is commonly called a slipped disc and curs most frequently in the lumbar region.

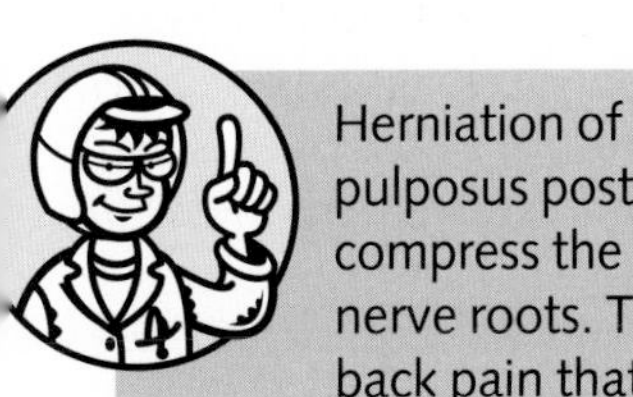

Herniation of the nucleus pulposus posterolaterally can compress the L5 and/or the S1 nerve roots. This causes lower back pain that radiates down the back of the lower limb.

oints of the vertebral arches

hese are between the articular processes on the rtebral arches and are called facet (zygopophyseal) ints. They are plane synovial joints and their ientation affects the movements that can take ace at the different vertebral levels.

igaments of the vertebral column

hese are described in Fig. 8.5.

lovement of the vertebral column

lexion, extension, lateral flexion, and rotation are ossible. Mobility results from the compression and asticity of the intervertebral discs (see Fig. 8.6). lovements of the atlanto-occipital and atlanto-axial ints are discussed in Chapter 7.

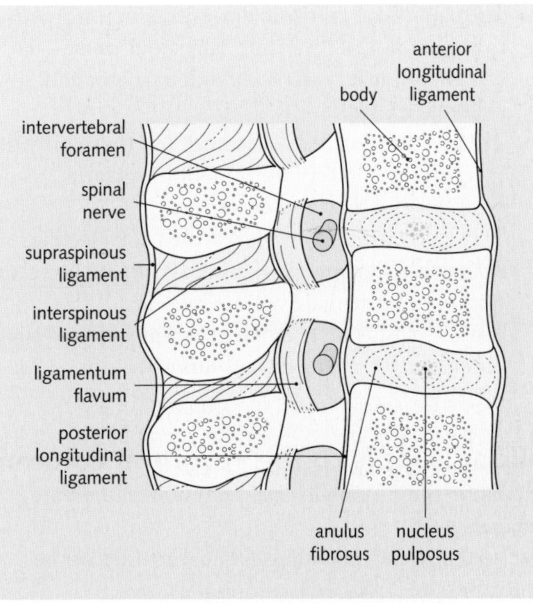

Fig. 8.4 Sagittal section of vertebrae showing the intervertebral discs and ligaments of the vertebral column.

Muscles of the vertebral column

An individual's body weight, for the greater part, is anterior to their vertebral column. To support this and move the vertebral column, there is a strong mass of muscle that runs longitudinally on the posterior aspect of the vertebrae. There are three main groups of back muscles:

Ligaments of the vertebral column

Ligament	Action
Anterior longitudinal	Strong band covering the anterior part of the vertebral bodies and the intervertebral discs running from the anterior tubercle of C1 vertebra to the sacrum; maintains stability of the intervertebral discs and prevents hyperextension of the vertebral column
Posterior longitudinal	Is attached to the posterior aspect of the intervertebral discs and posterior edges of the vertebral bodies from C2 vertebra to the sacrum; prevents hyperflexion of the vertebral column and posterior protrusion of the discs
Supraspinous	Accessory ligament uniting the tips of the spinous processes
Interspinous	Accessory ligaments uniting the spinous processes
Ligamenta flava	Help to preserve the curvature of the vertebral column and support the joints between the vertebral arches; unite the laminae of adjacent vertebrae

Fig. 8.5 Ligaments of the vertebral column.

- The superficial extrinsic muscles associated with the upper limb—trapezius, latissimus dorsi, levator scapulae, and rhomboideus minor and major (see Chapter 2).
- The intermediate extrinsic muscles, these provide accessory respiratory movements—serratus posterior and levatores costarum.
- The deep intrinsic muscles of the back (see Figs 8.7 and 8.8)—superficial vertical running (erector spinae), intermediate oblique running (transversospinalis), and deepest muscles (interspinales and intertransversarii).

Blood supply to the vertebral column

Spinal arteries supplying the vertebral column are branches of:

- Vertebral and ascending cervical arteries of the neck.
- Posterior intercostal arteries in the thoracic region.
- Subcostal and lumbar arteries in the abdomen.
- Iliolumbar and lateral sacral arteries in the sacrum.

The spinal artery branches enter the intervertebral foramina and divide into radicular arteries. These supply the anterior and posterior roots of the spinal nerves.

Spinal veins form plexuses inside (internal vertebral venous plexus) and outside (external vertebral venous plexus) the vertebral canal (see Fig. 8.9). Both plexuses have no valves.

Blood may return from the pelvis and abdomen to the heart via the vertebral venous plexuses and the superior vena cava. Abdominal and pelvic tumours may metastasize to the vertebrae in this way.

An aortic aneurysm can compress a segmental artery that gives rise to a spinal branch. This impairs blood supply to that spinal segment level and results in loss of sensation and voluntary movement below this level.

The suboccipital nerve has no cutaneous component supplying the back.

Back of the neck

At the back of the neck are muscles that connect the skull to the spine and pectoral girdle. A midline ligamentum nuchae is a vertical partition that separates the trapezius and sternocleidomastoid muscles of one side from those of the contralateral side. Beneath the deep intrinsic semispinalis capitus muscle is the suboccipital triangle.

Ligamentum nuchae

The ligamentum nuchae is a strong triangular fibroelastic ligament attaching superiorly to the external occipital protuberance and the cervical vertebrae spinous processes. It provides muscular attachments for trapezius and rhomboideus minor because the C3 to C5 spines are shorter than the other cervical spines.

Movements of the vertebral column

Vertebral region	Movements and accommodating factors	Limited movements and factors
Cervical	Flexion, extension, and lateral flexion occur because the intervertebral discs are thick compared to the vertebral bodies, facet joint capsules are loose, and the facet joints of C3 to C7 are horizontal, ovoid-shaped, and large	Lateral rotation is limited due to the shape of the articular processes of the facet joints of C3 to C7
Thoracic	Rotation and lateral flexion occur due to the oblique, nearly vertical shape of the facet joints and laminae	Flexion and extension are inhibited by facet joint shape, long spinous processes, the ribs, sternum, and thin intervertebral discs
Lumbar	Flexion, extension, and lateral flexion occur due to large intervertebral discs, and the shape of the facet joints	Rotation is prevented by the interlocking articular processes of the facet joints

Fig. 8.6 Movements of the vertebral column.

Deep intrinsic muscles of the back			
Muscle	**Origin**	**Insertion**	**Action**
		Deep layer	
Interspinales	Spinous process	Adjacent spinous process	Extension and rotation of vertebral column
Intertransversii	Transverse process	Adjacent transverse process	Lateral flexion and stabilize vertebral column
		Transversospinalis (intermediate layer)	
Rotatores	Transverse process	Lamina of vertebra above	Stabilize vertebrae
Multifidus	Sacrum, ilium, thoracic transverse processes, cervical articular processes	2nd to 4th spinous process above	Stabilize vertebrae
Semispinalis	Transverse processes C4 to T12	Spinous processes above and occipital bone	Rotate column, extend head, cervical and thoracic regions
		Erector spinae (superficial layer)	
Iliocostalis	Sacrum, iliac crest, lumbar spinous processes	Angles of ribs, cervical and thoracic transverse processes, mastoid process	Erector spinae muscles extend column and head, control flexion by lengthening their fibres, lateral flexion
Longissimus	Sacrum, iliac crest, lumbar spinous processes	Angles of ribs, cervical and thoracic transverse processes, mastoid process	
Spinalis	Spinous processes	Spinous processes	
Splenius	T1–T6 spinous processes	Mastoid process	

Fig. 8.7 Deep intrinsic muscles of the back.

Suboccipital region

This region is inferior to the occipital bone of the cranial base. Within this region is the suboccipital triangle, which contains the C1 (suboccipital) nerve and the vertebral artery before it enters the skull through the foramen magnum.

Four muscles are within this region (see Fig. 8.10), three of which form the boundaries of the suboccipital triangle (see Fig. 8.11):

- Rectus capitis posterior major (superomedially)
- Obliquus capitis superior (superolaterally)
- Obliquus capitis inferior (inferolaterally)

The spinal cord and meninges

The spinal cord lies in the vertebral canal. It commences just below the foramen magnum and it ends opposite L2 vertebra at the conus medullaris in adults (Fig. 8.12). In children the spinal cord can end as low as the L4 vertebra. Below the conus medullaris, the rootlets of the lumbar and sacral nerves form the cauda equina.

A cervical enlargement extends from the C4 to T1 spinal cord segments. Ventral rami from these segments form the brachial plexus. A lumbrosacral enlargement extends from the L2 to S3 spinal cord segments. Ventral rami from these form the lumbar and sacral plexuses.

Blood supply to the spinal cord

Anterior and posterior spinal arteries arise in the cranial cavity from the vertebral arteries or the inferior cerebellar artery. Anterior and posterior radicular branches of the spinal arteries reinforce the blood supply.

Venous blood joins venous plexuses on the surface of the cord. These communicate with the cranial

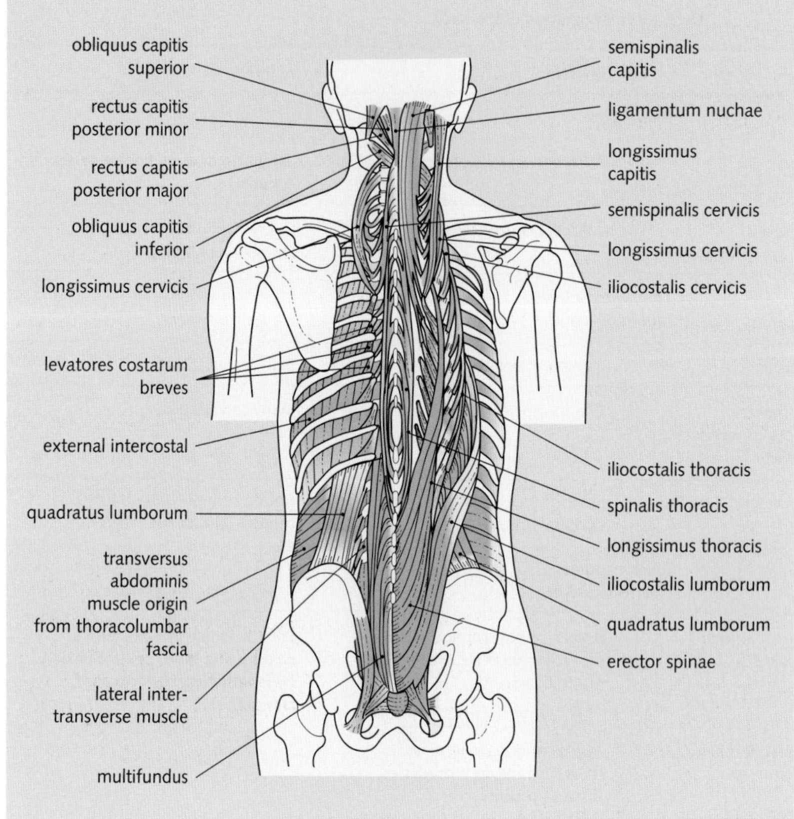

Fig. 8.8 Deep intrinsic muscles of the back. The right side shows the erector spinae components. Note longissimus cervicus has been moved laterally and semispinalis captis has been removed. (Adapted from *Gray's Anatomy*, 38th edn, edited by L H Bannister et al. Courtesy of Harcourt Brace and Co.)

veins and the venous sinuses of the skull and with the internal and external vertebral plexuses.

The spinal cord has no lymphatic vessels.

Spinal nerves

There are 31 pairs of spinal nerves. Each is composed of a dorsal and ventral root (see Fig. 1.9).

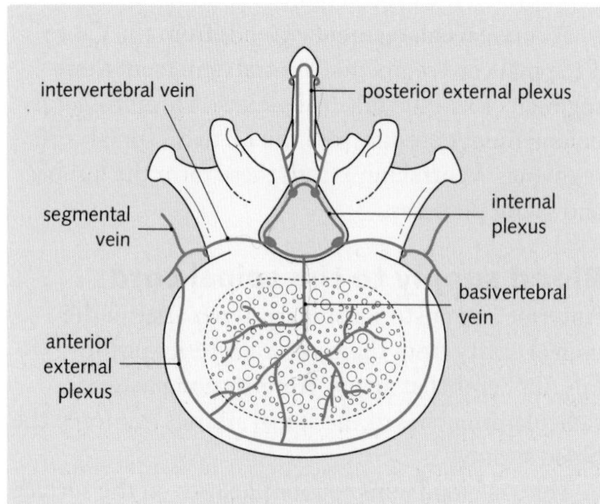

Fig. 8.9 Venous plexuses of the vertebral canal.

Note that there are seven cervical vertebrae but eight cervical nerves.

Spinal meninges and cerebrospinal fluid

The meninges and the cerebrospinal fluid (CSF) surround and protect the spinal cord.

Dura mater

The dura mater is tough fibrous membrane continuous with the dura of the brain. It is separated from the vertebral periosteum by the epidural space. It is attached to the foramen magnum superiorly and to the coccyx inferiorly by the filum terminale. The dura mater extends into, and adheres to, the periosteum of the intervertebral foramina. It continues along the spinal nerve roots investing them and eventually blending with the epineurium of the spinal nerves.

Muscles of the suboccipital region			
Muscle (nerve supply)	**Origin**	**Insertion**	**Action**
Rectus capitis posterior minor (suboccipital nerve)	Posterior tubercle of atlas (C1)	Inferior nuchal line	Extend head
Rectus capitis posterior major (suboccipital nerve)	Spinous process of axis (C2)	Inferior nuchal line	Extend and rotate head
Obliquus capitis inferior (suboccipital nerve)	Spinous process of axis (C2)	Transverse process of atlas (C1)	Rotate altas and hence head
Obliquus capitis superior (suboccipital nerve)	Transverse process of atlas (C1)	Occipital bone	Lateral flexion

Fig. 8.10 Muscles of the suboccipital region.

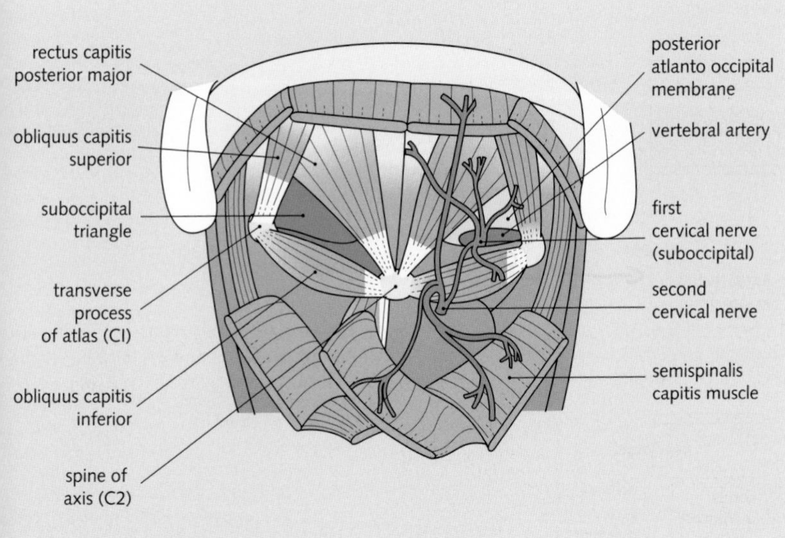

Fig. 8.11 The suboccipital region and contents. The right side shows the contents and the left side shows the boundaries of the suboccipital triangle. (Adapted from *Gray's Anatomy*, 38th edn, edited by L H Bannister et al. Harcourt Brace and Co.)

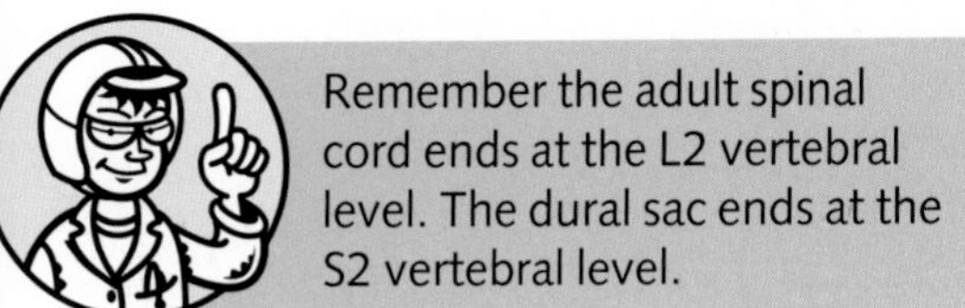

Remember the adult spinal cord ends at the L2 vertebral level. The dural sac ends at the S2 vertebral level.

Arachnoid mater

The arachnoid mater is a delicate avascular membrane that encloses the subarachnoid space. It may be separated from the dura by a potential space—the subdural space, e.g. in a subdural haematoma. Like the dura it covers the spinal nerve roots and spinal ganglia.

Subarachnoid space

The subarachnoid space lies between the arachnoid and the pia mater, and it contains the spinal cord, nerve roots, and ganglia. An enlargement of this space inferior to the conus medullaris is called the lumbar cistern. It contains the cauda equina and extends from the L2 vertebra to the S3 vertebra. It is a site where a lumbar puncture needle is inserted.

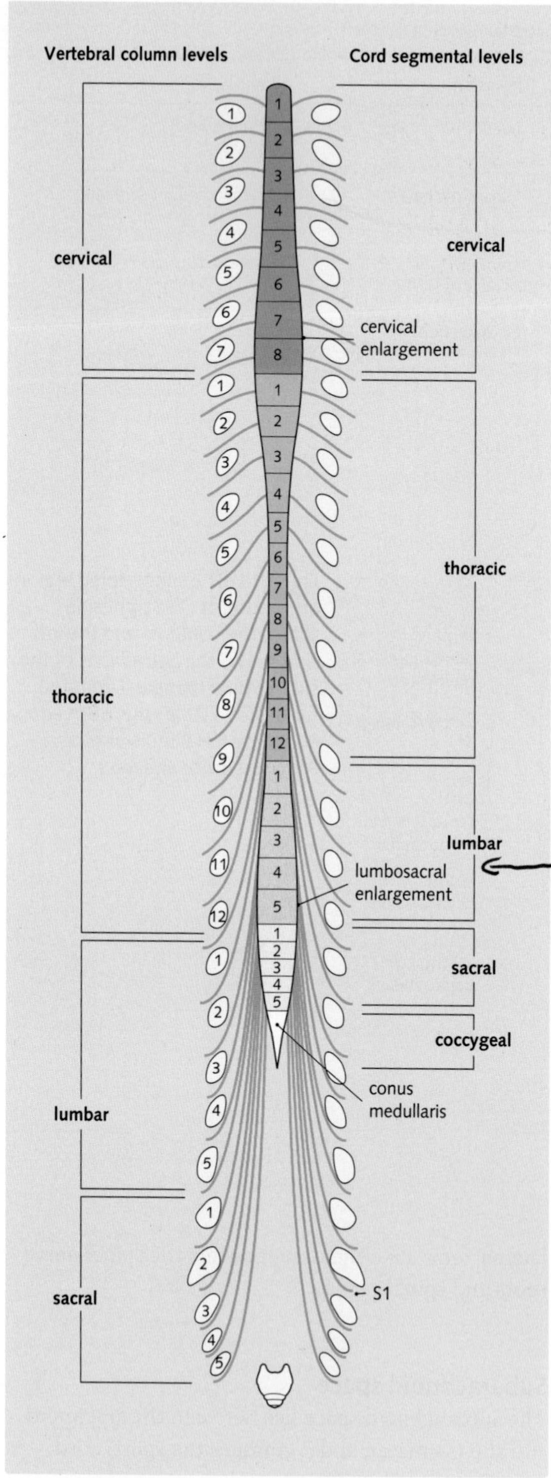

Fig. 8.12 Spinal cord, showing vertebral and segmental levels. (Adapted from *Anatomy as a Basis for Clinical Medicine*, by E C B Hall-Craggs. Courtesy of Williams & Wilkins.)

Pia mater

The pia mater is a finely vascular membrane that is closely adherent to the spinal cord. It covers the roots of the spinal nerves and the spinal ganglia. Below the conus medullaris the pia continues as the filum terminale. It pierces the dural sac at the S2 vertebral level to attach to the coccyx. The pia mater extends laterally as the denticulate ligament, which attaches to the dural sac's inner surface.

Lumbar puncture

The spinal cord ceases at the level of L2 vertebra, but the meninges continue well below this. By inserting a needle into the subarachnoid space below L2, a sample of CSF may be obtained without damaging the cord.

- Outline the surface anatomy of the back.
- Describe the cutaneous innervation of the back.
- Describe the features of individual vertebrae.
- Describe the joints of the vertebral column.
- What is an intervertebral disc and its function?
- Describe the ligaments of the vertebral column.
- Outline the movements of the vertebral column.
- Outline the blood supply to the vertebral column and the spinal cord.
- Outline the boundaries and contents of the suboccipital triangle.
- Outline the spinal cord and its surrounding meninges.

SELF-ASSESSMENT

Multiple-choice Questions

ndicate whether each answer is true or false.

Chapter 1—Basic Concepts of Anatomy

1. Concerning the anatomical position:

(a) The body is horizontal.
(b) The lower limbs are together.
(c) The palms of the hand face anteriorly.
(d) The feet point laterally.
(e) The head faces anteriorly with the eyes looking into the distance.

2. Concerning anatomical planes and terms of position:

(a) The median sagittal plane runs vertically through the midline of the body, from anterior to posterior.
(b) A parasagittal plane is perpendicular to the median sagittal plane.
(c) A coronal plane is perpendicular to the median sagittal plane.
(d) Proximal describes a structure as being closer to the trunk or an origin.
(e) Superficial describes a structure as being in front of another structure.

3. Regarding terms of movement:

(a) Abduction is a movement towards the median plane.
(b) Flexion is a movement in a sagittal plane that increases the angle of a joint.
(c) Eversion is a movement of turning the sole of the foot outwards.
(d) Abduction of the thumb is when the thumb moves anteriorly away from the palm.
(e) Extension of the thumb is when the thumb moves laterally away from the axis of digit abduction/adduction.

4. Regarding the skin:

(a) It has metabolic functions.
(b) It is composed of two layers.
(c) It is involved in thermoregulation.
(d) Its dermis contains a rich network of vessels and nerves.
(e) The hypodermis acts as a waterproof barrier.

5. Concerning bone:

(a) Long bones are involved in red blood cell synthesis.
(b) It is surrounded by periosteum.
(c) It has an outer spongy layer.
(d) It has two main blood supply sources.
(e) The metaphysis lies between the diaphysis and the epiphysis.

6. Regarding joints:

(a) Synovial joints lack a fibrous capsule.
(b) They receive their nerve supply from a motor nerve that supplies a muscle that moves that joint.
(c) Ligaments prevent excessive joint movement.
(d) Only ligaments and muscles stabilize joints.
(e) There are four types of joint.

7. Concerning muscle:

(a) A prime mover is a muscle responsible for the majority of a movement.
(b) Muscle fibres running parallel to the pull of a muscle allow great mobility.
(c) Multiple motor neurons in a motor unit allow precise movements.
(d) For a fixed muscle volume, shorter muscle fibres produce a greater contractile force.
(e) An aponeurosis is a flattened tendon.

8. Concerning the nervous system:

(a) It can be divided into somatic and autonomic parts.
(b) The head is supplied only by segmental spinal nerves.
(c) Sympathetic nerve fibres arise from the cranial and sacral parts of the central nervous system.
(d) The skeletal muscle motor fibres leave the spinal cord in the posterior root.
(e) The cell bodies of neurons in the central nervous system aggregate to form nuclei.

9. Concerning the cardiovascular and lymphatic systems:

(a) The left side of the heart pumps blood to the lungs (pulmonary circulation).
(b) All veins contain valves to prevent the backflow of blood.
(c) An anastomosis is a communication between two vessels.
(d) The lymphatic system removes excess tissue fluid.
(e) The lymphatics of the right side of the head, neck, and thorax drain into the thoracic duct.

Chapter 2—The Upper Limb

10. Regarding the bones of the upper limb:

(a) The scapula helps to increase the range of movement of the shoulder joint.
(b) The long head of the triceps muscle is attached proximally to the coracoid process.
(c) A surgical neck fracture of the humerus may result in injury to the axillary nerve.

(d) The glenoid cavity is surrounded by a rim of cartilage (the glenoid labrum), which is deficient inferiorly.
(e) The radial nerve runs in the spiral groove of the radius.

11. Regarding the muscles of the upper limb:

(a) The latissimus dorsi muscle is supplied by the long thoracic nerve.
(b) The pectoralis minor muscle divides the axillary artery into three parts.
(c) The subscapularis muscle is the only rotator cuff muscle to attach to the lesser tubercle of the humerus.
(d) The pectoralis major muscle is supplied by the medial pectoral nerve only.
(e) The supraspinatus muscle initiates abduction of the arm.

12. Concerning the joints of the upper limb:

(a) In the adult, they are all synovial joints (except the interosseous membrane if included as a joint).
(b) The shoulder joint is completely surrounded by the rotator cuff muscles.
(c) The shoulder joint relies greatly on the glenohumeral ligaments for stability.
(d) Abduction of the shoulder joint is performed by the deltoid and infraspinatus muscles.
(e) The sternoclavicular joint has an intra-articular disc.

13. Concerning the axilla:

(a) It contains no lymph nodes.
(b) The axillary artery commences at the outer border of the first rib.
(c) The axillary vein lies within the axillary sheath.
(d) It is bounded laterally by the intertubercular sulcus, and coracobrachialis and biceps brachii muscles.
(e) The trunks of the brachial plexus are found in the axilla.

14. Regarding the brachial plexus:

(a) It is formed by the posterior rami of C5–C8 and T1 spinal nerves.
(b) The divisions of the brachial plexus are behind the clavicle.
(c) The cords of the brachial plexus are named with respect to the axillary artery.
(d) The medial and lateral cords form the median nerve.
(e) The ulnar nerve supplies muscles in the arm, forearm, and hand.

15. Regarding the vessels of the upper limb:

(a) The axillary artery becomes the brachial artery at the inferior border of teres minor.
(b) The brachial artery is superficial throughout most of its course.
(c) The radial artery may be palpated in the anatomical snuffbox.
(d) The ulnar artery runs lateral to the ulnar nerve in the wrist region.
(e) The common interosseous artery arises from the ulnar artery just before it enters the hand.

16. Concerning nerves of the upper limb:

(a) Damage to the radial nerve results in sensory loss over the majority of the hand.
(b) The radial nerve supplies the extensor compartment of the arm.
(c) The radial nerve just below the elbow gives off the posterior cutaneous nerve of the forearm.
(d) The axillary nerve arises from the posterior cord of the brachial plexus.
(e) The musculocutaneous nerve is motor to the flexor muscles of the arm.

17. Concerning the cubital fossa:

(a) The supinator and brachialis muscles form the fossa' floor.
(b) The brachial artery usually divides in the fossa into radial and ulnar arteries.
(c) The biceps brachii tendon is found medial to the brachial artery.
(d) The radial nerve divides into its deep and superficial branches in the fossa.
(e) The median nerve lies lateral to the brachial artery.

18. Regarding the nerves of the upper limb:

(a) The median nerve is easily damaged as it crosses the wrist.
(b) The median nerve supplies all the muscles of the thenar eminence.
(c) The median and ulnar nerves have no branches in the axilla.
(d) The median and radial nerves supply flexor digitorum profundus muscles.
(e) The ulnar nerve is motor to all the interossei muscles of the hand.

19. Concerning the muscles of the upper limb:

(a) The supinator muscle is supplied by the median nerve.
(b) The median nerve and ulnar artery pass into the forearm between the origins of the flexor digitorum superficialis.
(c) The median nerve and ulnar artery pass into the forearm between the origins of the pronator teres.
(d) The ulnar nerve enters the forearm superficial to the heads of flexor carpi ulnaris muscle.
(e) The deltoid muscle is supplied by the axillary nerve.

20. Regarding the carpal tunnel:

(a) It is roofed by the extensor retinaculum.
(b) The flexor digitorum superficialis tendons enter the tunnel stacked in two rows.
(c) It transmits the ulnar nerve.
(d) Thenar and hypothenar eminence muscles arise from the flexor retinaculum.

(e) Compression of the nerve in the tunnel causes paralysis of the hypothenar eminence and the medial three and a half digits.

1. Concerning the intrinsic muscles of the hand:

(a) The dorsal interosseous muscles adduct the digits.
(b) The medial two lumbricals are supplied by the median nerve.
(c) The radial artery enters the palm between the two heads of adductor pollicis to form the superficial palmar arch.
(d) The opponens pollicis muscle inserts into the base of the proximal phalanx.
(e) The lumbrical muscles arise from the tendons of flexor digitorum superficialis.

Chapter 3—The Thorax

2. Regarding the surface anatomy of the great vessels:

(a) The aortic arch begins behind the manubriosternal joint, passing to the left and posteriorly.
(b) The left common carotid and subclavian arteries lie behind the manubrium.
(c) The brachiocephalic trunk bifurcates into the left common carotid and subclavian arteries behind the left sternoclavicular joint.
(d) The superior vena cava is formed behind the right first costal cartilage.
(e) The superior vena cava enters the heart behind the second costal cartilage.

3. Concerning the thoracic wall:

(a) The manubrium only articulates with the clavicles, sternum, and first and second costal cartilages.
(b) The body of the sternum articulates with the second to sixth costal cartilages.
(c) A typical rib articulates with the same numerical vertebral body and the vertebral body above.
(d) The eleventh and twelfth ribs do not articulate with the ribs above to form the costal margin and, therefore, they have no costal cartilage on their tips.
(e) A cervical rib can compress the subclavian artery and inferior trunk of the brachial plexus.

4. Regarding the thoracic wall and intercostal spaces:

(a) The thoracic wall is made of both bone and cartilage.
(b) The neurovascular bundle runs deep to the three muscle layers.
(c) The intercostal nerves are the posterior rami of the thoracic spinal nerves.
(d) Motor innervation to the intercostal muscles is from the phrenic nerve and the intercostal nerves.
(e) The intercostal arteries all arise from the aorta.

5. The diaphragm:

(a) Has a motor nerve supply from the phrenic nerve only.
(b) Transmits the aorta at the level of T10 vertebra.
(c) Transmits the inferior vena cava through the central tendon.
(d) Is pierced by the sympathetic trunks.
(e) Is the main muscle used in quiet respiration.

26. The pericardium:

(a) Is divided into serous and fibrous layers.
(b) Lies in the anterior and middle mediastinum.
(c) Consists of a fibrous layer that is attached to the central tendon of diaphragm, sternum, and great vessels.
(d) The nerve supply to the fibrous, parietal serous and visceral serous layers is by the phrenic nerve.
(e) The reflection of the serous pericardium around the pulmonary veins forms the transverse sinus.

27. Regarding the heart:

(a) It is completely surrounded by pericardium that is attached to the left dome of the diaphragm.
(b) It has the fossa ovalis lying on the interatrial wall.
(c) It is supplied by coronary arteries that arise from the aortic sinuses.
(d) The sinoatrial node is the normal cardiac pacemaker.
(e) The mitral valve lies in the left atrioventricular orifice.

28. Concerning the blood supply of the heart:

(a) The coronary arteries are the first branches of the aorta.
(b) The anterior cardiac vein drains into the coronary sinus.
(c) The great cardiac vein runs in the interventricular groove with the anterior interventricular artery.
(d) If the posterior interventricular artery arises from the circumflex artery, the heart is said to have right dominance.
(e) The cardiac veins drain into the right atrium (excluding the venae cordis minimae).

29. Regarding the thoracic aorta:

(a) The ascending aorta has no branches.
(b) The aortic arch is attached to the right pulmonary artery via the ligamentum arteriosum.
(c) The descending aorta commences at the level of the sternal angle.
(d) The descending aorta gives branches to the pericardium, oesophagus, and bronchi.
(e) The aorta leaves the thorax by passing through the central tendon of the diaphragm at the level of T12 vertebra.

30. Concerning thoracic nerves:

(a) The phrenic nerve passes anterior to the lung hilum.
(b) The phrenic nerve arises from anterior rami of C3–C5 spinal nerves.
(c) The vagus and phrenic nerves contribute to the formation of the pulmonary and oesophageal plexuses.
(d) Both right and left vagi leave the thorax lying anterior to the oesophagus.

(e) The thoracic sympathetic ganglia are concerned with supplying intrathoracic structures only.

31. The oesophagus:

(a) Is a continuation of the laryngopharynx at the level of C8 vertebra.
(b) Passes inferiorly behind the trachea.
(c) Has constrictions caused by the cricopharyngeus muscle, aorta, left main bronchus and diaphragm.
(d) Its lower part has a blood supply from oesophageal branches of the aorta.
(e) Exits the thorax by piercing the muscle of the diaphragm at T10.

32. Concerning the lungs and bronchi:

(a) The left main bronchus is shorter, wider, and more vertical than the right.
(b) The right lung has three lobes, and the left lung has two lobes.
(c) Bronchopulmonary segments are the functional units of the lungs.
(d) The apex of the lung lies below the clavicle.
(e) The visceral pleura has a rich somatic innervation, and it is very sensitive to pain.

33. Regarding the mechanics of the respiration:

(a) In quiet inspiration, the diaphragm contracts, increasing the vertical diameter of the thorax.
(b) In adult males, thoracic movement is greater than diaphragmatic movement.
(c) In forced inspiration, the twelfth rib is fixed by quadratus lumborum muscle allowing a more forceful diaphragmatic contraction.
(d) In quiet expiration, the intercostal muscles contract.
(e) In forced expiration, the abdominal wall muscles contract pushing abdominal viscera upwards and thus the diaphragm.

Chapter 4—The Abdomen

34. Concerning the surface anatomy of the abdomen:

(a) The upper border of the liver runs behind the fourth rib.
(b) The fundus of the gall bladder is marked by the linea semilunaris intersecting the transpyloric plane.
(c) The spleen lies deep to the right ninth, tenth, and eleventh ribs.
(d) The head of the pancreas lies in the transpyloric plane.
(e) The upper poles of the kidneys lie in front of the twelfth rib posteriorly.

35. Regarding the anterolateral abdominal wall:

(a) The external and internal oblique muscles both form the conjoint tendon.
(b) The rectus sheath, below the arcuate line, passes posterior to the rectus abdominis muscle.
(c) The inferior epigastric artery enters the rectus sheath and runs deep to the rectus abdominis muscle.
(d) The ilioinguinal nerve (L1) forms part of the nerve supply to all anterolateral abdominal wall muscles except external oblique.
(e) The right upper two lumbar veins join to form an ascending lumbar vein, which joins the azygos vein.

36. Regarding the inguinal canal:

(a) The superficial inguinal ring is an opening in the external oblique aponeurosis.
(b) The deep inguinal ring is a defect in the internal oblique muscle.
(c) The spermatic cord, with its three coverings, is formed at the deep inguinal ring.
(d) The ilioinguinal nerve enters the inguinal canal via the deep inguinal ring.
(e) The canal lies above and parallel to the inguinal ligament.

37. Concerning the peritoneum and the peritoneal cavity:

(a) The parietal layer lines the anterior and posterior abdominal walls, inferior surface of the diaphragm, and the pelvic cavity.
(b) The parietal layer lining the pelvis is supplied by the obturator nerve.
(c) The medial umbilical fold contains the inferior epigastric vessels.
(d) The greater sac is divided into two compartments by the transverse mesocolon.
(e) The epiploic foramen is bounded anteriorly by the portal vein, and posteriorly by the inferior vena cava.

38. The stomach:

(a) Is derived from the foregut.
(b) Has a fundus that lies above the level of entry of the oesophagus.
(c) Is not covered by peritoneum over the fundus.
(d) Has a blood supply mainly derived from the coeliac trunk.
(e) Drains mainly into the azygos system of veins.

39. Regarding the small intestine:

(a) The duodenum begins in the transpyloric plane.
(b) Lymph from the duodenum drains into the coeliac and superior mesenteric nodes.
(c) The jejunum has shorter vasa recta than the ileum.
(d) The ileum has a lesser amount of fat than the jejunum.
(e) The small intestine is supplied by the vagus nerve and sympathetic fibres of T9 and T10 spinal cord levels.

40. Concerning the colon:

(a) It is retroperitoneal throughout its course.
(b) It is supplied by the superior mesenteric artery only as far as the hepatic flexure.
(c) It is surrounded by a continuous layer of longitudinal and circular muscle.
(d) The sigmoid colon is closely related to the left ureter.

e) Appendices epiploicae are more frequent in the sigmoid and descending colon compared with the ascending colon.

1 Regarding the liver:

a) It is completely covered by peritoneum except over its anterior surface.
b) The hepatic artery, hepatic vein, and bile duct enter and leave the liver at the porta hepatis.
c) The fetal ductus venosus is represented by the ligamentum teres in the adult.
d) Venous blood from the liver enters the inferior vena cava via the portal vein.
e) It receives its arterial supply from a branch of the coeliac trunk.

2. Concerning the gall bladder:

a) The bile duct is formed by the union of the cystic duct with the common hepatic duct.
b) The bile duct unites with the main pancreatic duct to empty into the third part of the duodenum.
c) The gall bladder is usually supplied by a branch of the left hepatic artery.
d) The fundus of the gall bladder lies at the tip of the ninth costal cartilage.
e) The gall bladder stores and concentrates bile.

3. Regarding the kidneys and suprarenal blood vessels:

a) The right kidney is lower than the left kidney.
b) The ureter lies anterior to the renal artery and vein at the renal pelvis.
c) The kidneys are both retroperitoneal.
d) The suprarenal gland receives its blood from the suprarenal artery only.
e) The suprarenal gland is innervated by the abdominal sympathetic trunk.

4. Regarding the abdominal blood vessels:

a) The abdominal aorta ends at the level of L4 vertebra by dividing into the common iliac arteries.
b) The superior mesenteric artery arises from the aorta at the level of L2 vertebra.
c) The inferior mesenteric artery supplies the derivatives of the hindgut.
d) The inferior vena cava drains the two gonadal veins directly.
e) There are numerous anastomoses of blood vessels supplying the appendix.

5. Concerning the nerves of the abdomen:

a) The lumbar plexus is formed by the anterior rami of L1–L4 spinal nerves.
b) The femoral nerve and obturator nerve lie medial to psoas major.
c) The iliohypogastric and ilioinguinal nerves arise from L1 spinal nerve.
d) The genitofemoral nerve is involved in the cremasteric reflex.
(e) The obturator nerve is sensory to the medial aspect of the thigh.

Chapter 5—The Pelvis and Perineum

46. Concerning the bony pelvis:

(a) The greater or false pelvis lies below the pelvic brim.
(b) The pelvic inlet lies at 45 degrees to the outlet.
(c) The adult hip bone consists of two fused bones.
(d) In an articulated pelvis in an erect individual, the anterior inferior iliac spines and pubic symphysis lie in the same vertical plane.
(e) The female pubic arch is wider than the male pubic arch.

47. Concerning the rectum:

(a) It commences where the sigmoid mesocolon ends.
(b) The superior and middle rectal arteries supply the rectum.
(c) It usually has three permanent transverse folds.
(d) The rectovesical pouch in females separates the rectum from the bladder.
(e) The mucosa is non-keratinizing, stratified squamous epithelium.

48. Concerning the pelvic ureters:

(a) The ureters turn anteriorly towards the bladder at the level of the ischial spine.
(b) They enter the bladder horizontally.
(c) In the female the uterine artery passes inferior to the ureter.
(d) In the male no structure crosses superiorly over the ureter.
(e) The ureters cross the obturator vessels and nerve.

49. Regarding the pelvic viscera:

(a) The bladder is covered by peritoneum on its superior surface.
(b) The broad ligament of the uterus contains the uterine tubes and ovarian vessels.
(c) The ovary lies on the posterior surface of the broad ligament.
(d) The transverse cervical ligaments play a part in the stabilization of the uterus.
(e) The posterior fornix of the vagina is deeper than the anterior fornix.

50. Concerning the pelvic vessels:

(a) The ovarian artery arises from the internal iliac artery.
(b) The internal pudendal artery leaves the pelvis through the lesser sciatic foramen.
(c) The obturator artery is crossed laterally by the ureter.
(d) The uterine artery may anastomose with the ovarian and vaginal arteries.
(e) The uterine artery is closely related to the ureter just before it reaches the uterus.

51. Concerning the nerves of the pelvis:

(a) The sacral plexus is formed by the S1–S5 anterior rami only.
(b) The sacral sympathetic trunk contains four ganglia.
(c) The pelvic splanchnic nerves carry sympathetic nerve fibres.
(d) The pudendal nerve roots are S2, S3, and S4.
(e) The right and left hypogastric plexuses receive a right and left hypogastric nerve, respectively.

52. Regarding the perineum:

(a) It can be divided into an anterior (urogenital) triangle and a posterior (anal) triangle.
(b) The external anal sphincter is formed by two muscles.
(c) The anal triangle contains the two ischiorectal fossae.
(d) The urogenital triangle contains the perineal membrane, which lies above the pelvic floor.
(e) The pudendal canal transmits the pudendal nerve only.

53. Concerning the anal canal:

(a) It is supplied entirely by the autonomic nervous system.
(b) It commences at the anorectal junction.
(c) At the pectinate line, the epithelium becomes stratified squamous.
(d) The internal anal sphincter is composed of the longitudinal muscle layer.
(e) Voluntary contraction of the external sphincter may delay defaecation.

54. Regarding the urogenital triangle:

(a) It lies below the anterior part of the pelvic floor.
(b) The deep perineal space contains the sphincter urethrae and deep transverse perineal muscles.
(c) The perineal membrane is the inferior boundary of the deep perineal space.
(d) Lymphatics of the urogenital triangle drain into the common iliac nodes.
(e) The superficial perineal space is lined by a membranous fascia that is continuous with fascia of the anterior abdominal wall.

55. Regarding the external genitalia:

(a) The penis has a fascial covering that is an extension of the superficial fascia of the anterior abdominal wall.
(b) The penis receives its blood supply from the internal pudendal artery.
(c) The iliohypogastric nerves supply the labia majora anteriorly.
(d) The pudendal nerve does not supply the clitoris.
(e) Either side of the vagina there are two erectile masses called the bulb of the vestibule.

Chapter 6—The Lower Limb:

56. Concerning the bones of the lower limb:

(a) At puberty the three bones of the hip are separated by a Y-shaped cartilage in the acetabulum.
(b) The median sacral crest represents the fused spinal processes of the sacral vertebrae.
(c) The greater sciatic notch is formed by the posterior margin of the ischium.
(d) The greater trochanter of the femur gives attachment to the piriformis muscle.
(e) The obturator foramen transmits the obturator arter and nerve.

57. Concerning the gluteal region:

(a) The close fit of the femoral head in the acetabulum contributes to the stability of the hip joint.
(b) Abduction of the hip is performed by the gluteus medius and minimus muscles.
(c) The only structures to pass above the piriformis muscle from the greater sciatic foramen are the superior gluteal nerve and vessels.
(d) Cutaneous innervation is supplied by the gluteal branches of the posterior rami of lumbar and sacral nerves.
(e) The pudendal nerve exits the pelvis through the lesser sciatic foramen and re-enters through the greater sciatic foramen.

58. Regarding the femoral triangle:

(a) The femoral vein, artery, and nerve lie in the femora sheath.
(b) The femoral nerve lies most medially in the femoral triangle.
(c) The lateral border of the femoral triangle is formed by the lateral border of the sartorius muscle.
(d) The femoral canal lies medial to the femoral vein.
(e) At the apex of the femoral triangle, the femoral vessels pass into the adductor canal.

59. Regarding the vessels of the lower limb:

(a) The femoral artery is a direct continuation of the internal iliac artery as it passes below the inguinal ligament.
(b) The profunda femoris artery supplies the muscles of the medial and posterior compartments of the thigh
(c) The great saphenous vein drains into the femoral vein.
(d) The profunda femoris vein drains into the great saphenous vein.
(e) In increased venous return, the femoral vein can expand into the femoral canal.

60. Concerning nerves of the lower limb:

(a) The obturator nerve supplies the muscles of the adductor compartment of the thigh.
(b) The sciatic nerve leaves the pelvis via the greater sciatic foramen.

(c) The femoral nerve is formed by the anterior rami of the L3–L5 spinal nerves.
(d) The superior gluteal nerve supplies the gluteus maximus muscle.
(e) The sciatic nerve may be damaged by posterior dislocation of the hip joint.

61. Regarding the popliteal fossa:

(a) The popliteal artery is the deepest vessel of the popliteal fossa.
(b) The popliteal artery has no branches in the popliteal fossa.
(c) The small saphenous vein joins the popliteal vein in the popliteal fossa.
(d) The sciatic nerve divides into the tibial and superficial peroneal nerves in the popliteal fossa.
(e) The popliteus muscle locks the knee joint.

62. Concerning the knee joint:

(a) The knee joint is an articulation between the femur, tibia, and fibula.
(b) Only flexion and extension movements are possible at the knee joint.
(c) The cavity of the knee joint communicates with the suprapatella bursa.
(d) Tearing of the tibial (medial) collateral ligament can cause contaminant tearing of the medial meniscus.
(e) Extension of the knee joint is performed by the quadriceps femoris muscle.

63. Regarding the leg:

(a) The anterior compartment of the leg is supplied by the superficial peroneal nerve.
(b) The tibialis anterior muscle dorsiflexes and everts the foot.
(c) The anterior tibial artery is a branch of the posterior tibial artery.
(d) The posterior tibial artery may be palpated behind the lateral malleolus.
(e) The peroneus tertius muscle dorsiflexes and everts the foot.

64. Concerning the nerves of the lower limb:

(a) The tibial nerve ends by dividing into the medial and lateral plantar nerves.
(b) The saphenous nerve passes behind the medial malleolus to supply the medial side of the foot.
(c) Trauma to the common peroneal nerve results in foot inversion and footdrop.
(d) The tibial nerve supplies the anterior compartment muscles of the leg.
(e) The sural nerve is cutaneous only.

65. Concerning the dorsum of the foot:

(a) The dorsalis pedis artery is a continuation of the anterior tibial artery in the foot.
(b) The extensor digitorum longus tendons are joined by lumbrical and interossei muscle tendons to form an extensor expansion.
(c) Extensor hallucis brevis muscle inserts into the distal phalanx of the great toe.
(d) The superficial and deep peroneal nerves supply the dorsal skin of the foot.
(e) The extensor digitorum brevis is supplied by the deep peroneal nerve.

66. Regarding the leg:

(a) Muscles of the lateral compartment of the leg evert and plantar flex the foot.
(b) The posterior compartment muscles of the leg are divided into superficial and deep by the deep transverse crural fascia.
(c) The tibialis posterior tendon is an immediate posterior relation to the lateral malleolus.
(d) The gastrocnemius and plantaris muscles form the tendocalcaneus.
(e) The lateral compartment muscles of the leg are supplied by the superficial peroneal nerve.

67. Concerning the foot:

(a) The ankle joint is most stable in dorsiflexion.
(b) The axis of abduction and adduction passes through the third digit.
(c) Flexor accessorius (quadratus plantae) muscle action allows plantar flexion of the toes even if the foot is plantar flexed.
(d) Inversion and eversion movements occur at the subtalar joint.
(e) The lateral plantar arch formed by the calcaneus, cuboid, and lateral two metatarsals is higher than the medial arch.

Chapter 7—The Head and Neck

68. Concerning the skull:

(a) The sagittal suture is formed by the parietal bones articulating with the frontal bone.
(b) The bregma is where the lambdoid suture joins the sagittal suture.
(c) The lambda is where the coronal suture joins the lambdoid suture.
(d) The pterion is formed by the sphenoid, temporal, frontal, and parietal bones.
(e) The zygomatic arch is formed by the zygomatic process of the temporal bone and the temporal process of the zygomatic bone.

69. Regarding the skull foramina:

(a) The foramen rotundum transmits the mandibular (V_3) nerve.
(b) The vertebral arteries enter the skull via the foramen magnum.
(c) The external acoustic meatus transmits the facial and vestibulocochlear nerves.

(d) The middle meningeal artery passes through the foramen ovale.
(e) The jugular foramen transmits the CN IX, CN X and CN XI nerves.

70. Concerning the face and scalp:

(a) The buccal branch of the facial nerve is sensory to the buccal mucosa.
(b) The muscles of facial expression lie below the deep fascia of the face.
(c) The trigeminal nerve is sensory to the entire scalp region.
(d) The facial vein is formed by the union of the supraorbital and supratrochlear veins.
(e) The arterial supply to the face and scalp is from branches of the external carotid artery only.

71. Concerning the cranial cavity:

(a) The tentorium cerebelli forms the roof over the posterior cranial fossa.
(b) The sigmoid venous sinus becomes the internal jugular vein at the jugular foramen.
(c) Infection from the infraorbital region may spread to the cavernous sinus.
(d) The circulus arteriosus connects the internal and external carotid arteries.
(e) The internal carotid artery and the trochlear nerve run through the cavernous sinus.

72. Concerning the cranial cavity:

(a) The pituitary gland sits in the dorsum sellae.
(b) The superior sagittal sinus runs in the inferior border of the falx cerebri.
(c) The dura mater is supplied by the trigeminal nerve anteriorly.
(d) The falx cerebelli separates the two cerebral hemispheres.
(e) The internal carotid artery enters the middle cranial fossa via the foramen lacerum.

73. Regarding the cranial nerves:

(a) The optic nerve is accompanied by the ophthalmic artery through the optic canal.
(b) The cranial part of the accessory nerve is sensory to the mucosa of the pharynx.
(c) The trigeminal nerve is motor to the muscles of mastication.
(d) The oculomotor nerve transmits the parasympathetic fibres to the lacrimal gland.
(e) The facial nerve transmits taste fibres from the posterior third of the tongue.

74. Concerning the orbit:

(a) The central artery of the retina is an end artery.
(b) The superior rectus muscle is supplied by the trochlear nerve.
(c) Levator palpebrae superioris has a somatic and autonomic nerve supply.
(d) Inferior oblique moves the cornea inferiorly and laterally.
(e) The inferior orbital fissure communicates with the pterygopalatine fossa.

75. Concerning the orbit:

(a) The nasolacrimal duct empties into the middle meatus of the nasal cavity.
(b) The parasympathetic fibres of the lacrimal gland synapse in the pterygopalatine ganglion.
(c) The roof of the orbit is formed by the frontal bone.
(d) The abducent nerve supplies the medial rectus muscle.
(e) The long ciliary branch of the nasociliary nerve carries sympathetic fibres.

76. Regarding the salivary glands:

(a) The external carotid artery and the retromandibular vein lie in the substance of the parotid gland.
(b) The parotid gland duct opens in the oral cavity opposite the second premolar tooth.
(c) The facial nerve gives off its five terminal branches before it enters the parotid gland.
(d) The superficial and deep parts of the submandibular gland are continuous around the mylohyoid muscle.
(e) Paraesthesia to the tongue may result following surgery to the sublingual glands.

77. Regarding the temporal and infratemporal fossae:

(a) The deep temporal arteries are branches of the superficial temporal artery.
(b) The temporalis muscle attaches to the condyloid process of the mandible.
(c) To clinically test the masseter and temporalis the patient clenches their teeth.
(d) The temporomandibular joint is strengthened by the lateral temporomandibular ligament.
(e) Secretomotor fibres for the parotid gland synapse in the otic ganglion.

78. Concerning the ear and the vestibular apparatus:

(a) The external auditory meatus is one third bone and two thirds cartilage.
(b) The tympanic membrane's external surface is supplied by the auriculotemporal nerve.
(c) The tympanic cavity and nasopharynx are connected by the auditory tube.
(d) Tensor tympani increases the vibration of the tympanic membrane.
(e) The facial nerve exits the stylomastoid foramen of the temporal bone.

79. In the neck region:

(a) The external jugular vein is formed by the union of the posterior auricular vein and the posterior division of the retromandibular vein.
(b) The accessory nerve supplies the trapezius muscle.
(c) The digastric muscle is supplied by the facial nerve only.

(d) The thyrohyoid muscle is supplied by the recurrent laryngeal nerve.
(e) The superior thyroid artery arises from the posterior aspect of the aspect of the external carotid artery.

---

80. Concerning the neck:

(a) The investing layer of the deep fascia forms the sphenomandibular ligament.
(b) The cervical plexus is formed by the anterior rami of C1–C4 spinal nerves.
(c) The pretracheal fascia blends with the fibrous pericardium inferiorly.
(d) The internal carotid artery has only two branches in the neck.
(e) The deep cervical lymph nodes lie along the posterior aspect of the sternocleidomastoid.

---

81. Concerning the neck region:

(a) The investing layer of the deep cervical fascia splits to enclose the sternocleidomastoid and trapezius muscles.
(b) The pretracheal fascia extends laterally to form the axillary sheath.
(c) The brachial plexus roots emerge between the scalenus medius and scalenus posterior muscles.
(d) The platysma muscle is supplied by the ansa cervicalis.
(e) At the terminal part of the common carotid artery and the origin of the internal carotid artery there is a carotid sinus.

---

82. Concerning the neck:

(a) The infrahyoid muscles are all supplied by the ansa cervicalis.
(b) The hypoglossal nerve curves around the external carotid artery.
(c) There are three sympathetic ganglia in the neck.
(d) The internal jugular vein unites with the subclavian vein behind the sternoclavicular joint.
(e) The vagus supplies the stylopharyngeus muscle.

---

83. Regarding the neck:

(a) The common carotid artery divides at the upper border of the thyroid cartilage.
(b) The maxillary and superficial temporal arteries are the terminal branches of the external carotid artery.
(c) The deep cervical nodes eventually drain into the jugular trunk.
(d) The external laryngeal nerve is sensory to the mucosa above the vocal folds.
(e) The glossopharyngeal nerve is motor to the tongue.

---

84. Regarding the pharynx:

(a) The pharyngeal plexus supplies all the muscles of the pharynx.
(b) The pharynx is divided into the nasopharynx, oropharynx and laryngopharynx.
(c) The pharynx finishes at the C6 vertebra.
(d) During swallowing, the soft palate is elevated.
(e) Mucosa can herniate between the cricopharyngeus and the thyropharyngeus muscles.

---

85. Concerning the pharynx:

(a) The posterior third of the tongue forms the anterior wall of the oropharynx.
(b) Tongue mucosa is reflected on to the epiglottis.
(c) The pharyngobasilar fascia separates pharyngeal mucosa and muscle.
(d) Either side of the laryngopharynx is a groove—the piriform recess.
(e) The nasopharnyx has a nerve supply by the maxillary nerve (V_2).

---

86. Regarding the nose:

(a) The nasal septum is completely cartilaginous.
(b) The nasal floor is formed by the hard palate.
(c) The sphenoethmoidal recess has the sphenoid sinus opening into it.
(d) The CN I supplies sensory fibres to part of the lateral wall of the nasal cavity.
(e) The vestibule is a common site for epistaxis (nose-bleed).

---

87. Regarding the nasal cavity and the pterygopalatine fossa:

(a) The nasal cavity warms and removes dust particles from inspired air.
(b) Three conchae divide the lateral nasal wall into three meatus.
(c) The pterygopalatine fossa communicates with the middle cranial fossa.
(d) The nerve of the pterygoid canal is formed by the deep petrosal nerve and the greater petrosal nerve.
(e) The lesser palatine nerve supplies the soft palate.

---

88. Concerning the bones of the skull and face:

(a) The lateral pterygoid muscle is attached to the coronoid process of the mandible.
(b) The maxilla take part in formation of the hard palate.
(c) The lingual nerve may be damaged during removal of a mandibular third molar tooth.
(d) The temporal bone transmits CN VII (facial nerve) and CN VIII (vestibulocochlear nerve).
(e) The coronoid process articulates with the mandibular fossa of the temporal bone.

---

89. Concerning the oral cavity:

(a) Sensation to the anterior two thirds of the tongue is supplied by the lingual nerve.
(b) The vallate papillae lie just posterior to the sulcus terminalis.
(c) The hard palate is formed by the horizontal process of the palatine bones alone.
(d) The extrinsic muscles of the tongue are supplied by the lingual nerve.

(e) The foramen caecum on the tongue is the embryological remnant marking the site of the upper end of the thyroglossal duct.

90. Concerning the oral cavity:

(a) The intrinsic muscles of the tongue are all supplied by the hypoglossal nerve.
(b) The chorda tympani supplies taste fibres to all of the tongue.
(c) The lymphatic supply to the tongue drains into the deep cervical lymph nodes.
(d) There are between ten and twelve vallate papillae.
(e) The palatine tonsil lies between the palatoglossal and palatopharyngeal arches.

91. Concerning the floor of the mouth and the soft palate:

(a) The submandibular gland is supplied by the facial and lingual arteries.
(b) Parasympathetic fibres to the submandibular gland originate in the inferior salivary nucleus.
(c) The sublingual gland is supplied by the lingual nerve.
(d) Tensor veli palatini is supplied by the pharyngeal plexus.
(e) The pharyngeal plexus is formed by the glossopharyngeal and vagus nerves.

92. Concerning the larynx:

(a) The sensory nerve supply to the mucosa of the larynx above the level of the vocal folds is from the recurrent laryngeal nerve.
(b) The cricothyroid muscle is supplied by the internal laryngeal nerve.
(c) The epiglottis is composed of elastic cartilage.
(d) The vocal ligament is the thickened upper edge of the cricothyroid membrane.
(e) The posterior cricoarytenoid muscle abducts the vocal folds.

93. Concerning the intrinsic muscles of the larynx:

(a) The recurrent laryngeal nerve supplies all the muscles except cricothyroid and is sensory to the mucosa below the vocal folds.
(b) The lateral cricoarytenoid muscle causes widening of the rima glottis.
(c) The cricothyroid muscle tenses the vocal folds.
(d) The vocalis muscle relaxes the vocal ligament.
(e) The transverse arytenoid muscle is the only unpaired intrinsic muscle of the larynx.

94. Regarding the larynx:

(a) The thyrohyoid membrane is pierced by the external laryngeal nerve.
(b) The inferior border of the quadrangular membrane forms the vestibular ligament.
(c) The laryngeal inlet is bounded by the thyroid cartilage, aryepiglottic and interarytenoid folds.
(d) The joints of the larynx are synovial.
(e) There is no membrane connecting the cricoid cartilage and trachea.

95. The thyroid gland:

(a) The thyroid gland is an exocrine gland that regulates metabolism.
(b) The thyroid has two lobes connected by an isthmus.
(c) The superior thyroid artery runs with the internal laryngeal nerve.
(d) The recurrent laryngeal nerve runs with the inferior thyroid artery.
(e) All thyroid veins drain into the internal jugular vein.

Chapter 8—The Back

96. Concerning the back:

(a) There are eight cervical vertebrae.
(b) The sacral vertebrae are the only fused vertebrae in the vertebral column.
(c) Cervical and lumbar curvatures of the vertebral column begin to appear after birth.
(d) Thoracic and sacrococcygeal curvatures are anteriorly concave.
(e) The C7 vertebra's spine is the highest easily palpable spinous process.

97. Concerning the vertebral column:

(a) There are 33 vertebrae.
(b) Individual vertebrae articulate with each other via articular facet joints only.
(c) The vertebral column supports the weight of the upper body.
(d) Herniation of a lumbar intervertebral disc's nucleus pulposus can cause lower back and lower limb pain.
(e) The anterior longitudinal ligament prevents hyperextension.

98. Regarding the movements of the vertebral column:

(a) Movements can occur due to the elasticity of the intervertebral discs.
(b) Rotation occurs at the atlanto-occipital joint.
(c) Flexion, extension, and lateral rotation occur in the cervical region.
(d) The ribs and sternum limit flexion and extension in the thoracic region.
(e) There is free rotation in the lumbar region.

99. Concerning the spinal cord:

(a) The spinal cord commences below the foramen magnum, and it ends at the L2 vertebra in adults.
(b) There are eight cervical nerves.
(c) The spinal cord receives its blood supply from branches of the vertebral and superior cerebellar and segmental arteries.
(d) The spinal cord has no lymphatic vessels.
(e) The cauda equina is composed of sacral nerves.

100. Concerning the meninges:

(a) The epidural space is between the dura mater and the vertebral periosteum.
(b) The dura mater attaches to the coccyx inferiorly.
(c) The arachnoid mater is separated from the dura mater by the potential subarachnoid space.
(d) The lumbar cistern is an enlargement of the subarachnoid space.
(e) The pia mater continues below the conus medullaris as the filum terminale.

Short-answer Questions

1. Briefly describe the boundaries and contents of the axilla.
2. Write short notes on the biceps brachii muscle.
3. Describe the flexor retinaculum of the wrist, including its bony attachments, and the contents of the carpal tunnel.
4. List the muscles, vessels, and nerves of a typical intercostal space.
5. Write short notes on the diaphragm.
6. Describe the anatomy of the oesophagus.
7. Describe the pleurae.
8. Briefly describe the surface anatomy of the abdomen.
9. Briefly describe the inguinal canal.
10. Illustrate the components of the rectus sheath.
11. Describe the origin, course, and distribution of the obturator nerve.
12. Write short notes on the rectum.
13. List the lymphatic drainage of the pelvis.
14. Describe the boundaries and contents of the femoral triangle.
15. Briefly describe the hip joint.
16. A hockey player is hit on the side of the head by an opponent's stick. He is dazed and is substituted by the manager. At the end of the match the player is confused, and he has a severe headache. On the way to hospital the player losses consciousness where he is diagnosed as having an extradural haemorrhage.

    What important bony landmark has been hit, and what does it mark?

    Where is the extradural haemorrhage located?

    What structure causes the headache?
17. A 63-year-old man sees his general practitioner, complaining of a swelling in his left cheek in front of his ear. He adds that his face on the left is weak and that the left corner of his mouth and eye had 'dropped'. The man claimed he dribbled from his mouth during eating and that his left eye is dry and painful. The general practitioner told the man that he probably had a parotid tumour. What is the origin, course, and distribution of the nerve affected?
18. Draw a diagram to illustrate the components and branches of the cervical plexus.
19. Describe the innervation of the mucosa of the tongue.
20. Describe the thyroid gland and its blood supply.

Essay Questions

1. Discuss the descriptive anatomical terms used in anatomy and clinical practice.
2. Discuss the anatomy of the shoulder joint, including the muscles responsible for its movement.
3. Describe the origin, formation, and branches of the brachial plexus.
4. Discuss the muscles affected by damage to the ulnar nerve at the elbow.
5. Discuss the divisions of the thoracic cavity and their contents.
6. Describe the anatomy of the heart, including its nerve and blood supply.
7. Discuss the muscles and movements involved in respiration.
8. Describe the anatomy of the peritoneum, including the peritoneal folds and sacs.
9. Describe the course and distribution of the abdominal aorta.
10. Summarize the somatic and autonomic nerves of the abdominal wall.
11. Summarize the anatomy of the bony pelvis, including the joints and pelvic floor.
12. Describe the anatomy of the female reproductive tract.
13. Discuss the anatomy of the urogenital triangle, including deep and superficial perineal spaces.
14. Discuss the anatomy of the gluteal region.
15. Discuss the anatomy of the knee joint.
16. Discuss the muscles of the leg compartments, including actions, and blood and nerve supply.
17. Describe the arrangement of the cranial meninges, including cranial venous sinuses.
18. Summarize the boundaries and contents of the orbital cavity.
19. Summarize the boundaries and contents of the neck.
20. Describe the anatomy of the larynx.

MCQ Answers

1. (a) False—the body is vertical.
   (b) True—the lower limbs are together with the feet slightly apart.
   (c) True—the palms are turned forwards.
   (d) False—the feet point anteriorly.
   (e) True—the head and eyes are directed forwards.

2. (a) True—this plane runs vertically through the middle of the body.
   (b) False—a parasagittal plane runs parallel to the median sagittal plane.
   (c) True—a coronal plane is at 90 degrees to the median sagittal plane.
   (d) True—this term is used in contrasting positions e.g. the arm is proximal to the forearm.
   (e) False—it describes a structure as closer to the body surface e.g. skin is superficial to muscle.

3. (a) False—abduction is away from the median sagittal plane.
   (b) False—flexion is in a sagittal plane, but it reduces the angle of a joint.
   (c) True—the foot sole is turned away from the median sagittal plane.
   (d) True—remember that thumb movements are at right angles to the rest of the body.
   (e) True—remember that thumb movements are at right angles to the rest of the body.

4. (a) True—vitamin D synthesis occurs in the skin.
   (b) False—there are three layers: the epidermis, dermis and hypodermis.
   (c) True—sweat glands in the hypodermis regulate thermoregulation.
   (d) True—it contains many arteries, veins, and lymphatic vessels.
   (e) False—the epidermis is waterproof; the hypodermis acts as a shock-absorbing layer.

5. (a) True—this occurs in red marrow found in the medullary cavity of the long bones.
   (b) True—this connective tissue membrane surrounds the bone.
   (c) False—the spongy (cancellous) layer lies deep to the outer compact bone layer.
   (d) True—a major nutrient artery and vessels in the periosteum supply the bone.
   (e) True—this growing portion of bone lies between the epiphysis and the diaphysis.

6. (a) False—this joint type has a fibrous capsule that covers articulating bone surfaces.
   (b) True—this is the definition of Hilton's law.
   (c) True—ligaments are pulled tight in excessive joint movements.
   (d) False—ligaments, muscles, and bony contours all contribute to joint stability.
   (e) True—primary and secondary cartilaginous, fibrous and synovial joints.

7. (a) True—prime movers are the main active muscle for a specific movement.
   (b) True—this muscle exerts its force to keep articulating surfaces together.
   (c) False—a single motor neuron motor unit allows precise movements.
   (d) True—force generated is related to the cross-sectional area of muscles fibres.
   (e) True—an aponeurosis unites one muscle to another or muscle to bone.

8. (a) True—somatic nerves control voluntary movements; autonomic nerves regulate involuntary activities, e.g. heart rate.
   (b) False—the head is also supplied by the cranial nerves.
   (c) False—sympathetic nerves arise from the thoracic and lumbar parts of the central nervous system.
   (d) False—motor fibres exit through the anterior root of the spinal cord.
   (e) True—however, in the peripheral nervous system cell body aggregations form ganglia.

9. (a) False—the right side pumps blood to lungs; left side pumps blood to rest of body.
   (b) False—some veins lack valves, e.g. pelvic, head, and neck veins.
   (c) True—these communications can provide an alternative route for blood flow in normal route obstructions.
   (d) True—other functions include fat transportation and an immunological role against infection.
   (e) False—right side drains into the right lymphatic duct; left side drains into the thoracic duct.

10. (a) True—the scapula rotates so that the upper limb can be lifted above the shoulder.
    (b) False—the long head of triceps brachii attaches to the infraglenoid tubercle of the glenoid cavity.
    (c) True—the axillary nerve 'winds' around this neck. Damage paralyses deltoid and teres minor muscles.
    (d) False—the glenoid labrum completely encircles the glenoid cavity deepening the cavity.

(e) False—the radial nerve runs in the spiral groove of the humerus.

11. (a) False—latissimus dorsi is supplied by the thoracodorsal nerve.
(b) True—first part is medial, second part behind, and the third part lateral to the muscle.
(c) True—the other rotator cuff muscles attach to the greater tubercle of the humerus.
(d) False—it is supplied by the lateral and medial pectoral nerves.
(e) True—it initiates the first 15 degrees of abduction.

12. (a) True—the synovial joints have a fibrous capsule lined by a synovial membrane.
(b) False—the rotator cuff muscles are deficient inferiorly.
(c) False—the glenoid labrum and rotator cuff muscles provide greater stability.
(d) False—the supraspinatus and deltoid muscles perform abduction.
(e) True—an articular disc 'sits' in the cavity of the joint dividing it into two.

13. (a) False—axillary lymph nodes are present and drain the upper limb and breast.
(b) True—it is the continuation of the subclavian artery.
(c) False—the vein lies outside the sheath so it can expand on increased venous return.
(d) True—the sulcus is anteriorly on the humerus; the muscles are shoulder joint flexors.
(e) False—trunks are found in the posterior triangle of the neck; the axilla contains the cords.

14. (a) False—anterior rami of the C5–C8 and T1 spinal nerves.
(b) True—posterior divisions of upper and lower trunks join the middle trunk; anterior division of middle trunk joins the upper trunk.
(c) True—lateral cord is lateral, medial cord is medial, and posterior cord is posterior to the artery.
(d) True—the lateral root of the lateral cord joins the medial root of the medial cord.
(e) False—the ulnar nerve supplies some forearm and intrinsic hand muscles.

15. (a) False—the axillary artery finishes at the inferior border of teres major.
(b) True—the brachial artery can be palpated easily under biceps brachii medially.
(c) True—the radial artery passes through the snuffbox before passing between the first dorsal interosseous heads.
(d) True—they pass superficial to the flexor retinaculum to reach the hand.
(e) False—the common interosseous artery arises from the ulnar artery proximally in the forearm.

16. (a) False—radial nerve supplies the dorsal lateral three and a half digits.
(b) True—branches arise in the axilla therefore nerve damage in the spiral groove doesn't affect arm muscles.
(c) False—the nerve arises in the arm above the elbow.
(d) True—it is one of the cord's terminal branches. The other is the radial nerve.
(e) True—it pierces coracobrachialis. After supplying the muscles becomes the lateral cutaneous nerve of the forearm.

17. (a) True—supinator supinates the forearm; brachialis flexes the elbow.
(b) True—however, sometimes it may divide in the arm and a superficial ulnar artery arises.
(c) False—the biceps brachii tendon is lateral to the brachial artery.
(d) True—the radial nerve divides anterior to the lateral epicondyle.
(e) False—the median nerve lies medially to the brachial artery.

18. (a) True—the median nerve lies superficially in the carpal tunnel.
(b) True—a recurrent (muscular) branch supplies the thenar eminence.
(c) True—median nerve branches arise in arm and forearm; ulnar nerve branches arise in the forearm.
(d) False—median nerve supplies the muscle laterally; ulnar nerve supplies the muscle medially.
(e) True—the ulnar nerve supplies all intrinsic hand muscles except lateral two lumbricals and thenar eminence.

19. (a) False—the radial nerve supplies the supinator muscle.
(b) True—in the forearm the ulnar nerve is joined by the ulnar artery.
(c) True—pronator teres forms the medial border of the cubital fossa.
(d) False—the ulnar nerve passes deep between the two heads of the flexor carpi ulnaris muscle.
(e) True—the nerve supplies the muscle, teres minor and skin of the upper arm.

20. (a) False—the tunnel is roofed by the flexor retinaculum.
(b) True—middle and ring finger tendons lie upon the index and little finger tendons.
(c) False—it transmits the median nerve.
(d) True—the eminences also arise from the bones of the distal carpal rows.

(e) False—thenar eminence is paralysed and sensory loss in the lateral three and a half digits.

21. (a) False—dorsal interossei abduct the digits towards the middle digit.
(b) False—the medial two lumbricals are supplied by the ulnar nerve.
(c) False—it passes between the two adductor pollicis muscle heads and forms the deep palmar arch.
(d) False—the muscle inserts into the first metacarpal bone.
(e) False—the lumbrical muscles arise from the tendons of flexor digitorum profundus.

22. (a) True—the aorta passes posteriorly and to the left of the vertebral column.
(b) True—these arteries arise from the arch of the aorta directly.
(c) False—the trunk divides into right common carotid and subclavian arteries behind the right sternoclavicular joint.
(d) True—the left and right brachiocephalic veins unite to form the superior vena cava.
(e) False—the superior vena cava enters the heart behind the right third costal cartilage.

23. (a) True—only the upper part of the second costal cartilage articulates with the manubrium.
(b) True—but also the seventh.
(c) True—there are upper and lower articular facets on the rib head separated by a crest.
(d) False—these ribs do have costal cartilages but no articulation with the costal margin.
(e) True—causes ischaemic pain and paraesthesia (numbness) of the medial border of the forearm.

24. (a) True—the wall is composed of twelve pairs of ribs, thoracic vertebrae, and costal cartilages.
(b) False—the neurovascular bundle runs between the innermost and internal intercostal muscles.
(c) False—intercostal nerves arise from the anterior rami of thoracic spinal nerves.
(d) False—the phrenic nerve does not supply the intercostal muscles.
(e) False—the anterior intercostal arteries arise from the internal thoracic artery.

25. (a) True—the phrenic nerve arises from C3–C5 spinal segments.
(b) False—the aorta passes behind the diaphragm at the T12 vertebra.
(c) True—the inferior vena cava passes through the tendon at T8 vertebral level.
(d) False—the trunks pass behind the diaphragm.
(e) True—it is the main muscle of quiet respiration.

26. (a) True—it is connected to the sternum by weak sternopericardial ligaments.
(b) False—it lies in the middle mediastinum only.
(c) True—the serous layer is within the fibrous layer.
(d) False—the phrenic nerve supplies the fibrous and parietal serous layers only.
(e) False—this serous reflection forms the oblique sinus. Reflection around great vessels forms the transverse sinus.

27. (a) False—it attaches to the central tendon, as well as sternum and great vessels.
(b) True—this is the remnant of the foramen ovale in the fetal heart—an atrial communication.
(c) True—these are the first aortic branches.
(d) True—sympathetic and vagus nerves influence the pacemaker.
(e) True—this valve consists of two cusps, and it separates the left atrium and ventricle.

28. (a) True—arise from left and right aortic sinuses.
(b) False—it drains directly into the right atrium.
(c) True—the great cardiac vein ends in the coronary sinus.
(d) False—the heart has left dominance; the circumflex artery arises from the left coronary artery.
(e) True—the venae cordis minimae empty into the heart chambers.

29. (a) False—coronary arteries arise from the ascending aorta.
(b) False—it is attached between the pulmonary trunk and aortic arch.
(c) True—the sternal angle is at the level of T4 vertebral level.
(d) True—the descending aorta runs to the left of the vertebral level. These branches are small.
(e) False—it leaves the thorax behind the diaphragm.

30. (a) True—the phrenic nerve passes anteriorly and the vagus nerve passes posteriorly to the lung hilum.
(b) True—the nerve is the sole motor supply to the diaphragm, therefore, damage ipsilaterally paralyses it.
(c) False—only the vagus nerve contributes to these plexuses.
(d) False—the right nerve passes posteriorly and the left nerve anteriorly to the oesophagus.
(e) False—the thoracic sympathetic ganglia supply abdominal structures, e.g. stomach or liver through splanchnic nerves.

31. (a) False—the oesophagus is a continuation at C6 vertebral level of the laryngopharynx.
(b) True—the oesophagus is a posterior relation to the trachea.

(c) True—the cricopharyngeus muscle forms an upper oesophageal sphincter.
(d) False—the left gastric artery has oesophageal branches that supply the lower oesophagus.
(e) True—the oesophagus pierces the diaphragmic muscle at T10 vertebral level.

32. (a) False—this right main bronchus is shorter and wider, and it is prone to inhaled objects lodging in it.
(b) True—right lobes are separated by oblique and horizontal fissures, left lobes by an oblique fissure.
(c) True—there are ten segments in each lung.
(d) False—the apex lies 1 cm above the clavicle.
(e) False—it has autonomic nerve supply, and it is insensitive to pain.

33. (a) True—as the diaphragm contracts it lowers, thus increasing the vertical diameter.
(b) False—in adult males diaphragmatic movement is greater than thoracic. This is opposite for females.
(c) True—scalene muscles elevate ribs, intercostal muscles contract forcefully, and erector spinae arch back, contributing to inspiration.
(d) False—the intercostal muscles relax and it is entirely due to lung elastic recoil.
(e) True—the muscle reinforces the elastic recoil.

34. (a) False—the upper liver border runs behind the fifth ribs.
(b) True—transpyloric plane intersects the linea semilunaris at the ninth costal cartilage and gall bladder fundus.
(c) False—it lies deep to the left ninth to eleventh ribs.
(d) False—the pancreatic head lies to the right of the L2 vertebral level.
(e) True—the twelfth rib protects the upper pole of the kidney.

35. (a) False—the internal oblique and transversus abdominis muscles form the conjoint tendon.
(b) False—the rectus sheath passes anterior to the rectus abdominis muscle below the arcuate line.
(c) True—the artery runs superiorly and anastomoses with the superior epigastric artery.
(d) True—the external oblique is supplied by T7–T12 nerves only.
(e) True—right ascending lumbar vein joins the right subcostal vein to empty into the azygos vein.

36. (a) True—a medial crus attaches to the pubic crest; a lateral crus attaches to the pubic tubercle.
(b) False—it is a defect in the transversalis fascia.
(c) False—the coverings are formed as it passes through the inguinal canal.
(d) False—it enters the inguinal canal from the side.
(e) True—the inguinal canal lies above the medial half of the inguinal ligament.

37. (a) True—it is attached to these walls by extraperitoneal tissue.
(b) True—parietal peritoneum irritation can cause referred pain to the area supplied by the obturator nerve.
(c) False—medial fold contains remnants of the umbilical arteries; inferior epigastric vessels are in the lateral fold.
(d) True—the superior compartment lies above the mesocolon, the inferior compartment below.
(e) True—the foramen is a communication between the lesser and greater sacs.

38. (a) True—the oesophagus, liver, and gall bladder are also foregut derivatives.
(b) True—the fundus is usually full of gas.
(c) False—the lesser omentum splits into two 'layers' at the lesser curvature to completely cover the stomach.
(d) True—left gastric artery from coeliac trunk; right gastric artery from the common hepatic artery.
(e) False—it drains mainly into the portal vein.

39. (a) True—the transpyloric plane runs horizontally at the L1 vertebral level.
(b) True—remember that lymphatic supply follows the arterial supply.
(c) False—the jejunum has longer vasa recta.
(d) False—the ileum has a greater amount of fat, which encroaches upon the small intestine.
(e) True—the vagal input increases secretion and peristaltic activity; sympathetic input decreases these activities.

40. (a) False—the transverse colon has a mesentery (the transverse mesocolon).
(b) False—this artery supplies as far as the splenic (left colic) flexure.
(c) False—only the sigmoid colon has a complete layer; the rest has three bands—the taeniae coli.
(d) True—the left ureter is a close relation of the sigmoid colon just before it enters the pelvis.
(e) True—appendices epipolicae are distended pouches of fat, which help distinguish large from small intestine.

41. (a) False—posteriorly on the liver there is the bare area, which lacks peritoneum.
(b) False—portal vein enters the porta hepatis; hepatic vein empties into the inferior vena cava directly.
(c) False—ductus venosum is represented by the ligamentum venosum. Ligamentum teres is the umbilical vein remnant.

(d) False—three hepatic veins drain the liver into the inferior vena cava.
(e) True—a common hepatic artery from the coeliac trunk divided into left and right hepatic arteries.

42.(a) True—left and right hepatic ducts form the common hepatic duct, which joins the cystic duct.
(b) False—the ducts unite to drain into the second part of the duodenum.
(c) False—the right hepatic artery usually gives rise to the cystic artery.
(d) True—it is marked by the transpyloric plane intersecting the linea semilunaris.
(e) True—it concentrates the bile produced by the liver and stores it until it is needed.

43. (a) True—the right lobe of the liver causes the right kidney to be lower.
(b) False—the ureter is the most posterior structure of the renal pelvis.
(c) True—the peritoneum lies anterior to the kidneys.
(d) False—the gland receives its blood supply from the inferior phrenic, aorta directly and renal arteries.
(e) False—the thoracic lesser splanchnic nerve supplies the gland directly without synapsing.

44. (a) True—the common iliac arteries then divide to supply the pelvis and lower limb.
(b) False—the superior mesenteric artery arises at the L1 vertebral level from the aorta.
(c) True—the artery supplies the descending and sigmoid colon and the rectum superiorly.
(d) False—right gonadal vein joins the inferior vena cava; the left joins the left renal vein.
(e) False—the appendix is supplied by an end artery of the posterior caecal artery.

45. (a) True—the plexus is formed within the substance of the psoas major muscle.
(b) False—the femoral nerve is lateral to the psoas major muscle.
(c) True—the ilioinguinal nerve is a collateral branch of the iliohypogastric nerve.
(d) True—stimulation of the femoral branch causes the genital branch to produce contraction of the cremasteric muscle.
(e) True—the nerve also supplies the adductor muscle compartment of the thigh.

46. (a) False—the greater pelvis lies above the pelvic brim.
(b) True—the inlet is fifty to sixty degrees to the horizontal, and the outlet is fifteen degrees.
(c) False—the hip consists of three bones: ilium, ischium, and pubis.
(d) False—anterior superior iliac spine and pubic symphysis lie in the vertical plane.
(e) True—this wide angle allows a fetal head to pass through it.

47. (a) True—the sigmoid mesocolon ends at the third sacral segment.
(b) True—superior rectal artery (an inferior mesenteric artery branch); middle rectal artery (an internal iliac artery branch).
(c) True—these folds are composed of circular muscle and mucosa.
(d) False—the rectovesical pouch is in the male.
(e) False—the mucosa is columnar epithelium.

48. (a) True—they turn anteriorly to run along the pelvic floor.
(b) False—they enter the bladder obliquely; this prevents reflux of urine.
(c) False—the uterine artery is superior to the ureter.
(d) False—the ductus deferens passes superior to the ureter.
(e) True—the ureter crosses these structures as it passes down the pelvic wall.

49. (a) True—only the superior surface is covered by peritoneum.
(b) True—the broad ligament is a double fold of peritoneum.
(c) True—the ovary is suspended from the broad ligament by the mesovarium.
(d) True—this ligament, with the uterosacral ligament and the pelvic floor, prevents the uterus prolapsing.
(e) True—the posterior fornix is closely related to the rectouterine pouch.

50. (a) False—the ovarian artery is a branch of the abdominal aorta.
(b) False—the internal pudendal artery leaves the pelvis through the greater sciatic foramen.
(c) False—the obturator artery is crossed medially by the ureter.
(d) True—all are internal iliac artery branches except the ovarian artery.
(e) True—the uterine artery crosses the ureter superiorly.

51. (a) False—the plexus is formed by L4, L5, and S1–S5 anterior rami.
(b) True—from these ganglia sacral splanchnic nerves arise which carry sympathetic fibres.
(c) False—the pelvic splanchnic nerves carry parasympathetic nerves.
(d) True—remember: 'S2, 3, 4 keeps your guts off the floor'.
(e) True—the hypogastric nerves arise from the superior hypogastric plexus.

52. (a) True—the posterior triangle is larger than the anterior triangle.
(b) False—the sphincter is formed by three muscles: deep, superficial, and subcutaneous.
(c) True—the fossae are separated by the anal canal, anococcygeal ligament, and perineal body.
(d) False—the perineal membrane lies below the pelvic floor.
(e) False—the canal transmits the pudendal nerve and internal pudendal vessels.

53. (a) False—the external sphincter is supplied by the pudendal nerve (inferior rectal branch).
(b) True—the anorectal junction is where puborectalis 'slings' around the rectum.
(c) True—the epithelium changes from columnar epithelium just below the anal valves.
(d) False—the internal anal sphincter is composed of circular muscle, and it stops at the white line.
(e) True—an individual has conscious control over the external anal sphincter.

54. (a) True—the urogenital triangle contains the roots of the external genitalia.
(b) True—other contents include pudendal vessels, dorsal nerve of penis (clitoris), urethra, and bulbourethral glands (male).
(c) True—the perineal membrane is a strong fascial sheet.
(d) False—the lymphatic vessels drain into the superficial inguinal lymph nodes.
(e) True—accumulated fluid, e.g. urine, in the space can, therefore, track up the abdominal wall.

55. (a) True—the fascia separates superficial and deep dorsal veins.
(b) True—the artery has artery of the bulb, deep artery, and dorsal artery of the penis branches.
(c) False—the ilioinguinal nerves supply the anterior third of the labia majora.
(d) False—the pudendal nerve does supply the clitoris through its dorsal branch.
(e) True—it is the female equivalent of the male corpus spongiosum.

56. (a) True—hyaline cartilage separates the ilium, ischium, and pubis bones, which fuse at 19 years of age.
(b) True—the median crest represents the upper four sacral spinous process, the fifth lacks this process.
(c) False—the notch is formed by the posterior margin of the ischium and the ilium superiorly.
(d) True—the piriformis muscle attaches to the apex of the trochanter.
(e) True—the obturator foramen, closed by fascia, has a hiatus (hole) superiorly that transmits these structures.

57. (a) True—the depth of the acetabulum is increased by an acetabular labrum, increasing stability further.
(b) True—these muscles stabilize the pelvis in walking.
(c) True—the nerve arises from L4, L5, and S1. The artery arises from the internal iliac artery.
(d) True—gluteal (or clunial) nerves arise from the thigh's posterior cutaneous nerve, L1–3 and S1–3 posterior rami.
(e) False—the pudendal nerve exits via the greater sciatic foramen and enters the lesser sciatic foramen.

58. (a) False—the femoral artery and vein are within the femoral sheath.
(b) False—the femoral nerve is a lateral structure. Remember the mnemonic N-A-V-Y from lateral to medial.
(c) False—the medial border of the sartorius muscle forms the lateral border of the femoral triangle.
(d) True—the femoral canal begins superiorly at the femoral ring.
(e) True—adductor canal boundaries are: sartorius (anteromedially), vastus medialis (laterally), and adductor longus and magnus (posteriorly).

59. (a) False—the femoral artery is a direct continuation of the external iliac artery.
(b) True—the profunda femoris artery's four perforating branches that supply adductor magnus and the hamstring muscles.
(c) True—the great saphenous vein pierces the saphenous opening's cribriform fascia before joining the femoral vein.
(d) False—the profunda femoris vein joins the femoral vein just below the inguinal ligament.
(e) True—the femoral canal contains a few lymphatic vessels and loose connective tissue.

60. (a) True—the obturator nerve supplies the adductor muscles, gracilis, and hip and knee joints.
(b) True—the sciatic nerve exits the greater sciatic foramen usually below piriformis.
(c) False—the femoral nerve arises from the anterior rami of L2–L4.
(d) False—the inferior gluteal (L5, S1, S2) nerve supplies gluteus maximus.
(e) True—the sciatic nerve becomes compressed paralysing the hamstrings and muscles supplied distal to knee joint.

61. (a) True—the popliteal artery lies adjacent to the femur's distal end, and fractures may damage it.
(b) False—the popliteal artery has genicular, muscular, and sural branches.

(c) True—the small saphenous vein pierces the popliteal fascia to join the popliteal vein.
(d) False—the sciatic nerve divides into the tibial and common peroneal nerves.
(e) False—the popliteus muscle unlocks the knee joint by lateral rotation of the femur.

62. (a) False—the knee joint is an articulation between the femur and the tibia.
(b) False—popliteus laterally rotates the femur on the tibia. Rotation can occur in the flexed knee.
(c) True—other bursae that communicate with the joint are popliteus, and gastronemius bursae.
(d) True—the medial meniscus is attached to the tibial collateral ligament.
(e) True—the quadriceps femoris tendon has a seasmoid bone (patella) within it.

63. (a) False—the anterior compartment muscles are supplied by the deep peroneal nerve.
(b) False—the tibialis anterior muscle dorsiflexes and inverts the foot.
(c) False—the anterior tibial artery is a branch of the popliteal artery.
(d) False—the posterior tibial artery can be palpated behind the medial malleolus.
(e) True—peroneus tertius arises from the fibula and inserts into the fifth metatarsal.

64. (a) True—the tibial nerve divides just below the flexor retinaculum.
(b) False—the saphenous nerve, the femoral nerve's terminal branch, passes anterior to the medial malleolus.
(c) True—common peroneal nerve branches supply muscles whose actions produce eversion and dorsiflexion.
(d) False—the tibial nerve supplies the posterior compartment.
(e) True—the sural nerve supplies the skin on the posterior and lateral parts of the leg.

65. (a) True—the dorsalis pedis artery can be palpated between the extensor hallucis and digitorum longus tendons.
(b) True—the foot has an extensor expansion similar to the expansion found in the hand.
(c) False—extensor hallucis brevis inserts into the proximal phalanx of the great toe (hallux).
(d) True—the deep peroneal nerve supplies the cleft between the first and second digits.
(e) True—extensor digitorum brevis dorsiflexes the middle three toes.

66. (a) True—peroneus longus and peroneus brevis muscles evert and plantar flex the foot.
(b) True—the superficial muscles (triceps surae) act as a venous calf pump for the deep veins.
(c) False—the tibialis posterior tendon is the first to wind around the medial malleolus.
(d) False—gastrocnemius and soleus muscles join to form the tendocalcaneus.
(e) True—the common peroneal nerve divides into superficial and deep nerves at the fibula neck.

67. (a) True—the talar articular surface is wider anteriorly. In dorsiflexion this is driven between the malleoli.
(b) False—the axis of abduction/adduction runs through the second digit.
(c) True—flexor accessorius pulls and removes the 'slack' of the flexor digitorum longus tendons in plantarflexion.
(d) True—the subtalar joint is between the inferior surface of talus and superior surface of calcaneum
(e) False—the medial longitudinal arch is higher than the lateral arch. Arches act as shock absorbers.

68. (a) False—the sagittal suture is formed by the parietal bones articulating together.
(b) False—the coronal suture joins the sagittal suture at the bregma.
(c) False—the sagittal suture joins the lambdoid suture at the lambda.
(d) True—the pterion overlies the anterior branches of middle meningeal artery.
(e) True—temporalis passes deep to it.

69. (a) False—the foramen rotundum transmits the V_2 (maxillary) nerve.
(b) True—the arteries join to form the basilar artery.
(c) False—the internal acoustic meatus transmits the facial and vestibulocochlear nerves.
(d) False—the middle meningeal artery passes through the foramen spinosum.
(e) True—also the sigmoid sinus empties into the superior bulb of the internal jugular vein.

70. (a) False—the buccal branch of the facial nerve is motor to the facial muscles.
(b) False—the facial muscles are superficial, and they attach to the overlying skin.
(c) False—the greater occipital and lesser occipital cervical nerves supply the scalp posteriorly.
(d) True—the facial vein eventually empties into the internal jugular vein.
(e) False—the internal carotid artery, through supraorbital and supratrochlear branches, also contributes to the supply.

71. (a) True—the tentorium cerebelli separates the occipital lobes of the brain from the cerebellar hemispheres.

(b) True—the sigmoid sinus also receives the superior petrosal sinus.
(c) True—this is due to the veins of the head and neck having no valves.
(d) False—the circulus arteriosus connects the internal carotid and vertebral arteries.
(e) False—the internal carotid artery and the abducens nerve run through the cavernous sinus.

72. (a) False—the pituitary sits in the sella turcica.
(b) False—the superior sagittal sinus runs in the superior border of the falx cerebri.
(c) True—the maxillary (V_2) and mandibular (V_3) nerves have meningeal branches.
(d) False—the falx cerebelli separates the cerebellar hemispheres.
(e) True—the internal carotid artery enters the carotid foramen and exits the foramen lacerum.

73. (a) True—the ophthalmic artery is a branch of the internal carotid artery.
(b) False—CN XI (cranial part) is motor to pharyngeal and laryngeal muscles via the vagus nerve.
(c) True—the muscles are supplied by the mandibular branch of the trigeminal nerve.
(d) False—the oculomotor nerve transmits parasympathetic fibres that supply the sphincter pupillae and ciliary eye muscles
(e) False—the facial nerve transmits taste fibres for the anterior two thirds of the tongue.

74. (a) True—obstruction of the central retinal artery by an embolus results in blindness.
(b) False—the oculomotor nerve supplies the superior rectus muscle; the trochlear nerve supplies superior oblique.
(c) True—the sympathetic supply travels in the oculomotor nerve and innervates the smooth muscle part.
(d) False—the inferior oblique muscle causes the eye to look superiorly and laterally.
(e) True—the inferior orbital fissure connects the orbit with the pterygopalatine fossa.

75. (a) False—the nasolacrimal duct empties into the inferior meatus of the nasal cavity.
(b) True—the facial nerve's greater petrosal branch carries the parasympathetic nerve fibres to the ganglion.
(c) True—the orbital part of the frontal bone forms the roof of the orbit.
(d) False—the abducent nerve supplies the lateral rectus muscle.
(e) True—the long ciliary nerve carries sympathetic fibres to the dilator pupillae muscle of the eye.

76. (a) True—the vein is superficial to the artery, but both are deep to the facial nerve.
(b) False—the parotid duct opens opposite the upper second molar tooth.
(c) False—the facial nerve divides into its five terminal branches within the parotid gland.
(d) True—the superficial part is larger than the deep part of the gland.
(e) True—this is due to the close relationship of the lingual nerve to the sublingual gland.

77. (a) False—the deep temporal arteries are branches of the maxillary artery.
(b) False—the temporalis muscle attaches to the coronoid process of the mandible.
(c) True—this tenses the muscles, which can then be palpated.
(d) True—the temporomandibular ligament prevents posterior dislocation.
(e) True—the lesser petrosal nerve carries the fibres to the otic ganglion.

78. (a) False—the outer third is cartilage and the inner two thirds are bone.
(b) True—the tympanic branch of the glossopharyngeal nerve supplies the internal surface of the tympanic membrane.
(c) True—this connection allows equalization of pressure in the middle ear.
(d) False—tensor tympani dampens the vibration of the tympanic membrane protecting it from loud noises.
(e) True—the facial nerve enters the internal acoustic meatus and exits the stylomastoid foramen.

79. (a) True—The external jugular vein descends over sternocleidomastoid and enters the subclavian vein.
(b) True—the accessory nerve appears in the posterior triangle from behind the posterior border of sternocleidomastoid.
(c) False—the nerve to mylohyoid supplies the anterior belly of digastric; the facial nerve supplies the posterior belly.
(d) False—the thyrohyoid muscle is supplied by the C1 fibres traveling in the hypoglossal nerve.
(e) False—the superior thyroid artery arises from the anterior aspect of the external carotid artery.

80. (a) False—the thickening of the investing cervical fascia forms the stylomandibular ligament.
(b) True—the nerves form loops and off these loops arise the branches of the cervical plexus.
(c) True—laterally the pretracheal fascia blends with the carotid sheath also.
(d) False—the internal carotid artery has no branches in the neck.

(e) False—the deep cervical lymph nodes lie along the internal jugular vein within the carotid sheath.

81. (a) True—the investing fascia also encloses the parotid gland.
(b) False—the prevertebral fascia extends laterally to form the axillary sheath.
(c) False—the brachial plexus roots emerge between the scalenus anterior and scalenus medius muscles.
(d) False—the platysma muscle is supplied by the cervical branch of the facial nerve.
(e) True—baroreceptors in the carotid sinus monitor arterial pressure.

82. (a) False—the thyrohyoid muscle is supplied by the C1 fibres carried in the hypoglossal nerve.
(b) False—the hypoglossal nerve curves around the occipital artery of the external carotid.
(c) True—superior, middle, and inferior ganglia. The inferior ganglion usually fuses with the first thoracic ganglion.
(d) True—the union of the veins forms the brachiocephalic vein.
(e) False—the glossopharyngeal nerve supplies stylopharyngeus muscle.

83. (a) True—the common carotid divides into external and internal carotid arteries.
(b) True—the external carotid artery divides within the parotid gland.
(c) True—the jugular trunk empties into the (left) thoracic duct or the right lymphatic duct.
(d) False—the external laryngeal nerve is motor to cricothyroid muscle.
(e) False—the hypoglossal nerve supplies the tongue musculature except palatoglossus.

84. (a) False—the glossopharyngeal nerve supplies stylopharyngeus.
(b) True—the pharynx is divided into nasopharynx, oropharynx, and laryngopharynx.
(c) True—the pharynx finishes at the cricoid cartilage.
(d) True—the soft palate forms a seal so that food cannot enter the nasopharynx.
(e) True—the bulge of mucosa forms a pharyngeal pouch (Killian's dehiscence).

85. (a) True—the tongue surface has an irregular appearance due to the lingual tonsil.
(b) True—they form the lateral and median glossoepiglottic folds; between the folds lies a depression—the vallecula
(c) True—the fascia blends with the periosteum of the skull above.
(d) True—a food bolus passes off the epiglottis, which covers the laryngeal inlet, into the recess.
(e) True—the pharyngeal branch of the pterygopalatine ganglion supplies the nasopharynx.

86. (a) False—septal cartilage, perpendicular plate of the ethmoid bone, and vomer bone form the nasal septum.
(b) True—palatine process of maxilla and horizontal plate of the palatine bone form the hard palate.
(c) True—the sphenoid sinus sits below the pituitary fossa.
(d) True—lateral olfactory nerves supply the lateral nasal wall superiorly.
(e) True—it is where the sphenopalatine artery joins the septal branch of the superior labial artery.

87. (a) True—nasal hairs trap and remove the dust. A vascular plexus humidifies the inspired air.
(b) True—three conchae separate the sphenoethmoidal recess, superior meatus, middle meatus, and inferior meatus.
(c) True—pterygopalatine fossa communicates via the foramen rotundum and pterygoid canal with the middle cranial fossa.
(d) True—the deep petrosal nerve carries sympathetic fibres; the greater petrosal nerve carries parasympathetic fibres.
(e) True—the lesser palatine nerve has sensory and taste fibres of soft palate.

88. (a) False—the lateral pterygoid muscle attaches to the condyloid process.
(b) True—the hard palate is formed by the maxilla and palatine bones.
(c) True—the lingual nerve runs close to the third molar tooth.
(d) True—the internal acoustic meatus transmits both nerves. The stylomastoid foramen transmits the facial nerve.
(e) False—the condyloid process articulates with the mandibular fossa of the temporal bone.

89. (a) True—lingual nerve supplies the anterior two thirds; the glossopharyngeal nerve the posterior third of the tongue.
(b) False—the vallate papillae lie anterior to the sulcus terminalis.
(c) False—the maxilla and palatine bones form the hard palate.
(d) False—the hypoglossal nerve supplies the extrinsic muscles except palatoglossus, supplied by pharyngeal plexus.
(e) True—thyroid gland descends, attached by thyroglossal duct to foramen caecum. The duct usually disappears.

90. (a) True—the intrinsic muscles run in three bundles, transverse, longitudinally, and vertically.
(b) False—chorda tympani supplied the anterior two thirds with taste fibres.
(c) True—a carcinoma may spread to neck structures through the lymphatics.
(d) True—the vallate papillae are on the oral part of the tongue.
(e) True—the palatoglossus muscle forms the palatoglossal arch; the palatopharyngeus muscle forms the palatopharyngeal arch.

91. (a) True—the glandular branch of the facial and the sublingual branch of the lingual arteries contribute.
(b) False—the parasympathetic fibres originate in the superior salivary nucleus.
(c) False—the sublingual gland is supplied by the submandibular ganglion.
(d) False—the medial pterygoid nerve supplies the tensor veli palatini muscle.
(e) True—the glossopharyngeal supplies sensory fibres; the vagus nerve carries motor fibres from the accessory nerve.

92. (a) False—the internal laryngeal nerve supplies the mucosa above the vocal folds.
(b) False—the external laryngeal nerve supplies the cricothyroid muscle.
(c) True—the epiglottis covers the laryngeal inlet as a food bolus passes towards the oesophagus.
(d) True—the vocal ligament and cricothyroid membrane together are known as the conus elastus.
(e) True—this abduction action opens the rima glottis into a diamond shape.

93. (a) True—in nerve damage the ipsilateral vocal fold is adducted and the voice is hoarse.
(b) False—the lateral cricoarytenoid muscle adducts the vocal cords and thus closes the rima glottis.
(c) True—cricothyroid tilts the cricoid cartilage, and hence the arytenoid cartilages, thus tensing the vocal cords.
(d) False—vocalis tenses the vocal ligament anteriorly and relaxes it posteriorly.
(e) True—the transverse arytenoid adducts the vocal ligaments closing the rima glottis.

94. (a) False—the thyrohyoid membrane is pierced by the internal laryngeal nerve.
(b) True—the vestibular ligament is superior to the vocal ligament, separated by a ventricle.
(c) False—the laryngeal inlet is bounded by the epiglottis, aryepiglottic, and interarytenoid folds.
(d) True—the joints have a lax capsule, which allows freer movement.
(e) False—the cricotracheal membrane connects the cricoid cartilage to the first tracheal ring.

95. (a) False—the thyroid is an endocrine gland, and it regulates metabolism through thyroxine.
(b) True—the isthmus lies opposite the second and third tracheal ring cartilages.
(c) False—the superior thyroid artery runs with the external laryngeal nerve.
(d) True—the inferior thyroid artery arises from the thyrocervical trunk of the subclavian artery.
(e) False—the inferior thyroid vein joins left brachiocephalic vein.

96. (a) False—there are seven cervical vertebrae.
(b) False—the five sacral vertebrae and four coccygeal vertebrae are fused.
(c) False—cervical and lumbar curvatures begin to appear before birth.
(d) True—the thoracic curvature results from the wedge shaped vertebral bodies.
(e) True—the C7 vertebral process is called the 'vertebra prominens'.

97. (a) True—eight cervical, twelve thoracic, five lumbar, five sacral, and four coccygeal vertebrae.
(b) False—vertebrae articulate with each other via articular facets and intervertebral discs.
(c) True—it supports the body weight, transferring it to the lower limbs. It protects the spinal cord.
(d) True—the herniation traps the L4, L5 nerve root and causes lower back and limb pain.
(e) True—the anterior longitudinal ligament prevents hyperextension.

98. (a) True—the intervertebral discs are compressed during vertebral column movements.
(b) False—rotation occurs at the atlanto-axial joint. Flexion and extension at the atlanto-occipital joint.
(c) True—thick intervertebral discs, a lax capsule, and horizontal facets accommodate these movements.
(d) True—flexion pushes the ribs together. In extension, rib movement is limited by their attachments.
(e) False—rotation is prevented in the lumbar region by interlocking articular facets.

99. (a) True—the spinal cord is shorter than the vertebral column, and it ends in the conus medullaris.
(b) True—there are eight cervical nerves but seven cervical vertebrae.
(c) False—blood supply is from vertebral, inferior posterior cerebellar, and segmental arteries.
(d) True—the central nervous system lacks a lymphatic supply.
(e) False—the lumbar and sacral nerves make up the cauda equina.

100. (a) True—into this space anaesthetic agent can be introduced, producing an epidural block, e.g. in childbirth.

(b) True—dura mater attaches to the coccyx via the filum terminale.

(c) False—the subdural space is between the dura mater and the subarachnoid mater.

(d) True—this space is below the conus medullaris, and it is the site of a lumbar puncture.

(e) True—the pia mater attaches to the coccyx via the filum terminale.

SAQ Answers

1. The axilla is a space between the arm and the thoracic wall that is bounded anteriorly and posteriorly by axillary folds. Superiorly, the axilla communicates via its apex with the posterior triangle of the neck, and it contains the neurovascular structures for the upper limb.
The boundaries of the axilla are:
   - apex—the superior border of the scapula, the clavicle and the outer border of the first rib.
   - anterior wall—pectoralis major, pectoralis minor muscles, and clavipectoral fascia (these form the anterior axillary fold).
   - lateral wall—intertubercular groove.
   - posterior wall—teres major, latissimus dorsi, and subscapularis muscles (these form the posterior axillary fold).
   - medial wall—serratus anterior, upper four ribs, and intercostal muscles.
   - floor—axillary fascia and skin.

   The contents of the axilla are the axillary artery, axillary vein, lymph nodes, and the three brachial plexus cords. These structures enter or exit through the apex of the axilla. The axillary artery is a continuation of the subclavian artery from the outer border of the first rib, and it finishes at the lower border of the teres major muscle. The artery is divided into three parts by the pectoralis minor muscle. The first part is medial to the muscle and has one branch. The second part behind the muscle and has two branches. The third part lateral to the muscle and has three branches. The axillary vein is formed by joining of the brachial veins and basilic vein at the lower border of teres major muscle, and it becomes the subclavian vein at the outer border of the first rib. The tributaries are the same as the arterial branches. The cords of the brachial plexus run with the axillary artery in the axillary sheath. The axillary sheath is derived from the prevertebral fascia of the deep cervical fascia of the neck. The cords divide into branches within the axilla that supply the upper limb mostly. The axillary vein lies outside the axillary sheath and this allows it to expand in increased venous return.

   Within the fatty tissue of the axilla are five groups of axillary lymph nodes. These lymph nodes drain the upper limb and the majority of the breast. The lymph nodes empty into a subclavian lymph trunk that empties into the thoracic duct on the left or the right lymphatic duct.

2. Biceps brachii is a muscle of the anterior (or flexor) compartment of the arm. As the name denotes, the muscle has two heads. The long head originates from the supraglenoid tubercle of the scapula. The long head tendon passes through the capsule of the shoulder joint and is extrasynovial, but it is surrounded by a double-layered tubular sheath (an extension of the shoulder joint's synovial membrane). It emerges from the capsule of the shoulder joint in the intertubercular groove of the humerus from under the transverse ligament. This transverse ligament runs between the greater and lesser tubercles of the humerus. The short head arises from the apex of the coracoid process of the scapula.

   The two muscle bellies fuse in the lower part of the arm and give rise to a tendon that is attached to the tuberosity of the radius. The tendon, at the level of the elbow joint, gives rise to a bicipital aponeurosis that crosses and forms the roof of the cubital fossa. The bicipital aponeurosis fuses with the deep fascia of the forearm.

   The musculocutaneous nerve, a branch of the lateral cord of the brachial plexus, supplies the muscle. The muscle is a flexor of the shoulder joint and a powerful flexor of the elbow joint. Moreover, the muscle can also supinate the forearm if the elbow is flexed.

3. The flexor retinaculum is a strong thickening of the deep fascia that serves to bind the long flexor tendons of the forearm against the surface of the wrist. It is attached to the pisiform bone and hook of hamate medially, and the trapezium and scaphoid bones laterally. This osseofibrous tunnel formed by the flexor retinaculum and the carpal bones is known as the carpal tunnel.

   Through the carpal tunnel pass the four tendons of flexor digitorum superficialis, the four tendons of flexor digitorum profundus, the tendon of flexor pollicis longus, and the median nerve. As the tendons of flexor digitorum superficialis enter the carpal tunnel, the index finger and little finger tendons lie upon the middle finger and ring finger tendons. At the distal row of carpal bones, these four tendons all lie in the same plane. A synovial sheath surrounds all the tendons within the carpal tunnel. The median nerve is the most superficial structure in the carpal tunnel, and it supplies the thenar eminence, lateral two lumbrical muscles, and the skin of the lateral three and a half digits.

   The intrinsic muscles of the thenar and hypothenar eminences take origin from the flexor retinaculum.

   Compression of the median nerve in the carpal tunnel can occur, and this is known as carpal tunnel syndrome, a common clinical condition. This compression paralyses the thenar eminence muscles and the thumb takes an adducted position since adductor pollicis has no antagonist to its action. Paralysis of the two lateral lumbrical muscles can be seen when the patient is asked to make a fist and the two lateral fingers lag behind the two medial digits. The lateral three and a half digits have paraesthesia (numbness). The skin of the palm is unaffected due to the palmar cutaneous branch of the median nerve arising before the nerve enters the carpal tunnel and passing over the flexor retinaculum. The compression may be relieved by the flexor retinaculum being cut, removing the tension upon the median nerve.

4. The muscles of the intercostal space are the external intercostal, internal intercostal, and innermost intercostal. They are all supplied by the intercostal

nerves. Their functions are to elevate and depress the ribs during respiration.

The vessels of an intercostal space are the posterior and anterior intercostal arteries. Posterior intercostal arteries of the first and second spaces originate from the superior intercostal artery—a branch of the costocervical trunk of the subclavian artery. The posterior intercostal arteries of the lower nine spaces originate from the thoracic aorta. The anterior intercostal arteries in the upper six spaces originate from the internal thoracic artery. The anterior intercostal arteries for spaces seven to nine originate from the musculophrenic artery—a branch of the internal thoracic artery. Each artery supplies the skin, muscles, and parietal pleura.

The intercostal nerves are the anterior rami of the thoracic spinal nerves. The first six nerves are distributed within their intercostal spaces. They supply the skin, intercostal muscles, and parietal pleura. The lower five nerves (T7–T11) and the subcostal nerve (T12) also pass to the anterior abdominal wall, where they supply the skin, anterior abdominal wall muscles, and parietal peritoneum.

The nerves and vessels are protected by the costal groove of the ribs. This neuromuscular bundle runs between the second (internal intercostal muscle) and third (innermost intercostal muscle) layers.

5. The diaphragm is a thin sheet of muscle that separates the thoracic and abdominal cavities. In profile the diaphragm is an inverted 'J-shape' with two domes. The right dome is higher than the left. The diaphragm is the primary muscle of inspiration and on contraction it lowers. This action is seen in male respiration; however, in female inspiration there is greater movement of the thoracic wall.

   The diaphragm arises from crura. The right crus arises from the upper three lumbar vertebral bodies and intervening intervertebral discs, and fibres of the right crus form a sling around the oesophageal opening in the diaphragm. The left crus arises from the upper two lumbar bodies and intervening intervertebral disc. These crura pass superiorly into a central tendon. The tendinous fibres of the medial edge of the crura unite with each other in front of the T12 vertebra and form the median arcuate ligament. The medial arcuate ligament is a thickening of psoas fascia, and the lateral arcuate ligament is a thickening of the quadratus lumborum fascia. Muscle fibres arise from the medial and lateral arcuate ligaments, the twelfth rib, and the costal margin and adjacent rib as far as the seventh rib. The muscle fibres then pass upwards and insert into a central tendon.

   There are three large openings in the diaphragm. An aortic opening is opposite T12 vertebra behind the median arcuate ligament. It transmits the aorta, azygos vein, hemiazygos vein, and thoracic duct. The oesophageal opening is opposite T10, and it is within the muscle of the diaphragm. It transmits the oesophagus, vagus nerves, and oesophageal branches of the left gastric vessels. The vena caval opening is opposite the T8 vertebra in the central tendon. It transmits the inferior vena cava and the right phrenic nerve. Other structures pass around the diaphragm. Splanchnic nerves and the sympathetic trunk pass behind the diaphragm. Lymphatics and the superior epigastric vessels pass anteriorly to the diaphragm. The lower six intercostal neurovascular bundles pass laterally to diaphragm.

   The motor nerve supply to the diaphragm is phrenic nerve (C3–C5). The sensory nerve supply is the phrenic nerve centrally and the lower five intercostal and subcostal nerves peripherally. The blood supply is from the musculophrenic branch and superior epigastric branch of the internal thoracic artery. It receives phrenic branches from the thoracic and abdominal aorta.

6. The oesophagus is a muscular tube that transmits swallowed food to the stomach. The oesophagus commences at the cricoid cartilage (C6 vertebral level), and it is a continuation of the laryngopharynx. The cricopharyngeus muscle forms an upper oesophageal sphincter, and it is the narrowest part of the oesophagus. The oesophagus descends through the thorax behind the trachea until the trachea bifurcates at the T4 vertebral level. The oesophagus is then crossed by the aortic arch and the left main bronchus, which form a constriction of the oesophagus. The oesophagus leaves the thorax by passing through the muscle of the diaphragm at T10 vertebral level. Here, fibres of the right crus of the diaphragm form a sling around the oesophagus and again form a constriction. The muscle in the wall of the oesophagus changes gradually from entirely striated superiorly to entirely smooth muscle in the abdomen.

   The upper oesophagus receives its blood supply from the inferior thyroid artery—a branch of the thyrocervical trunk of the subclavian artery. The middle oesophagus receives oesophageal branches of the thoracic aorta. The lower oesophagus receives oesophageal branches of the left gastric artery. The venous drainage of the oesophagus follows the arteries. The upper oesophagus drains into the brachiocephalic vein, the middle part drains into the azygos vein and the lower part drains into the left gastric vein and eventually to the portal vein.

   The nerve supply to the upper part is from the recurrent laryngeal nerve and sympathetic nerves of the middle cervical ganglion. The lower part is supplied by the sympathetic trunk and great splanchnic nerve. The parasympathetic supply arises from the vagus nerves and forms anterior and posterior oesophageal plexuses. Lymphatic drainage is to the deep cervical lymph nodes superiorly, tracheobronchial nodes in the middle, and prearotic nodes inferiorly.

   An important anastomosis exists between the portal and systemic systems at the lower end of the oesophagus. In cases of portal obstruction the lower veins can dilate and become varicose. If left untreated the veins can rupture.

7. The pleurae consist of an outer parietal and an inner visceral layer separated by a potential space filled with a small amount of serous fluid. These two layers are a sheet of mesothelium (flat cells) that covers the lungs. They reduce friction as the lungs expand and deflate in respiration with the thoracic wall and contents, e.g. pericardium.

The parietal pleura lines the thoracic wall (costal pleura), the upper surface of the diaphragm (diaphragmatic pleura), and the lateral aspect of the mediastinum (mediastinal pleura). At the thoracic inlet it arches over the lung as the cervical pleura, 1 cm above the clavicle. The mediastinal pleura forms a cuff around the lung hilum (root) and it reflects back on itself to form the visceral pleural layer. The visceral layer completely invests the lung surface, extending into the fissures of the lung.

The cuff of pleura around the lung hilum is 'too big' and it hangs below the lung hilum as an empty fold called the pulmonary ligament. The parietal pleura is reflected off the diaphragm onto the thoracic wall, and a recess is formed that is neither filled with lung nor visceral pleura except on deep inspiration. This is the costodiaphragmatic recess. A similar recess is formed between the thoracic wall and the mediastinum known as the costomediastinal recess.

The visceral pleura has an autonomic supply derived from the pulmonary plexus and it is insensitive to pain. However, the parietal pleura has a somatic supply. The costal pleura is supplied segmentally by the intercostal nerves. The mediastinal and diaphragmatic pleura have a phrenic nerve supply. The parietal pleura has a blood supply from the internal thoracic and intercostal arteries. The visceral pleura has a blood supply from the bronchial arteries.

In diaphragmatic parietal pleura irritation, referred pain to the shoulder can occur because the phrenic nerve innervates the parietal pleura.

8. To facilitate description of abdominal pain and/or swellings the abdomen is divided into four quadrants by vertical and horizontal lines through the umbilicus. However, a more accurate division of the abdomen into nine regions is also used. These nine regions are formed by two vertical lines that correspond to the midclavicular lines, and two horizontal lines. One horizontal line corresponds to the transpyloric plane (L1 vertebral level) and the second runs between the two tubercles of the iliac crest (intertubercular plane).

The inferior border of the liver extends from the right 10th costal cartilage in the midaxillary line to the left 5th rib in the midclavicular line. The liver's upper border runs between the left and right 5th ribs; both points are in the midclavicular line. The right border of the liver runs from the 5th right rib to the 10th costal cartilage.

The fundus of the gall bladder lies deep to where the linea semilunaris intersects the costal margin in the transpyloric plane. At this point the fundus lies behind the 9th costal cartilage.

The spleen lies deep to the 9, 10, and 11th ribs on the left. It is not palpable unless it is enlarged, at which point the spleen extends inferiorly and anteriorly along the 10th rib to below the costal margin.

The head of the pancreas lies in the C-shaped concavity of the duodenum at the level of the L2 vertebra. The neck of the pancreas lies at the level of the L1 vertebra in the transpyloric plane. The pancreas continues to the left curving upwards towards the hilum of the spleen.

The hilum of the kidney lies in the transpyloric plane, 5 cm from the midline. The upper poles of the kidneys lie anterior to the 12th rib. The right kidney is lower than the left because of the presence of the liver, but they both lie roughly opposite the first three lumbar vertebrae.

The ureters begin at the hilum of the kidney in the transpyloric plane. They run inferiorly on the anterior surface of the psoas major muscle, in front of the tips of the lumbar vertebrae transverse processes (as seen on a urogram), to the sacroiliac joint to enter the pelvis.

9. The inguinal canal is an oblique slit, approximately 6 cm long. The canal runs parallel and superiorly to the medial half of the inguinal ligament, commencing at the deep inguinal ring and ending at the superficial inguinal ring. In males the inguinal canal transmits the spermatic cord and ilioinguinal nerve. In females the round ligament and ilioinguinal nerve form the canal contents.

The anterior wall of the inguinal canal is formed by the external oblique aponeurosis. The inguinal ligament forms the floor, which is the lower free edge of the external oblique aponeurosis. The floor is reinforced medially by the lacunar ligament, which extends from the inguinal ligament to the pectinate line of the superior pubic ramus. The internal oblique and transversus abdominis muscles form the roof laterally. The posterior wall is formed by the transversalis fascia laterally and by the conjoint tendon medially.

The deep ring is an opening in the transversalis fascia bounded laterally by the arching fibres of transversus abdominis and the inguinal ligament and medially by the transversalis fascia. The superficial ring is a triangular slit in the external oblique aponeurosis. The fibres of the aponeurosis split into two crura. A medial crus attaches to the pubic crest and a lateral crus attaches to the pubic tubercle.

Structures entering at the deep inguinal ring gain the three coverings of the spermatic cord as they pass through the inguinal canal. Only when these structures exit the canal at the superficial ring are they completely invested in these three layers derived from the anterior abdominal wall. The ilioinguinal nerve lies outside the spermatic cord because it does not enter the canal at the deep inguinal ring. It enters the canal from the side as the nerve runs between the external oblique and internal oblique layers. However, it does leave the canal through the superficial ring.

An indirect hernia is a clinical condition in which one of the contents of the abdomen enters the inguinal canal and may descend into the scrotum or labia majus. This type of hernia is less common in females because the female inguinal canal transmits only the round ligament.

10. See Fig. 4.5.

11. The obturator nerve (L2–L4) arises from the lumbar plexus. The nerve runs down the medial side of the psoas major muscle and across the bifurcation of the common iliac vessels. In the pelvis, it supplies the parietal peritoneum lining the pelvis. On entering the obturator canal with the obturator vessels, the nerve splits into anterior and posterior divisions.

The posterior division supplies and pierces the obturator externus muscle and then runs posterior to the adductor brevis muscle. The posterior division then supplies the adductor part of adductor magnus (the hamstring part is supplied by the tibial component of the sciatic nerve). Finally the posterior division sends a branch to innervate the knee joint.

The anterior division passes anterior to the obturator externus muscle and supplies the hip joint. The anterior division then descends anterior to the adductor brevis muscle and supplies adductor brevis, adductor longus, gracilis, and pectineus muscles. This division continues and supplies the skin over the medial aspect of the thigh.

In the pelvis the obturator nerve crosses the ovary as it sits between the bifurcations of the common iliac vessels. In ovarian pathology referred pain to the medial aspect of the thigh can occur.

12. The rectum is 12 cm long and commences at the level of the third sacral segment as a continuation of the sigmoid colon. The rectum has no mesentery and three convex curves: two to the right (upper and lower) and one to the left (middle curve). The three curves correspond to transverse folds that project into the rectal lumen. These transverse folds are composed of circular muscle and mucus membrane. The lowest curve dilates into the rectal ampulla. The rectum ends at the anorectal junction where the puborectalis muscle slings around the rectum and pulls the anorectal junction anteriorly. The rectum has a complete layer of circular and longitudinal muscle. The longitudinal muscle in the form of taeniae coli that run over the sigmoid colon expand and join together to invest the rectum in a complete layer. There are no sacculations or appendices epiploicae.

Peritoneum covers the upper third of the rectum anteriorly and laterally. The middle third is covered anteriorly, and the lower third lies below the level of the peritoneum. The peritoneum is reflected onto the bladder in males or vagina in females and it forms the rectovesical pouch or rectouterine pouch (of Douglas), respectively. In these pouches, which are the lowest in the peritoneal cavity, are small intestine and sigmoid colon. Posteriorly, the rectum is related to the sacrum, coccyx, and pelvic floor.

The rectum receives its blood supply from the superior rectal artery (a branch of the inferior mesenteric artery), the middle rectal artery (a branch of the internal iliac artery), and it receives anasotomoses from the inferior rectal artery (a branch of the internal pudendal artery). A rectal venous plexus drains into the inferior mesenteric vein via the superior rectal vein and thus into the portal venous system. The plexus also drains into the internal iliac vein via middle rectal vein and thus into the systemic venous circulation.

The nerve supply is from the sympathetic hypogastric plexus and the parasympathetic pelvic splanchnic nerves. The parasympathetic nerves are motor to the rectal muscles. The lymphatic drainage accompanies the arteries and drains into the preaortic nodes and internal iliac nodes.

13. See Fig. 5.21.

14. The femoral triangle is formed in the thigh. Its boundaries are:
   - the inguinal ligament, superiorly.
   - the medial aspect of sartorius, laterally.
   - the medial aspect of adductor longus, medially.
   - iliopsoas, adductor longus and pectineus muscles form the floor.
   - the fascia lata and cribiform fascia form the roof.

   The femoral triangle contains, from lateral to medial, the femoral nerve, femoral artery, and the femoral vein. Medial to the femoral vein is the femoral canal. The femoral artery, femoral vein, and the femoral canal are enclosed within the femoral sheath. This sheath is derived from the transversalis fascia anteriorly, and from the iliopsoas fascia posteriorly. The femoral nerve is not in the femoral sheath because it lies outside the fascia, which forms the sheath. The femoral artery has a profunda femoris branch that arises at the apex of triangle, which then passes behind the adductor longus muscle. At the apex of the triangle the femoral vessels pass into the adductor canal.

   The femoral canal is bounded superiorly by the femoral ring. This is formed by the inguinal ligament, femoral vein, pectinate line, and lacunar ligament. In the canal is loose connective tissue in which lymphatic vessels run. Since the canal is medial to the femoral vein, it provides a dead space into which the vein can expand in increased venous return from the lower limb.

   A femoral hernia is when one of the contents of the abdomen enters the femoral canal through the femoral ring.

15. The hip joint is a synovial ball and socket joint in which the head of the femur articulates with the acetabulum of the hip bone. The joint is highly stable and it allows a high degree in range of movement. The stability is achieved by the shape of the joint surfaces and enhanced by the presence of a rim of fibrocartilage—the acetabular labrum. The acetabular and femoral articular surfaces are covered by hyaline cartilage. The great range of mobility results from the femoral neck being narrower than the diameter of the femoral head. This relationship allows considerable movement in all directions before the neck impinges upon the acetabular labrum.

   The capsule of the hip joint, which is loose but strong, is attached to the labrum and the femoral neck. Three ligaments that pass from the pelvic bone to the neck of the femur reinforce the capsule. The ligaments are:
   - pubofemoral—prevents excessive hyperextension and abduction.
   - iliofemoral—prevents excessive lateral rotation and hyperextension.
   - ischiofemoral—prevents medial rotation and hyperextension.

   In flexion of the hip joint the head of the femur moves about a transverse axis passing through both acetabula, and it causes the shaft to swing anteriorly. The muscles of the anterior compartment of the thigh

accomplish this. Extension of the hip joint is the opposite and the femoral shaft swings posteriorly. Muscles of the gluteal region and the posterior compartment of the thigh accomplish this. Adduction of the hip joint causes the head of the femur to move in the acetabulum about an anteroposterior axis and causes the femoral neck and shaft to swing medially. The adductor muscles cause this movement. Abduction performed by the gluteus medius and minimus is the opposite action but through the same axis and the thigh moves laterally. Medial rotation involves the rotation of the femoral head in the acetabulum about a vertical axis that passes through the femoral head and medial condyle. The neck of the femur swings anteriorly. The gluteus medius and minimus and tensor fasciae latae are responsible for this movement. Lateral rotation in the same axis causes the femoral neck to swing posteriorly. The obturator internus, piriformis, and quadratus femoris muscles contribute to this action.

The blood supply is from the medial circumflex, obturator, superior and inferior gluteal arteries. The nerve supply is governed by Hilton's law: 'the motor nerve to a muscle tends to give a branch of supply to the joint which the muscle moves and another branch to the skin over the joint'. For the hip joint, the femoral, obturator, sciatic, and superior gluteal nerves all innervate it.

16. The player has been struck in the temporal fossa, which is bounded by the temporal bone, zygomatic bone, and zygomatic arch. Here, the temporal bone is very thin and susceptible to fracturing. The bony landmark that has been struck is the pterion. The pterion is an area consisting of an H-shaped pattern of sutures formed by the articulation of the greater wing of the sphenoid bone, frontal, temporal, and parietal bones. The pterion is important because the anterior branch of the middle meningeal artery passes deep to the pterion in a groove.

The fracturing of the pterion has caused a tear in this branch of the middle meningeal artery, which has caused an extradural haemorrhage between the dura mater and periosteum of the cranium. The middle meningeal artery is a branch of the maxillary artery and enters the cranium through the foramen spinosum. A middle meningeal vein may also be torn and exits through the foramen spinosum to join the pterygoid venous plexus.

As the blood accumulates in the extradural space, the dura mater, which is strong, resists expansion. This is conveyed by pain receptors in the dura mater that cause the players headache. The dura mater is supplied by the trigeminal nerve in the anterior and middle cranial fossae. The dura mater of the posterior cranial fossa is supplied by the upper three cervical nerves and meningeal branches of the vagus and hypoglossal nerves.

17. The affected nerve is the facial nerve (CN VII) as it runs through the parotid gland. The facial nerve arises from the junction of the pons and medulla of the brainstem. The nerve has two divisions: a larger motor root that supplies the muscles of facial expression, and a sensory root (nervus intermedius) that carries parasympathetic and taste fibres.

The two roots of the facial nerve enter the internal acoustic meatus of the petrous bone and join together. The facial nerve then enters the facial canal, which passes anterior to the semicircular canals and above the vestibule of the inner ear. At this medial wall of the middle ear the nerve turns posteroinferiorly and the bend forms the geniculate ganglion. This ganglion contains the cell bodies of the taste fibres. From the geniculate ganglion the greater petrosal nerve originates and passes anteriorly carrying lacrimal gland parasympathetic secretomotor fibres. The greater petrosal nerve joins the deep petrosal nerve in the foramen lacerum to form the nerve of the pterygoid canal, which synapses at the pterygopalatine ganglion. The facial nerve, now running posteroinferiorly in the facial canal, passes under the lateral semicircular canal but above the promontory and it curves inferiorly at the posterior wall of the middle ear to leave through the stylomastoid foramen. Before the facial nerve exits the facial canal, it gives off its nerve to stapedius and chorda tympani branches. The chorda tympani runs across the internal surface of tympanic membrane before exiting through the petrotympanic fissure to join the lingual nerve. It carries taste fibres and parasympathetic secretomotor fibres for the submandibular and sublingual glands. Parasympathetic secretomotor fibres synpase in the submandibular ganglion.

The facial nerve exits the stylomastoid foramen and gives off its posterior auricular branch, which supplies the occipital belly of the occipitofrontalis. A muscular nerve is the next branch, and this splits into two supplying the posterior belly of digastric and stylohyoid muscles. The facial nerve now enters the parotid gland where it divides into five terminal branches superficial to the retromandibular vein. These branches (temporal, zygomatic, buccal, mandibular, and cervical nerves) supply the muscles of facial expression.

18. See Fig. 7.50.

19. The mucosa of the tongue has sensory fibres that supply the special sense of taste, as well as sensations of pain, temperature, and touch. Parasympathetic secretomotor fibres also supply the salivary glands of the tongue.

For the anterior two thirds of the tongue, fibres of general sensation are carried to the trigeminal ganglion via the lingual nerve. Taste fibres travel in the lingual nerve, then the chorda tympani, then in the facial nerve to the geniculate ganglion of the facial nerve. In the geniculate ganglion are the cell bodies of these taste fibres. Parasympathetic secretomotor fibres from the superior salivary nucleus run in the facial nerve then the chorda tympani. The chorda tympani runs with the lingual nerve and then it joins the submandibular ganglion. In this ganglion the secretomotor fibres synapse and the postganglionic fibres then join the lingual nerve and supply the glands of the anterior two thirds of the tongue.

For the posterior third of the tongue and the vallate papillae, the sensation of general sensation, pain, and taste are conveyed by the lingual branch of the glossopharyngeal nerve. Parasympathetic secretomotor fibres travel in the glossopharyngeal nerve to supply mucosal glands on the posterior third of the tongue.

Sympathetic nerves from the superior cervical ganglion reach the tongue by travelling with the lingual artery.

20. The thyroid gland is an endocrine organ that regulates the rate of metabolism by secretion of the hormone thyroxine. The gland consists of two lateral lobes that are connected together by an isthmus. The isthmus lies in front of, and adheres to, the second to fourth tracheal cartilaginous rings. A small glandular projection often arises from the isthmus. This pyramidal lobe is to the left of the midline, and it is attached to the hyoid bone. It represents the development of the glandular tissue from the caudal end of the thyroglossal duct. The thyroid gland is enclosed within its own capsule and within the pretracheal fascia of the deep cervical fascia of the neck.

    The blood supply is from the superior and inferior thyroid arteries. The superior thyroid artery arises from the anterior aspect of the external carotid artery and enters the upper pole of the lateral lobe. With the superior thyroid artery runs the external laryngeal nerve, which supplies the cricothyroid muscle. The inferior thyroid artery arises from the thyrocervical trunk of the subclavian artery. The inferior thyroid artery enters the lower pole of the lateral lobe. The recurrent laryngeal nerve runs with the inferior thyroid artery and the nerve supplies all the intrinsic laryngeal muscles (except cricothyroid) and the mucosa below the vocal cords. These two thyroid arteries anastomose with each other. The venous return is via three pairs of veins, which are superior, middle, and inferior thyroid veins. The superior and middle thyroid veins join the internal jugular vein, and the inferior thyroid veins join together and empty into the left brachiocephalic vein. In 3% of individuals a thyroid ima artery enters the isthmus after arising from either the brachiocephalic trunk or aortic arch.

    The relationship of the thyroid arteries to laryngeal nerves is important to surgeons performing a complete or partial removal of the thyroid gland since the arteries need to be ligated. If the laryngeal nerves are damaged, the voice becomes weak and hoarse since the intrinsic muscles are paralysed.

Index

Specific arteries, muscles, nerves and veins are listed under their individual full names, not grouped under general entries

G